# CSR, Sustainability, Ethics & Governance

**Series Editors**

Samuel O. Idowu, London Metropolitan University, London, UK

René Schmidpeter, BFH, Bern, Switzerland

In recent years the discussion concerning the relation between business and society has made immense strides. This has in turn led to a broad academic and practical discussion on innovative management concepts, such as Corporate Social Responsibility, Corporate Governance and Sustainability Management. This series offers a comprehensive overview of the latest theoretical and empirical research and provides sound concepts for sustainable business strategies. In order to do so, it combines the insights of leading researchers and thinkers in the fields of management theory and the social sciences and from all over the world, thus contributing to the interdisciplinary and intercultural discussion on the role of business in society. The underlying intention of this series is to help solve the world's most challenging problems by developing new management concepts that create value for business and society alike. In order to support those managers, researchers and students who are pursuing sustainable business approaches for our common future, the series offers them access to cutting-edge management approaches. CSR, Sustainability, Ethics and Governance is accepted by the Norwegian Register for Scientific Journals, Series and Publishers, maintained and operated by the Norwegian Social Science Data Services (NSD).

Milan Todorovic • Samuel O. Idowu
Silvia Puiu
Editors

# Sustainability and Social Responsibility of the Media and in the Media

## Media Perception and Environmental Impact in the 21st Century

Springer

*Editors*
Milan Todorovic
Guildhall School of Business and Law
London Metropolitan University
London, UK

Samuel O. Idowu
Guildhall School of Business and Law
London Metropolitan University
London, UK

Silvia Puiu
University of Craiova
Craiova, Romania

ISSN 2196-7075 ISSN 2196-7083 (electronic)
CSR, Sustainability, Ethics & Governance
ISBN 978-3-032-00085-9 ISBN 978-3-032-00086-6 (eBook)
https://doi.org/10.1007/978-3-032-00086-6



This Springer imprint is published by the registered company Springer Nature Switzerland AG
The registered company address is: Gewerbestrasse 11, 6330 Cham, Switzerland

# Foreword

It is a great honour and privilege to write this foreword for this book: *Sustainability and Social Responsibility in the Media, and of the Media: Media Perception and Environmental Impact in the twenty-first Century.* I am grateful for the invitation of the Editors Professor Silvia Puiu, Professor Samuel O Idowu, and Dr Milan Todorovic, and I congratulate them for the piece. The book is organized into two parts and 12 chapters. The contributors in each chapter have extensive research, academic, and industry experience and have produced several research outputs of pronounced quality.

The global perception and understanding of what sustainable development is about has grown significantly across the global village where we all exist. This is not surprising, and it is also not because they are told to do so nor is it because their socioeconomic and political orders have forced them to be attentive to what Agenda 2030 is asking all global citizens to do. It is simply because of the growing understanding of the negative consequences of the absence of sustainability and social responsibility in various sectors of the developed and developing economies. Research shows that there are significantly growing voices expressing the exigent need for industries such as media and others to align their operational practices with the ethos of sustainable development and social responsibility.

The past, current, and expected unstable world political and socioeconomic climate compels industries to align their strategies with a sustainable development and socially responsible roadmap for limiting the negative impacts on financial and economic volatilities. In our world today, sustainability has transformed from a roundtable discussion to a solid and significantly indispensable factor to determine the success of not only businesses but also the wellbeing of the human nature and all other living organisms. This universal success can only be real and perceived not only by focusing on one or a few sectors, but also all businesses, industries, institutions, and other sectors with public and private connotations. One of these industries that has a dominant role in the social fabric of every nation, economic growth, development, and political landscapes is the media industry. The extension of the sustainability landscape towards the media industry has become one of the focal points in sustainability and social responsibility research and development. To inform readers, policymakers, decision-makers, and the general population, this book is of overriding importance in creating a forum and research niche to contribute to the knowledge sphere of sustainability and social responsibility in the media industry to investigate its extent, perception, and the environmental impact in the twenty-first century.

The media industry is very diverse. There is no sector in the economy that is not linked directly and/or indirectly to the media industry. The extensive reach of the Internet, the ever-growing use of artificial intelligence, the continuous changes perceived in the education sector, and their environmental impacts have been underlined by various researchers.

This book is unique by its nature due to its particular focus on specific aspects of the media industry such as the sustainability challenges in the industry, impact and responsibilities, and the responsible media dialectics with a specific focus on communities, sustainability, education, and ethical voices. The book through its contributors should deliver current insights into the sustainability practices of the media industry and its social responsibilities. It is my strong belief that this book provides a broad understanding of the sustainability practices of the industry and enhances our view of how industry practices can be aligned with what the sector's sustainability and social responsibility practices should be.

I hope global readers of the new book would agree with some of the sentiments I have advanced above about this new addition to what is currently available in the areas discussed in the chapters of the book.

Strategic Management Department
Guildhall School of Business and Law,
London Metropolitan University,
London, UK
June 2025

Eyob Mulat-Weldemeskel

# Preface

What makes sustainability and social responsibility in the media and of the media? A versatile examination of the notion is required: The media "universe" has become infinitely fragmented—yet it remains dominated by very large economic interests: the multitude of platforms, devices, formats, and technologies spread across "verticals" is ordered by a set of changing and evolving community rules. The discourses of modern media are just as diverse as human thought and knowledge ever was, and the contest for trust-[post]truth-fact-authenticity is excited, passionate, and engaging, dangerously so at times.

The gadgets we use appear safe, clean, and highly energy efficient, yet the ICT giants that power the digital media consume enormous amounts of power and disrupt everyday life. Media modes have an impact on individuals and societies through increasingly interactive traits. In a community-oriented meaning of what is "sustainable", the powerful engines of social consent can perpetuate discord and community alike. In such contexts, the word "environment" attains a sociocultural meaning wider than the physical world itself. Therefore, the debate opens to a broad variety of questions, answered by more questions on a diverse spectrum of subjects. These relate to the notions of "media" across various corporate, autonomous, and community contexts of beliefs, proprietorship, authorship, originality, trustworthiness, and agency in economic, social, and environmental sustainability. Much of this is defined by technology and the forceful advances of artificial intelligence.

This book directs our gaze at the notions of sustainability and social responsibility in media contexts from a broad set of emerging, non-aligned, and diverse angles, including those of health, lifestyle, creativity, morality, and human agency within the context of interactive and increasingly intelligent media. This book is far from uniform. We aim to widen the debate on what do CSR and sustainability really mean in the third decade of the twenty-first century.

This book presents a variety of perspectives, theories, and methods from diverse locales:

- Environmental issues and mitigation of their impact through modern tools of media persuasion
- Representation and resistance from psychological, sociocultural, and behavioural perspectives
- Community-building, cultures, identities, and values adopted by media users
- Authenticity, journalism, freedom of expression, and freedom of engagement
- Post-truth and media power in contemporary contexts
- The impact of Internet use, digitization, and devices on mental health
- Technological progress, artificial intelligence, user experiences, policies, and media CSR
- Market effects on ethics, responsibility, and accountability of media stakeholders

London, UK — Milan Todorovic
London, UK — Samuel O. Idowu
Craiova, Romania — Silvia Puiu
June 2025

# Acknowledgements

We wish to thank everyone who made this book possible. Of course, that means starting with all the authors who put considerable skill, labour, effort, and faith into it. Their work was demanding and time-consuming, and it required a great deal of labour, confidence, conviction, commitment, and dedication on their part. Not least, patience—for which we are especially grateful.

This project is among the first of its kind, which makes it both rewarding and challenging in equal measure. Therefore, Dear Authors: Thank you for your creativity, imagination, and, of course, superb expertise and research! We also wish to thank Dr Eyob Mulat-Weldemeskel of London Metropolitan University for preparing such a fine Foreword for us.

Thank you so much, everybody.

We wish to extend our deep appreciation to our Publishers at Springer, including Barbara Bethke, for having the vision that this book will indeed fly and giving us a shot at it, Parthiban Gujilan Kannan, for his endurance, trust, and selfless support, Anette Weiss, and Ruth Milewski, and all the Springer staff whom I have not met as yet but who indeed made this happen.

To our friends and loved ones, both present and departed, who shone as the beacons of inspiration and good spirit—especially at those times when the light at the end of the tunnel felt like an approaching train and not the blaze of triumph.

Last not least, to take the advice of a much-accomplished colleague and mentor: If a few slips did happen, it was never the intention, as this joint effort was truly rewarding.

# Contents

# About the Editors and Contributors

## About the Editors

**Milan Todorovic, PhD** is Senior Lecturer in Creative Industries at London Metropolitan University. He is also the Course Leader in Music Business and several legacy curricula in Media and Music Industries. He is Lead Doctoral Supervisor of top experts in their fields, such as Media Value + Supply Chains powered by AI, Leadership Management in Art Music and Culture, etc. Todorovic created unique study programmes, e.g. Business Modelling for Machine Learning in Streaming Media, and several strategic curricula. His scientific and artistic background combines with incisive analysis for multidisciplinary projects. He designs novel research methodologies converging sociocultural and scientific work. In a career before academia, he revolutionized a European capital's underground scene spanning electronic music, daring events, organized chaos, digital film, video and **Art**. A renaissance man equally at ease with creative writing, laser physics and strategy, his films and artworks were shown globally. He advises art + music festivals on rebranding and strategic modelling, contributes to European Capital of Culture projects, and delivers consultancy for leading change. Milan Todorovic operates under artistic aliases for ***Sonic Painting*** projects he had singularly developed. A classically trained Fine Art painter, producer, and videographer, he is both versatile and clandestine in this field. Academic work forms part of his socially conscious mission. He wrote an influential multidisciplinary monograph on Creative Industry Strategies, to which he is preparing a sequel. To Milan all this forms a logically integrated continuum as he is a polymathic synaesthete whose innovative projects tend to break boundaries.

**Samuel O. Idowu, PhD** is Professor of Accounting and Corporate Social Responsibility at the Guildhall School of Business and Law, London Metropolitan University, where he is currently the course leader for the MSc Corporate Social Responsibility & Sustainability and Advanced Diploma in Professional Development (ADPD) Corporate Social Responsibility & Sustainability. Samuel is also Professor of CSR and Sustainability at Nanjing University of Finance & Economics, China.

He is a fellow member of the Chartered Governance Institute, a fellow of the Royal Society of Arts, a Liveryman of the Worshipful Company of Chartered Governance Institute, and a named freeman of the City of London. He is the CEO and President of the Global Corporate Governance Institute an international network of CSR scholars. Samuel has published over one hundred articles in both professional and academic journals and contributed chapters in several edited books and is the Editor-in-Chief of three major global reference books by Springer—the *Encyclopedia of Corporate Social Responsibility* (ECSR), the *Dictionary of Corporate Social Responsibility* (DCSR), and the *Encyclopaedia of Sustainable Management* (ESM), and he is Series Editor for Springer's CSR, Sustainability, Ethics and Governance books. Samuel is the Editor-in-Chief of the *International Journal of Corporate Social Responsibility* and the *American Journal of Economics and Business Administration.* Samuel has been in academia for more than 30 years winning one of the Highly Commended Awards of Emerald Literati Network Awards for Excellence in 2008 and 2014. In 2010, one of his edited books was placed in 18th position out of 40 top Sustainability books by Cambridge University Programme for Sustainability Leadership, and in 2016, one of his books won the outstanding Business Reference Book of the year of the American Library Association. In 2018, he won a CSR Leadership Award in Cologne, Germany, and in 2019, he won the 101 Most Impactful CSR Leaders Award in Mumbai, India. Samuel is on the Editorial Advisory Boards of the *International Journal of Business Administration.* He has been researching in the field of CSR since 1983 and has attended and presented papers at several national and international conferences and workshops on CSR. Samuel has made a number of keynote speeches at international conferences and workshops and written the *foreword* to a number of leading books in the field of CSR and Sustainable Development. And he has examined on a few PhD theses in the UK, Australia, South Africa, the Netherlands, and New Zealand.

**Silvia Puiu, PhD** is Associate Professor PhD Habil. at the University of Craiova, Faculty of Economics and Business Administration. She has 18 years of teaching and conducting research related to corporate social responsibility, management, and marketing, and in recent years, she has developed a high interest in studying topics related to the psychology of human resources. During 2018 and 2025, she was the president of a nongovernmental organization focused on nonformal education, and starting in 2024, she started offering workshops on emotional intelligence after she got her certification as a Personal Development Coach. At the end of 2024, she also got her certification as a Career Coach. In the last 10 years, she edited a magazine for students, a project started with passion and love for her students and her profession. Silvia Puiu appreciates that being a teacher and a mentor is the best thing that has happened to her, not able to imagine a different career path for her. In 2024, she started a new journey as a student at the Faculty of Psychology and Education Sciences in Brasov, Romania. Psychology has been a passion for her since high school, and recently, she realized that this path can help her become a better teacher for her students. Silvia Puiu published numerous research articles in journals indexed in Web of Science and also contributed to numerous Springer books on the vast domain of CSR. She also published three personal books, in part autobiographies and in part psychological and philosophical short stories, one of them published as an audiobook in which she used her voice.

## Contributors

**Olga Amarie** Department of World Languages and Cultures, Georgia Southern University, Statesboro, GA, USA

**Predrag Đ. Bajić** Faculty of Sport, University "Union – Nikola Tesla", Belgrade, Serbia

**Cecilia Chirieleison** Department of Political Science, University of Perugia, Perugia, Italy

**Christopher Cohen** London, UK

**Germina Alina Cosma** University of Craiova, Craiova, Romania

**Bernardo Cruz** Hei-Lab: Digital Human-environment Interaction Labs, Universidade Lusófona, Lisbon, Portugal

**Suzana Demyen** Department of Business Administration—Reşita, Faculty of Economics and Business Administration, Babeş—Bolyai University, Resita, Romania

**Samuel Domingos** Hei-Lab: Digital Human-environment Interaction Labs, Universidade Lusófona, Lisbon, Portugal

**Ana Rita Farias** Hei-Lab: Digital Human-environment Interaction Labs, Universidade Lusófona, Lisbon, Portugal

**Carmela Gulluscio** Department of Law and Economics, University of Rome UnitelmaSapienza, Rome, Italy

**Samuel O. Idowu** Guildhall School of Business and Law, London Metropolitan University, London, UK

**Zoran B. Jevtović** Department of Communicology and Journalism, Faculty of Philosophy, University of Niš, Niš, Serbia

**Adebimpe Adesua Lincoln** Robert Kennedy College, Zurich, Switzerland

**Christian J. Lincoln** Atlantic Wharf, Cardiff Bay, UK

**Ana Loureiro** Hei-Lab: Digital Human-environment Interaction Labs, Universidade Lusófona, Lisbon, Portugal

**Sílvia Luís** Hei-Lab: Digital Human-environment Interaction Labs, Universidade Lusófona, Lisbon, Portugal

**Eliana Portugal** Hei-Lab: Digital Human-environment Interaction Labs, Universidade Lusófona, Lisbon, Portugal

**Catarina Possidónio** Hei-Lab: Digital Human-environment Interaction Labs, Universidade Lusófona, Lisbon, Portugal

**Silvia Puiu** University of Craiova, Craiova, Romania

**Gabrielle Round** University of Salford, Salford, UK

**Sam Sarpong** University of Central Lancashire, Lancashire, UK

**Georgiana Camelia Stănescu** Faculty of Letters, University of Craiova, Craiova, Romania

**Amalia Raluca Stepan** University of Craiova, Craiova, Romania

**Adrian-Costinel Tănase** Department of Business Administration—Reşita, Faculty of Economics and Business Administration, Babeş—Bolyai University, Resita, Romania

**Florenţa-Diana Tănase** Faculty of Political, Administrative and Communication Sciences, Babeş—Bolyai University, Resita, Romania

**Milan Todorovic** Guildhall School of Business and Law, London Metropolitan University, London, UK

**Martina Topić-Rutherford** Department of Advertising and PR, College of Communication and Information Sciences, The University of Alabama, Tuscaloosa, AL, USA

**Teresa Turzo** Department of Law and Economics, University of Rome UnitelmaSapienza, Rome, Italy

**Mihaela Tinca Udristioiu** Department of Physics, Faculty of Sciences, University of Craiova, Craiova, Romania

**Mihaela Zăvăleanu** University of Craiova, Craiova, Romania

**Marija Geiger Zeman** Ivo Pilar Institute of Social Sciences, Zagreb, Croatia

# List of Figures

# List of Tables

**Artificial Intelligence and its Implications for Media Sustainability in Romania**



**A Rigid Dichotomy: Artificial Intelligence and Ethical Responsibility: Comparative Assessment of EU, UK, and US AI Policy Framework**



**The Media Impact on the Economic News Consumer**



**"They Don't Promote It, in the Capitalist Society that Promotes the Meat Industry": Echo Chambers and Media Mistrust Among Women Vegans in the UK and Croatia: A Contribution to the Debate on the Social Responsibility of the Media**

# Sustainability and Social Responsibility in the Media, and of the Media: An Introduction

**Milan Todorovic, Samuel O. Idowu, and Silvia Puiu**

**Abstract** This chapter introduces the diverse, ethically oriented book aimed at a broad audience of practitioners, researchers, scholars and advanced students alike. The breadth of topics, already significant at this stage, commands an inclusive and purposeful approach. The notions of media competence in cultural terms, fused with multifaceted sustainability and corporate social responsibility demand such a take on the subject.

Should this volume see new editions, they will indeed move with the times and technologies, media formats and methodologies that shall emerge.

This introductory chapter serves primarily as an overview.

It introduces the reader to the broad complexity of media-related subjects within a variety of contexts. Some contributions have been designed to focus on technology. More than technology, however, at the heart of this book are the human communication needs, the demand for insight and entertainment, the community-driven uses of media, as are indeed the evolving meanings of the terms 'medium' and 'media'.

This chapter concentrates on a diverse, yet integrated set of foci presented in the book.

The multitude of authors' perspectives explore a variety of subjects in media ethics, sustainability, pragmatism and more, including: applied ontological analyses, legal and policy angles, community-oriented discourses in health and customs alike, various notions of fake news, and debatable forms of content and modes of interaction.

Lastly, we call for an exploration of the perimeters of technology not yet prepared for media use: Ones that may in the future cause an extensive and novel

M. Todorovic (✉) · S. O. Idowu
Guildhall School of Business and Law, London Metropolitan University, London, UK
e-mail: m.todorovic@londonmet.ac.uk; s.idowu@londonmet.ac.uk

S. Puiu
University of Craiova, Craiova, Romania
e-mail: silvia.puiu@edu.ucv.ro



M. Todorovic et al. (eds.), *Sustainability and Social Responsibility of the Media and in the Media*, CSR, Sustainability, Ethics & Governance,
https://doi.org/10.1007/978-3-032-00086-6_1

deep-seated transformation, likely to be disruptive. These may soon trigger new ethical debates around policies fit for processes as disrupting as those accompanying the advent of the Web or the current commanding impact of AI. Media, in the broadest sense, may again evolve, qualitatively so. It is therefore argued that we may soon need to direct our gaze to new technologies emerging from the advances in sciences of the human mind.

**Keywords** Corporate social responsibility (CSR) · Media · Sustainability · Social marketing · Public discourse · Information and communication technology (ICT) · Artificial intelligence (AI) · Fake news · Authenticity · Creativity · Labour · Behaviour · Psychology · Mental health · Media-induced anxiety · Practical philosophy · Storage · Mediums · Environmentally conscious filmmaking · Good practice · Regulatory framework · Policy · User-generated content (UGC) · Citizen journalism · Reality television/litigation hybrids · Gamification of newsworthy reporting · Duty of care of the media · (Problematic practices in) economic news reporting · (Communal) holidays · Festive traditions · Leisure · Educational campaigns · Effective communication · Stimulating sustainable practices · Media-enabled persuasion · Health · Vegans · Women · STEM · STEAM · Media texts/narratives · Media channels · HBO · Netflix · COVID-19 · Climate change · Learner motivation · Neuroimaging · Thoughts into words

This book addresses the growing need to fill a gap in literature for media practitioners, theorists, academics and consultants alike. While the notions of Sustainability and Corporate Social Responsibility (CSR) in the Media—and for the Media—are not an obvious match, this indeed is a subject that warrants scholarly and specialist practitioner attention for a number of reasons.

If we were to look at the two notions in a literal sense, it would not be immediately clear just how important this link is. Responsible behaviour by media organisations is easier to grasp; sustainability requires a bit more thought, however. This, of course, if one is familiar with the nuanced, and depth-aspects of what sustainability represents. The evident aspect of it from the point of view of media theory and practice, methods and means, is the one where media serve the role of an educational and awareness-raising tool. Not unlike the purpose that Social Marketing plays in enlightening the public to good social causes such as health and public safety, utilising a variety of media channels, methods and effective narratives.

Some of the 12 chapters in this book are entirely geared towards that purpose, the one of educating the public on matters of environmental sustainability primarily. Equally, however, there are less discernible albeit significantly impactful aspects of sustainability that relate to social (including cultural) and economic aspects, impacts and corresponding narratives that media can engage to the benefit of many. Such are the aspects of media practice that relate to understanding of how the notions of truth, fact, and compassionate approach to public discourse are construed to public benefit—or otherwise. The claim to truthfulness is a bold one and represents a key factor in policy making and dissemination, reception and perception of ‘the value of values' and identity constructs and practices at the heart of diversity, equity and opportunity for all. Blurring the lines and disregarding transparency in any of those aspects of our increasingly mediated lives can have undesirable consequences which may then become detectable and tangible in everyday lives of many, through politics, resistance to cultural change, empowerment and other facets of positive social action.

Equally, humanity is becoming increasingly aware of its ever-greater dependence on sophisticated technologies rooted in information and communication technology (ICT) sectors, and the latest, ongoing mix of challenges and opportunities is epitomised in the disruptive power of the innovations in Artificial Intelligence (AI). The growing industry is seen by some as an apocalyptic force that will destroy jobs and subsume much of the demand for human labour; equally, many will appreciate the AI and its contribution to humanity as an incredibly powerful tool for the enhancement of human lives and communities.

While it is extraordinarily difficult to evaluate the effects of current developments and predicting the impact of science and technology bolstered by large commercial investment facilitated by state support and government incentives, it is clear to everyone that AI is a powerful catalyst of change. The notions of authenticity, creativity, labour and the very concept of human agency are either subverted or boosted by the emerging technology, depending on viewpoint.

AI is a tool that is seen as deus ex machina by some; and yet, neglecting the role of human agency both on the level of civil society would be a reductionist oversimplification. Every technology known throughout human history was, one, a means to an end depending on who employs it and for what purpose, and two, something that was eventually co-opted into everyday life, however disruptive it is. The very medium of the Internet, which is attaining ever more tangible features, had caused unprecedented disruption in recent decades, and is something that was at first seen by many as a fad that may be ignored. Far from being a purely disruptive force, it created solutions to problems that were not even perceived to be in existence a generation ago. While it made some professions obsolete, the Luddite approach was never a good response to innovation.

Therefore, more than one third of this book is dedicated to the impact of AI as it emerges into the media sphere, and through it, into aspects of our lives worldwide. Equally, authors of other treatises here have focused on a more established role of the media in terms of creative outputs and how they serve a social purpose to engage, educate and enhance the human condition, and indeed the conditions of people's lives. The educational aspect of media is well-represented in this book, and so is its community form, especially in the context of diversity, dialectic relationships between social groupings, their media clusters and major commercial enterprises ever so often facilitating such clustering, and the policy angle that connects the media and its social, economic and environmental impact with myriad manifestation of interactive digital technologies within and beyond the AI context.

To focus solely on disruption would be limiting, as much innovation and social transformation is incremental and less dramatic than that. Hence the focus on the seemingly mundane—such as leisure time, or on the established media formats less interactive than the ones mentioned above: those of established content industries we normally associate with the traditional arts such as music or drama.

The behavioural, psychological and social aspects of the role of media thus include it as a means for educating and enlightening the public, as stated above in the context akin to that of social marketing—enabling the public to understand and appreciate, through the media, notions of environmental sustainability and with it a responsible socially engaged behaviour that promotes a healthy physical, biological environment for all. Equally, the same tools can be used for social contests between the groups of involved individuals whose lifestyle choices and gendered identities can stand at odds with the consensus-promoting media narratives and where food is more than an element of sustenance but is politically charged as a source of meaning, at times such meaning being construed around deep environmental commitments.

This book crosses geographical, ideological and industry boundaries. From responsible use of media tools and formats in high-end creative industries, to the formulation of day-to-day existence under regulatory regimes; from the mental health impact through the disruptive technical characteristics of cognitively-engaging media to the novel blurring of boundaries and reality that failed to deliver in virtual reality but is likely to create unprecedented convergences between the disruptive but familiar worlds of gamification and the enhanced capacities for an AI-powered daydreaming of the future. Cultural conditions are at times crystallised in material form through policy and regulation yet the challenge of media-enhancement of daily conduct tend to surpass geographic regions and societal norms of traditional civilisation distinction. Humanity being unified by the narrative of an emerging force greater than itself may be far-fetched but only so much.

## 1 Overview

This book comes at a time when it is too soon to tell exactly how the new forms of Corporate Social Responsibility may evolve. The dozen chapters presented here, innovative and probing though they are, approaching the subject from various angles, disciplines and methodological traditions, are just a starting point for a debate to unfold in the future. Just as important as the known CSR and Sustainability definitions are at this stage, the interactive, corporate, digital-physical/intangible-tangible equivalences are, the additional self-producing, automated learning component of the novel intelligent media will pose new questions and challenges, forcing us to dream up new solutions qualitatively different from the ones that the Web brought us a generation ago, but just as novel in relative terms, and just as exciting if a bit startling.

The book is divided into two sections: One that addresses the technologies primarily, from those recognised for decades in the realm of culture, media and creative sectors. The other, that deals with social, psychological, educational, cultural and indeed ethical questions and multiple directions in which solutions are being sought.

## 2 The Sequence of Chapters

***The opening chapter*** by Mihaela Zăvăleanu, Amalia Raluca Stepan, Germina Alina Cosma, Mihaela Tinca Udriştioiu addresses the well-authenticated notion of mental health issues that affect media users in the current increasingly interactive age. It looks at strategic means and logics that can be developed to pre-empt the most undesirable by-products of current interactive media such as cyber aggression, as well as similar documented digital technology issues while formulating positive, policy-regulatory frameworks to facilitate best practice in utilisation of digital media innovation as the field evolves at an ever-greater pace. There is clearly a need for more such research to be done to connect the reported increase in the nature and scope of health impacts, varied forms of media-induced anxiety and emotional instability among interactive digital media users, and this chapter's authors should be commended for looking at the strategies that need to be adopted to utilise the media to users' advantage, rather than place them in a position of emotionally damaging *technostress*.

In Chap. 2, Christopher Cohen critically approaches *Medium Theory* from the perspective of a practical philosophy underpinned by comprehensive multidisciplinary fieldwork. Focusing on the *changing materiality* of content storage/play formats, in this case that of music, the author examines '*the clutter*', both metaphorical *and* physical—(let us note, and not lose sight that in Physics, *digital is physical*). These equivalences, clearly including the storage of electronic files which by definition *cannot be infinite*—and the superseding futures of content. Moreover, the environmental impact of such *mediums* is yet to be ascertained. Based on rich empirical data gathered by different means and varied paradigms, extracted from wide-ranging samples, this chapter aims to uncover the fundament of the distinction between different technologies that store, play and distribute content not only on the level of functionality and pragmatic account but also regarding the more fundamental state of embedded work, as stated above, a form of practical philosophy, *a teleology of form*.

In Chap. 3, Carmela Gulluscio, Cecilia Chirieleison, Teresa Turzo present an environmental analysis of the film industry across different nation states and jurisdictions of the West, and the industry policies designed to reduce undesired, primarily carbon-focused effects of film production. They draw on the utilisation of locally-regionally or nationally contextualised best practice for environmentally conscious filmmaking. This, together with the dissemination of good practice across industry sectors and regulatory areas. This, with the caveat that context is once more everything and that articulating any international film sustainability standards poses a challenge in practical and methodological terms, therefore inherently. The resulting work contains a detailed comparative taxonomy of good practice, and it identifies the unique features of each regulatory framework, some of those being impossible to replicate for a number of cultural, economic and physical reasons alike.

The take on AI in Georgiana Camelia Stănescu's Chap. 4 is an analytical cautionary tale, drawing correlations between the disruptive challenges of Artificial Intelligence in the media, such as those in the newsroom and new production. The author's warning prompts us to think and reflect about the pitfalls of low-cost convenience of this mode of content creation and the slippery slope of fake news and similar undesired effects of technological convenience. 'Too good to be true' spring to mind. A seemingly elegant solution that comes at an incalculable cost? Suggesting that we cannot delegate news production and dissemination to bots may raise ever more profound questions about the mediation of truth and interpretation of fact, especially beyond the realm of news. The previous technological revolution started around text, then text attained a new meaning, and if intelligent automation of news is the new text, then the author's caution surpasses the fake newsrooms and mass-customisation of Plato's caves.

A judiciously incisive comparative analysis of different regulatory systems for AI in Chap. 5 is both refined and troubling in equal measure. Its nuanced evaluation of all the ethical and practical effects of the global lack of preparedness for the challenges that the exponential pace of AI/machine learning development poses on a non-harmonised, overstretched and rapidly ageing regulation that cannot catch up with, let alone match the ever-converging AI/media industries' rapid disruptive transformation. As a well-argued call for action, it seems to correspond with the observations made elsewhere as to the increasingly obsolete position of public-sector bodies, regulations and their respective responses to the R&D leaps and bounds in Big Tech inscrutable to the public, lawmakers and legislators alike, the scientific modelling of strategies of which are not disclosed to citizen-elected bodies the world over. Its warnings resemble those raised nearly a generation ago, concerning Net Neutrality or the disruption to copyright, and may soon be reflected in the thought mechanics akin to the much delayed the kneejerk reactions of many a government regarding the (un)regulated access to social media and smart devices among minors, with, in this instance, far more at stake due to the sweeping impact of AI/machine learning on every aspect of human creation, labour and skill.

Predrag Bajić and Zoran Jevtović in Chap. 6 expand on related subject matter by raising concerns not dissimilar to those in Stănescu's Chap. 4. In the chapter concluding this, ***Part One*** of the book the two authors examine the removal of attested filtering and quality assurance processes of news production and analysis with the advent of AI. Not too long ago, during the dawning of social networks and video streaming UGC, a concept of 'citizen journalism' was hailed as an exciting and proactive initiative-taking development—until the darker, opaquer and menacing aspects of distributed human agency in media content creation, dissemination and execution had raised its spectre best known to the mainstream public through the well-recounted woes of social media moderation and deeply troubling, unhinging content. Bajić and Jevtović go a step beyond. They match the problematising nature of the said analogous effect augmented by AI in this instance—exploring journalism's role in generating historical records in contemporary settings, which, now in increasingly shortened timeframes and without adequate gatekeeping and filtering processes may lead to an unpredictable impact on numerous levels of post-modern knowledge-generation. Thus a fundamental shift on such conceptual basis cannot be considered without the potential for tangible real-world impact.

***Part Two*** of this selection of works, opens with Chap. 7 by Sam Sarpong. This work, suitably commencing the socially oriented section of the book, and that which addresses the more pronounced communal impacts of media practice is a highly critical piece that uncovers a disquietingly successful Ghanian media format centred on the circumvention of due process through a reality television/litigation hybrid, made popular, and indeed possible through the specific socioeconomic conditions of the Ghanaian context. Sarpong brings to the table the elements of technology, media formats loosely defined as Reality TV with the infotainment component concerningly tangible in its outcomes, and the blurring of boundaries between documentary content, gamification of newsworthy reporting, and regulatory issues side-stepped due to economic challenges facing the litigants. In *Sustainability* terms, as stated elsewhere in this Introduction, *the Social and Economic dimensions of Sustainability*—often overlooked by the general public—are at the heart of the matter, and directly, critically so, connected to the CSR role and the neglected public *duty of care of the media*, now seldom performed. Regulatory implications abound, together with core critical questions this author posits.

Chapter 8, by Suzana Demyen, Adrian-Costinel Tănase and Florența-Diana Tănase, addresses the context of unreliable fact-checking news media procedures—or lack thereof, combined with low levels of literacy among media audiences and thus resulting in high incidence of fake news. This chapter particularly focuses on the problematic practices in economic news reporting. Such news attracts less indignation than sensationalist content that grabs major headlines, and the authors argue that its corrosive nature is no less damaging. Not dissimilar from the previous chapter where *the real impact is a reality impact* of a poorly regulated infotainment show format which has practical legal implications, the damaging effects on financial resources of the viewing public are both preventable and disturbingly, perchance lastingly impactful. Once more, the detrimental effects are tangible: the unreliable media contents disseminated with minimal regulation focus on varied components of the news reporting on economic developments to an undiscerning public. A cautionary analysis of the brand of fake news that is seldom looked at in public discourse—as it lacks the outrageous flair of 'big politics'—yet it has an effect on audiences that the authors argue causes factual damage to real lives in a less detectable way; being induced into poor economic choices is quite as real as other, more glamorised hazardous behaviours.

In Chap. 9, the author Olga Amarie provides quite a contrasting reflexive analysis, one that on the surface level of reading has an air of nostalgia. Still, even though it appears to focus on the decidedly positive context of holidays, this chapter does indeed encompass a critical take on the role of contemporary media, specifically the Internet-driven digital media system and its socially atomising characteristics. Calling for a greater attention to holidays and celebratory events in general, many notably missing and omitted from the public discourse of today, the author compares, relates, conveys and recounts various festive traditions from around the world within contemporary as well as historic, traditional contexts. The unifying, communal nature of shared celebrations in this text are seen as rather more than a form of leisure as an absence of work: The communal and social components of holiday-making, contrasted with the isolating and socially divergent features of Internet-based media. It is a hopefully intoned, contemplative text, and equally although subtly, connoting a placid 'call to arms' against the widespread mores of digitally facilitated seclusion.

Chapter 10 is authored by Ana Rita Farias, Bernardo Cruz, Samuel Domingos, Catarina Possidónio, Ana Loureiro, Eliana Portugal and Sílvia Luís. It focuses on the most effective modes of media-enabled persuasion through effective communication designed to elicit a set of meaningful responses to consistent educational campaigns. The said campaigns are conceived, intended and calculated to stimulate sustainable practices among the publics, particularly in environmental contexts. Therefore, within the best-known sphere of Sustainability in public discourse. Special attention is paid to targeted communications, subtle yet preset incentives, tactics for building audience and recipient motivation to act in environmentally conscious ways, and consistent strategies for overcoming attitude-behaviour gaps. Designed for practical application and construed around overcoming the publics' cognitive barriers by refined, delicate and understated encouragements based on behavioural economics, this study looks at pragmatic empirical tools for effective and [mass]customised communication, opt-ins and other public-facing tactics for utilising advanced psychology tools to promote climate change awareness.

In Chap. 11, authored by Martina Topić, Marija Geiger Zeman and Gabrielle Round, a study was conducted with the participation of 48 respondents sampled across two countries (20 in Croatia, 28 in the UK). Two-tier thematic analysis was used; participants were vegan women who were asked to recount their experience of vegan-related topics in the mainstream media from their respective sets of news information or features' sources. It was found that in both societies the participants had felt there exists a bias against their lifestyle/ethics/health-oriented choices. The analysis based on carefully constructed method, had concluded that there was an '*Othering*' at play, sometimes elusive and at times more obvious. In the concluding section, remarks by the respondents cited highlight the notion that a common excuse for the prevalence of meat-promoting news articles had purported it to be preferable to plant-based diets on health grounds—a rather '*Bernaysian*' method. It was emphasised that personal preference of individual journalists still played a key part in the reporting and further findings were indicating a conceivable commercial interest involved in media narratives centred on veganism. Moreover, this study's findings maintain that the women interviewed preferred the safe familiarity of echo chambers which came as a result of their experiences of mainstream media partiality *vis a vis veganism.* This suggests that despite a changing set of public-facing conversations on the ethics and politics of food, the vegan community, in this case vegan women interviewed in two different societies still do feel *Othered.*

The concluding Chap. 12 authored by Mihaela Tinca Udriştioiu, looks at the ways to reinforce the relevance of STEM education through the media and disseminate it to the young audiences to enhance learning. Why STEM? Science Technology Engineering and Mathematics is far more than a policy mantra. It is indeed the evidence-based foundation of competitive education system fit for the very technologies examined in this book. It is not without irony that the most frequent consumers of networked platforms, and astute users of payable digital media dashboards that thrive on statistics and forecasting are often the same young people whose cultural resistance to STEM curricula appears passionate, immutable and vocal in equal measure (Todorovic, 2024).

The study articulates a set of policy recommendations that align educational content across the channels and formats through critical engagement with the subject matter and strategies to improve levels of media literacy. The study suggests that young audiences are overwhelmed by information content coming from all angles, thus finding it harder to discern about the quality of respective sources and synthesise the data in a meaningful way. This chapter further concludes that governments, technology firms and the media need to work in concert strategically to maximise the effectiveness of such messages and articulations. [Note—often the tech and media are the same entity—this too is not without irony.] Udriştioiu further explores a number of media texts, narratives and channels (HBO, Netflix and others), in which these messages and articulations are embedded as means of creating attractive content that facilitates critical thought key to STEM education, which in turn is being recognised across the globe as an essential foundational skillset for the competitive twenty-first century workforce. Scientific insight as a way of thinking and broad audience interest in it, it is suggested, was further facilitated through the COVID-19 pandemic as scientific news had provided the opportunity to promote the relevance of such elucidation for clearly pragmatic reasons, thus additionally showing that interactive media gained in prominence as a means of curricular delivery during the many lockdowns of the day. Similar reasoning that improves audience and learner motivation stems from the set of climate change accounts and a similar approach to engagement as espoused in Chap. 10.

## *2.1 Summarising*

Thus, the content of this selection of chapters comes full circle. Starting with scientific orientation in the context of the digital media's impact on human cognition, covering a broad variety of topics that include traditional concepts of creativity applied in media contexts, moving to a critical evaluation of the irresistible power of media narratives and formats, especially in the settings of erosion of social equity thus covering the other key points of Sustainability and CSR for, and in the media.

Interactivity is clearly a theme that runs through much of this book, as is the augmented *opportunity to threat ratio* coming from a novel game-changer, machine learning/AI at a time when societies around the world have just about absorbed the previous two innovation cycles, those of the Web and interactive digitisation respectively. The form and presentation of the media had changed, as has the method of its capture and dissemination, but the purpose—to inform, influence, engage and control, among other things—remains—albeit the versatility, diversity of dimensions, themes and modes increases by splitting the narrowcasting of messages infinitesimally.

The variety of standpoints within channel proliferation is such that it hardly surprises the reader that mental health impact is detected; that both media literacy and focused attention among digital natives appear on the wane; that the blurring between game, reality and 'reality' increases, not least due to the impact of previous generations of media forms, formats and designs. Isolation, alienation and permeant messaging work together well to intensify market opportunities while effective communications of profound meaning around health, the environment, inclusion, diversity and education are gaining a platform: Yet there are no guarantees or silver bullets to ensure positive alignment of societal drives and the media.

The status quo in environmental terms appears impossible to sustain and various industries take note of that in their practices. What appeared *near fictional* in the sciences of 2015 (Todorovic, 2016) such as computer-enhanced mental imaging, is now reaching its own '2.0' moment:

In 2025, at the time of writing, MIT reports on '*Brain scans that translate thoughts into words*' (Williams, 2025) so if this book reaches a new edition in a few years' time, it may well consist of 24 or 36 chapters discussing the inter-intra-brain media that treat *Neuralink* as a crude tool of the past and 'hacking' may well become synonymous with the narratives once confined solely to the fanciful realms (Straiton, 2023) of cyberpunk, whereby safety and security of any '*New New Media*' may blur the boundaries of what it is to be human, sentient and autonomous of/from the increasingly empowered learning machines.

Regardless, the questions of ethics, power structures, equity, equality, diversity and socioeconomic sustainability will remain just as relevant as are the woes of retaining a shared, clean, healthy and safe tangible world within our changing social and physical geographies, regardless of the cognitive structures reshaped by gadgets and channels that are becoming ever more invasive into the human mind and body.

New, screenless devices are announced by AI giants at the time of writing so perhaps the subject of media may become obsolete as the medium may become the very body and the message may become literally—not metaphorically—equivalent with it, sooner than the fanciful wordplay here suggests.

We may soon require a qualitatively new upgrade to the ethical toolkit, just as was suggested in some of the more disquieting analyses presented in this selection of works.

## References

Straiton, J. (2023August 25). I know what you're thinking; can neuroimaging truly reveal our innermost thoughts?. *BioTechniques., 75(3),* pp. 81–83. Available from: https://www.tandfonline.com/doi/full/10.2144/btn-2023-0066

Todorovic, M. (2016). *Rethinking strategy for creative industry: Innovation and interaction.* Routledge Research.

Todorovic, M. (2024). 'Novel STEAM engines for the new industrial revolution: A new education paradigm'. *46 EBES Conference, Sapienza University 10/01/2024.*

Williams, R. (2025, May 1). *Brain scans can translate a person's thoughts into words.* MIT Technology Review: Biotechnology and Health. Available from: https://www.technologyreview.com/2023/05/01/1072471/brain-scans-can-translate-a-persons-thoughts-into-words/

**Milan Todorovic** is Senior Lecturer in Creative Industries at London Metropolitan University. He is also the Course Leader in Music Business and several legacy curricula in Media and Music Industries. He is Lead Doctoral Supervisor of top experts in their fields, such as Media Value + Supply Chains powered by AI, Leadership Management in Art Music and Culture, etc. Todorovic created unique study programmes, e.g. Business Modelling for Machine Learning in Streaming Media, and several strategic curricula. His scientific and artistic background combines with incisive analysis for multidisciplinary projects. He designs novel research methodologies converging sociocultural and scientific work. In a career before academia, he revolutionized a European capital's underground scene spanning electronic music, daring events, organized chaos, digital film, video and Fine Art. A renaissance man equally at ease with creative writing, laser physics and strategy, his films and artworks were shown globally. He advises art + music festivals on rebranding and strategic modelling, contributes to European Capital of Culture projects, and delivers consultancy for leading change. Milan Todorovic operates under artistic aliases for ***Sonic Painting*** projects he had singularly developed. A classically trained Fine Art painter, producer, and videographer, he is both versatile and clandestine in this field. Academic work forms part of his socially conscious mission. He wrote an influential multidisciplinary monograph on Creative Industry Strategies, to which he is preparing a sequel. To Milan all this forms a logically integrated continuum as he is a polymathic synaesthete whose innovative projects tend to break boundaries.

**Samuel O Idowu** is Professor of Accounting and Corporate Social Responsibility at the Guildhall School of Business and Law, London Metropolitan University, where he is currently the course leader for the MSc Corporate Social Responsibility & Sustainability and Advanced Diploma in Professional Development (ADPD) Corporate Social Responsibility & Sustainability. Samuel is also Professor of CSR and Sustainability at Nanjing University of Finance & Economics, China. He is a fellow member of the Chartered Governance Institute, a fellow of the Royal Society of Arts, a Liveryman of the Worshipful Company of Chartered Governance Institute, and a named freeman of the City of London. He is the CEO and President of the Global Corporate Governance Institute an international network of CSR scholars. Samuel has published over one hundred articles in both professional and academic journals and contributed chapters in several edited books and is the Editor-in-Chief of three major global reference books by Springer—the *Encyclopedia of Corporate Social Responsibility* (ECSR), the *Dictionary of Corporate Social Responsibility* (DCSR), and the *Encyclopaedia of Sustainable Management* (ESM), and he is Series Editor for Springer's CSR, Sustainability, Ethics and Governance books. Samuel is the Editor-in-Chief of the *International Journal of Corporate Social Responsibility* and the *American Journal of Economics and Business Administration.* Samuel has been in academia for more than 30 years winning one of the Highly Commended Awards of Emerald Literati Network Awards for Excellence in 2008 and 2014. In 2010, one of his edited books was placed in 18$^{th}$ position out of 40 top Sustainability books by Cambridge University Programme for Sustainability Leadership, and in 2016, one of his books won the outstanding Business Reference Book of the year of the American Library Association. In

2018, he won a CSR Leadership Award in Cologne, Germany, and in 2019, he won the 101 Most Impactful CSR Leaders Award in Mumbai, India. Samuel is on the Editorial Advisory Boards of the *International Journal of Business Administration*. He has been researching in the field of CSR since 1983 and has attended and presented papers at several national and international conferences and workshops on CSR. Samuel has made a number of keynote speeches at international conferences and workshops and written the *foreword* to a number of leading books in the field of CSR and Sustainable Development. And he has examined on a few PhD theses in the UK, Australia, South Africa, the Netherlands, and New Zealand.

**Silvia Puiu** is Associate Professor PhD Habil. at the University of Craiova, Faculty of Economics and Business Administration. She has 18 years of teaching and conducting research related to corporate social responsibility, management, and marketing, and in recent years, she has developed a high interest in studying topics related to the psychology of human resources. During 2018 and 2025, she was the president of a nongovernmental organization focused on nonformal education, and starting in 2024, she started offering workshops on emotional intelligence after she got her certification as a Personal Development Coach. At the end of 2024, she also got her certification as a Career Coach. In the last 10 years, she edited a magazine for students, a project started with passion and love for her students and her profession. Silvia Puiu appreciates that being a teacher and a mentor is the best thing that has happened to her, not able to imagine a different career path for her. In 2024, she started a new journey as a student at the Faculty of Psychology and Education Sciences in Brasov, Romania. Psychology has been a passion for her since high school, and recently, she realized that this path can help her become a better teacher for her students. Silvia Puiu published numerous research articles in journals indexed in Web of Science and also contributed to numerous Springer books on the vast domain of CSR. She also published three personal books, in part autobiographies and in part psychological and philosophical short stories, one of them published as an audiobook in which she used her voice.

# Media Technologies, Sustainability Challenges, Impact and Responsibility

# Impact of Internet Use, Digitalisation, Devices on Mental Health

**Mihaela Zăvăleanu, Amalia Raluca Stepan, Germina Alina Cosma, and Mihaela Tinca Udriştioiu**

**Abstract** Digital technology (DT), devices and Internet use are essential components of present life. These are used and will indeed find many uses in the future to facilitate many human activities. Consequently, many changes in human behaviour are expected, the most obvious and rapid being those reported after the pandemic lockdown.

A question that arises is whether people are prepared or aware of the risks related to DT use, especially those in the mental health (MH) area. Studies already talk about excessive use at the wrong age for the wrong purposes.

Many other possible questions and controversies will occur in the future as the human environments in the digital age will quantify the modification of human MH changes.

Obtaining only the beneficial effects and advantages of using DT is not easy. Is MH safety just at the user's personal level, or is it societal? The complexity of positive effects should be prevalent in long-term use. However, the novelty of behaviours generated by DT needs solid legal frameworks (cyber aggression, for example) that regulate these security changes to maintain, especially, the MH of users.

The prophylactic component was a good strategy, as observed and confirmed in the fight against other conditions. Awareness of risks and negative effects or disadvantages remains the main way to counteract the effects on MH and, at the same time, can enhance the positive effects of DT as an integral part of our lives in the era we live and will live.

M. Zăvăleanu (✉) · A. R. Stepan · G. A. Cosma
University of Craiova, Craiova, Romania
e-mail: mihaela.zavaleanu@edu.ucv.ro; amalia.stepan@edu.ucv.ro; germina.cosma@edu.ucv.ro

M. T. Udriştioiu
Department of Physics, Faculty of Sciences, University of Craiova, Craiova, Romania
e-mail: mtudristioiu@central.ucv.ro



M. Todorovic et al. (eds.), *Sustainability and Social Responsibility of the Media and in the Media*, CSR, Sustainability, Ethics & Governance,
https://doi.org/10.1007/978-3-032-00086-6_2

**Keywords** Addiction · Algorithm · Anxiety · Apps · Artificial intelligence · Culturally appropriate · Cyberbullying · Data security · Depression · Digitalisation · Digital literacy and strategy · Digital markers of health · Digital societies · Digital strategy · Digital technology · E-patients · Ethical standards · Excessive and long-term use · Emotional intelligence · Inappropriate content · Internet addiction · Internet use · Legal framework · Learning · Gore content · Medical ethics · Mental health · Prevention · Quality of life · Risk behaviour · Screen time · Stress · Technology · Technostress

## 1 Introduction

According to the World Health Organization website, ***mental health (MH)*** is represented by "*mental disorders and psychosocial disabilities as well as other mental states associated with significant distress, impairment in functioning, or risk of self-harm*", not just an absence of clinical symptoms or distress (Jenkins et al., 2011).

Good MH includes the ability to perceive, understand and interpret changes in the environment and adapt thoughts and behaviour (e.g. an ability to speak and communicate coherently with other individuals). The term "mental capital" is used to cover the conceptual understanding of what might be called the "*bank account of the mind*", comprising intellectual and emotional resources that can be built up, used, depleted or damaged deformed throughout life, and the *maintenance* of *good MH* is a "subtle process" that is influenced by multiple factors, existing in our life environment, acting from the moment of conception. We can control many of them, or we should learn to have control over or relate to them safely.

Research suggests that the human brain takes about 25 years to mature, as here differences can vary between men and women and between individuals. Although the human brain matures in size during adolescence, essential changes do not occur in the prefrontal cortex and other regions until the second decade of life. Brain maturation is influenced by heredity and environment, prenatal and postnatal insults, nutritional status, sleep patterns, pharmacotherapy and significant medical conditions (surgery) during early childhood.

In addition, physical, mental, economic and psychological stress, drug abuse (caffeine, nicotine and ethanol) and sex hormones, including oestrogen, progesterone and testosterone, influence the development and maturation of the adolescent brain (Arain et al., 2013).

During immaturity, the brain continually "learns" and "adapts-regulates" in response to the external (Lockwood Estrin & Bhavnani, 2019) and internal environment. The development and maturation of the brain's nervous structures must be especially stimulated in positive and supportive environments for healthy integrity of nervous system functioning and maintenance of MH. Neurological and/or psychiatric disorders occur when the maturation of neural circuits is disrupted, although, at the structural level, there are no pathological changes (Meyer & Lee, 2023; Tooley et al., 2021).

Factors influencing mental health are ***biological*** (genetic predisposition, diseases that alter blood circulation and substances that ensure proper functioning of brain cells), ***environmental*** *and* ***economic*** (family, friends, support persons, living conditions, living status, access to specialised treatments or like Moitra et al. underline in 2023 some global event like COVID-19 pandemic, geo-political events, and environmental change). It should also be taken into account how specific experiences arise during the period of greatest vulnerability, such as the period ***of growth and brain development*** (abuse, physical or psychological suffering or trauma, discrimination of any kind, stigmatisation). Some factors influencing MH are linked to the existence of health problems, such as chronic illnesses and the emotional changes they cause, chronic consumption of various substances (alcohol, tobacco, exposure to toxic or psychoactive substances, etc.), and how the person is exposed to these substances, inappropriate behaviours over long periods (overwork, failure to rest and sleep, lack of nutritional principles such as vitamins or minerals etc.) which can have an impact on MH. We can only deal with many of these personally, which can be considered an advantage or a disadvantage if we do not have the necessary information on health risks and proper resources.

Excepting the genetic predisposition, according to Morrison and Frank (2023), "public policy, social, environmental, or cultural norms" can modify factors regulating many life experiences with a certain degree of impact, which can exacerbate serious mental illness.

Comorbid mental disorders are a risk factor for decreased life expectancy for other conditions (Momen et al., 2022) and are the "substantial cause of death" worldwide (Walker et al., 2015).

In recent decades, neuroscience has focused on understanding the functioning and relationships between brain structures. Prospects for answering many questions in this field vary, Marsch (2021) considers "*digital health data-driven*" to be of great help in elucidating human behaviour through the emergence of "***digital markers of health/risk behaviour***" the ability to describe neural brain functions and structures have guided the development of neural computing technologies such as neuromorphic hardware and artificial intelligence (AI) that can be applied, including, to describe neural computational perturbations in mental health and or neurological disorders (Aimone & Parekh, 2023).

The Internet and digital devices have become a core part of the external environment for children, young adults and elderly, with different impacts on behaviours (individual and social). Human capacities like socialization, creativity, curiosity, critical thinking or metacognition have facilitated our species adaptation and survival.

In the late years, this new tools changes our reality facilitating many aspects of our life, being increasingly embraced and integrated more and more by all.

People decide to use Internet-enabled devices for quick contact with people they know, for social interactions in various forms (communication, information exchange, chat, web pages), for entertainment, to discover new information, entertainment, shopping, communication, leisure, in response to the characteristics of the human being to explore and discover new horizons.

It is also possible for people to want to do things virtually that they cannot do in real life (travel, experience other people's experiences, feel free, judge without being judged, experience freely without being constrained by other people's behaviours and feelings, feel a different kind of freedom in an environment they consider safe).

Nevertheless, many of these will arise as other discoveries present difficulties or challenges. Challenges due to new factors in people's lives manifest themselves positively and negatively. Possible changes in human behaviour may not yet be detectable—visible—knowable, perhaps only intuited by those who monitor the phenomenon on a small number of subjects.

However, a proactive expectation that risk awareness is more beneficial and safer, as has been proven suitable in therapeutic/medical approaches; *the principle of a* risk-aware *approach* seems the most plausible way to help treat TD-related conditions in the MH field.

## 2 The Impact of the Internet on Human Health

The Internet and its derivatives are ***a newly emerging*** but ubiquitous (***stressor***) ***factor***, especially overwhelmingly present in urban and highly developed areas.

Drawing a parallel to the problems caused by the 2019 pandemic, Ishfaq and Mengxing (2022) consider Internet use essential for survival. Many studies examine Internet use via digital devices (smartphones, computers) by age groups. Studies from various countries worldwide show that the number of people using the Internet is increasing.

It is intended to be used even more intensively through digitisation. The adaptation and use of this technology in all aspects of our lives will reveal both the positive and negative sides over time, especially the impact on human health. However, the complexity of the positive effects should be prevalent for a good MH of the DT user in the long run.

## 3 Medical Problems Associated with the Use of these Technologies

For those who overuse the Internet, it is called an addiction (*Internet addiction*—IA) and has been a recognised ***red flag*** since the advent of this technology.

There has also been an increase in the number of studies that have attempted to identify and quantify the adverse effects of its use. The occurrence is not only at the behavioural level. The negative effects can be physical and mental, similar to those associated with lack of physical activity.

Some of the behaviours associated with excessive and long-term use of the Internet or digital devices:

- Neglecting social responsibilities, relationships and typical age group activities.
- Withdrawal symptoms, social isolation.

- Mood shifts, sleep disturbances.
- Obsessive-compulsive disorder.
- Altered relationships with family members or social environment.
- Increased anxiety, depression, FOMO or "fear of missing out".
- Post-traumatic stress disorders.
- Suicidal ideation, death.
- Low concentration by attention deficits span etc (Fig. 1).

Among children and adolescents in particular, there is an addictive relationship between ***problematic Internet use*** and psychological and behavioural outcomes. The specialists underline some psychosocial symptoms like avoidance behaviour and aggression with an impact on life satisfaction due to the association of low physical activity behaviours (Pan, Lan, et al., 2024).

*Internet addiction is a new pattern of behaviour and is not (yet) a diagnosis*, but it is considered a risk factor that can be quantified by apps that can provide daily reports of digital device use. There are also assessment scales that quantify abusive use behaviours (e.g. "*Compulsive Internet Use Scale—CIUS*" developed by Meerkerk et al., 2009, "The Problematic and Risky Internet Use Screening Scale—PRIUSS" developed by Jelenchick et al., 2014, "Coping with Cyberbullying Questionnaire" by Mcguckin, 2015, "Scale on Coping with Cyber Bullying" by Peker et al., 2015, "Problematic use of Mobile phones scale" by Merlo et al., 2013). Harris et al. in 2020 find 78 validated scales related to the smartphone use.

Those working in the medical field should identify it as a risk factor (for the problems mentioned above) and apply therapeutic interventions according to this aspect as well to mitigate all health risks.

For example, a new behaviour, ***nomophobia***, can be developed unconsciously by any individual. The most important aspect of any addiction is the awareness of the person who is experiencing the specific symptoms of excessive or inappropriate, unhealthy use.

Bouna-Pyrrou et al. (2015) considered that a higher Internet use of ≥30 h/week spent on the Internet could be classified as pathological.

Heavy use of these devices is the other side of the coin regarding beneficial effects.

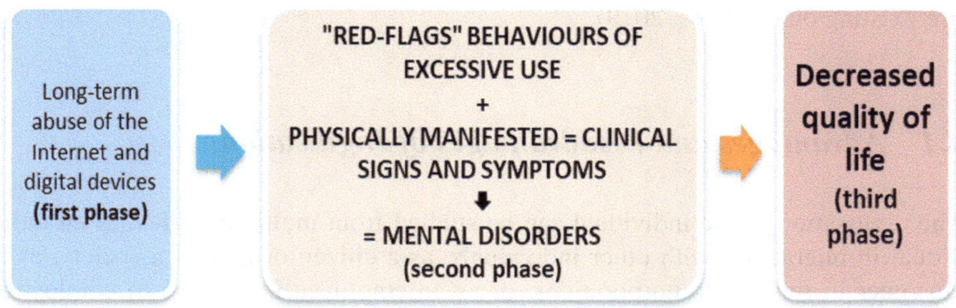


**Fig. 1** Long-term or intensive abuse of the use of the Internet or digital devices can lead to behaviours manifested in health

# 4 The Impact of Digitisation on Human Behaviour

It is clear that the exponential growth of digitisation has changed many human behaviours and activities irreversibly, and the changes are constantly changing as the pace of innovation is also particularly accelerated. It thus becomes necessary to understand the dynamics between technology use and society to maximise the positive effects of digitisation and minimise as many of its negative effects as possible (Serrano-Santoyo & Cabrera-Flores, 2018). Positive effects should be prevalent for good DT user MH in the long term. What is clear is that there are not only positive effects.

In the digital age we are in, what defines the human being and the set of values and behaviours, in parallel with the emergence of artificial intelligence (AI), is the subject brought up in many studies, media debates, etc. With the emergence of ***digital societies***, implicitly new emerging intra- and inter-human relationships, new systems of thinking, are and will be redefined, depending on the understanding and assimilation of these technological innovations (Nikitenko, 2019), especially as they start to be used by an even more significant number of individuals.

However, can we still talk about excessive use of the Internet and digital devices becoming a necessity and an unhealthy habit when children, teenagers and young learners are being guided by their teachers to use these more and more frequently to solve scholar-work, learn or study? For example, digital libraries are much more accessible and "easy to visit", just a click away. Books are digital, and with a device (tablet), you can carry many of them in a pocket. Knowledge travels through these new devices, and the knowledge bases are increasingly available.

The young generation in highly developed countries will no longer know how to do many activities without using AI, Internet, DD (digital device) and DT. They will be an integral part of all their life experiences and activities.

Concerning the impact of Information and Communication Technology on society in general and education in particular, some researchers point out that the focus is beginning to shift from initial optimism to the opposite pole, with social, financial, accessibility and affordability differences being cited through lack of literacy, availability and understanding of technology by all those involved in the same process, such as education, to the extent that the digital revolution does not solve previous needs, but instead brings other barriers, which can sometimes only exacerbate pre-existing ones (Hatos, 2019).

## *4.1 Advantages and Disadvantages of Digitisation*

The digitisation of the individual can be studied from multiple angles: as an individual in interaction with other individuals, as a citizen or part of a society, as a customer in relation to corporations, or as an employee in relation to employers (Matt et al., 2019).

We want to highlight the impact of the use of technological devices on the individual, especially in terms of the level of personal use.

These include hedonic technologies, social networks on different topics or subjects, virtual reality and many others that bring today more people together which generate new behaviours acquired as a result of digitisation, especially at the individual level. Researchers bring up the behavioural visibility that has arisen due to algorithms shaped by the emergence of artificial intelligence (Leonardi & Treem, 2020). Generative artificial intelligence is still a new field. Many look at it with significant reservations and high expectations. Others look at generative artificial intelligence even with fear.

The facets of the problem are multiple, and it is difficult to state a single outcome of this equation represented by the individual in association with the Internet, digital devices and mental health.

In association with digitisation, Turel, in 2018, in ***negative terms***, mentions the emergence of symptoms of overwork, decreased privacy and personal security, the emergence of negative emotions, addiction, increased stress levels, and in ***positive terms***, identifies increased motivation and quality of life, even public health, even activation of neural networks in various diseases (Turel et al., 2018).

To the extent that everyday activities and the use of technological devices tend to overlap as much as possible, there are effects ***even at times of non-use***, both positive (such as a sense of security) and negative (recording of personal data, cyberattacks) (Matt et al., 2019).

The recent editorial by Pan, Lan, et al. (2024); Pan, Teng, et al. (2024) discusses the acute need to define and understand the relationship between Internet use and "psychological well-being" at three levels for children: individual, family and school environment. So-called "*digital resilience*" depends on several factors statistically associated positively with digital literacy in school, and perhaps the unexpected result is that family monitoring and well-being do not produce the expected coping strategies. However, teachers must also be subject to interventions or extended studies regarding their own MH challenges related to the DT (Costardi et al., 2023). They have a double role: DT users and assure student support.

Digital literacy by knowledgeable people can reduce the risk of exposure to ***gore content***. Some studies show that as a result of these exposures, some changes in behaviours and social norms, such as decreased empathy and sensitivity to tragic events, fear and anxiety, hostility or behavioural violence. Huesmann (2007) explains these patterns by saying that "media" impacts children's learning, mainly by imitating observed behaviours in the environment.

Although previous studies have said that parents need to impose time limits to help boost children's ICT use outcomes to limit creativity and not increase anxiety levels (Milenkova & Manov, 2020), parents also have a role to play in setting an example and inspiring children to use devices (Vevera & Albescu, 2019). It is also parents who can be the first to identify addictive or withdrawal behaviour in a child who is unhealthily using a device if they are provided with appropriate information or education, to this end. If a child becomes inattentive, has low self-esteem, is hyperactive, tired, socialising less and is only concerned with using a device, they should be watched by adults for possible addiction. Digital parenting is also a key factor for "*promoting emotional intelligence and self-control as well as parenting qualities to reduce problematic Internet use in children and adolescents*" (Pan, Lan, et al., 2024).

Last but not least, we believe that adequately calibrated and culturally adapted studies on children's dependence on the Internet and digital devices can shed light on the exponential increase in recent times of ADHD-type pathologies, with their specific reactions, which may be signs of a distorted perception of reality and the social microclimate of the family or school, misinterpreted or superficially by medical staff.

Shikwambi & Angula wanted to identify how digitisation changes students' behaviour; 60% of the subjects they analysed positively appreciated the effects of using technology, while 40% of them disagreed, concluding the authors' recommendation to keep technology as up-to-date as possible so that young people's interest and performance will be some increased (Shikwambi & Angula, 2021).

Another sequence of the impact of digitisation that needs to be studied concerns older people, some of whom have tried to accommodate themselves to digital technologies and the Internet, becoming addicted users, and the effects on their health have not yet been sufficiently studied due to the relatively short time of exposure. While on the motor side addiction can pose problems, it is possible that from an MH perspective, we may have beneficial surprises in terms of maintaining mental and neurological function and preventing degenerative diseases (including Alzheimer's, and Parkinson's), as those who use the Internet more can benefit from exposure to stimulation and easy ways of therapy. Perhaps studies in this area will bring the expected clarification.

Goodday et al. (2023) highlighted the need for "*digital navigators*" to solve the problems of using digital technology in health, especially for the elderly.

Devices are very popular among children, and kids are online more than ever. Sometimes, a child starts using a digital device without being aware of the dangers in the virtual world, leaving a mark on the mind in a way similar to that of those in the real world.

There is talk of "***online predators, cyberbullying, inappropriate content or phishing attempts***" or other dangers that can lead to psychological trauma and MH damage in the face of traumatic experiences on the brain and healthy behaviours that are still forming in the child, which may be altered or modified in adult or older adults. Morea and Calvete (2022) discuss this topic and cognitive flexibility as a protective factor to prevent, mitigate and even treat the effects that occur at the psychological level following online aggression and trauma.

For this reason, awareness-raising interventions and rules of use at the end-user level (child-parent) presented by competent bodies (such as schools) can be key elements for healthy use, to the extent that such concerns should be often or more often the subject of analytical programmes in the education system.

## 5 Mental Health and Digital Technology

Zheng et al. (2023) point out that there has been a great emphasis on the positive effects of the Internet, but also those such as "*loneliness, sadness, and life* hardship need to be known". People need to be made aware and guided to use digital devices with Internet access, with all the facilities they offer that are very attractive or exciting, appropriately to reduce the negative effects and have only life-enhancing effects.

The benefits of use when mental problems occur are still found, according to Zheng et al. (2023), in only a third of users.

Digitisation and mental health issues such as anxiety and depression were the subject of big data **linguistic** research using Google Books Ngram Viewer on eight million books written in six languages, looking at the prevalence and associations between the three concepts.

In the last 50 years, the frequency of associated and correlated words like "digitisation, depression and anxiety" has increased (Teepe et al., 2023).

The review article by Cai et al. (2023) reviewed 223 studies that analysed data from 498,167 participants to determine the effects of Internet use on aspects of MH. As the authors also pointed out, the association of symptoms characteristic of those conditions, such as *depression or anxiety,* with various other mental health problems and overall altered well-being (Cai et al., 2023) can be a key factor in clinical diagnosis and treatment management. Xie et al. (2023) pointed out that "anxiety predicts internet addiction, which predicts depression". Management intervention would have better results, according to them, if it targeted anxiety, which affects the brain (functional and organic) and changes behaviours and human activity.

Among the new types of anxiety that people face, ***environmental anxiety has been*** described in recent literature, the effects of which could be countered by providing support services through digital technology (Yellowlees, 2022).

***Technostress*** is defined as stress caused by direct or indirect interaction with the use of ICT (*information and communication technology*). One age group that is more likely to experience more significant consequences of this type of stress is adolescents, who are in the process of self-discovery and defining their personality and personal identity. A quantitative and qualitative study conducted with the responses of 30 adolescents aged 10–17 has captured five factors as coping modalities used by them, including avoidance, use, conscious assimilation of knowledge, following the rules, and obtaining the expected satisfaction. The findings were that coping strategies increase with increasing intensity of the technostress experienced, but there are also situations where these strategies are insufficient (Schmidt et al., 2021).

Young people's mental health problems and their link to the digitalisation of education are the subject of another research review published in 2021. The study drew attention to deteriorating mental health among students as a result of taking courses exclusively online. Among the best practices that could help could be a combination of positive psychology concepts with high-tech applications, which could be made available to young people through apps on mobile devices or specific websites (Shilko et al., 2021).

Depending on how people use social media, it is recommended that interaction on digital social networks, especially through likes, comments, followers, media and posts, “should be done with caution due to *their significant negative effect* on mental health”, according to Beyari (2023).

## *5.1 Digitising Therapy for Mental Disorders*

The World Psychiatric Association and the World Health Organization are concerned about diversifying intervention modalities through digital means. A survey they conducted involved 145 partner entities and looked at the impact of digital health globally by identifying changes in mental health services. Such surveys underpin action plans aiming for better collaboration between professionals, greater coverage, and equal access to such services (Ramalho et al., 2023).

The digitisation of therapy has a positive effect on beliefs about mental health. It shows that despite the difficulty or impossibility of physically reaching a specialist, psychosocial support can still be offered, stress and depression levels can be reduced, and people’s quality of life increased by accessing these services, but in a digital format (Johar et al., 2023).

Technological novelties can solve growing psychological vulnerabilities globally (Sun et al., 2022). Digital interventions for subjects with mental disorders can improve access to specialist treatment (Lecce et al., 2023; Kalman et al., 2023) and the quality of specialist treatments and methods of study, assessment and monitoring (Kalman et al., 2023).

In the direction of access to mental health services in disadvantaged and poor areas, which are deprived even of qualified medical staff (for them, the equity of provision will increase), the importance of the presence of digital technologies is stressed, but the same problems, which digitisation also raises, remain in question about access to the Internet, openness of the people involved (patients and specialists), data security, medical ethics, language skills, level of education.

Thus, it is impossible to discuss replacing traditional medical services with digital ones. They could, however, facilitate access, through technology, to some services in some areas and to some patients, where otherwise, through physical, one-to-one presence, these consultations would not be possible (Balikuddembe & Reinhardt, 2020) or perhaps partly, in a hybrid way can reduce costs associated with the therapeutic process (and for the patient and social systems).

A team of authors has presented a best practice example of digitising mental health services through remote interventions for 200,000 people isolated following a humanitarian disaster in the northern province of Mozambique. They achieved good results by harnessing digital technology and delegating tasks, for example, to non-specialists who were thus able to provide mental health support and care (Mootz et al., 2024).

Stern et al. (2023) believe that digital psychiatry decreases stigma, is more convenient and readily available, and can produce an objective scientific knowledge base for prevention. It can be used without geographical boundaries, constraints, or limits and reduces infrastructure resources (Alenoghena et al., 2023), which are key factors for today's healthcare field.

Current studies underline the importance of the changes that have become necessary during the COVID-19 pandemic, which has also increased digitisation in the field of psychiatric services in the direction of online consultations and hybrid healthcare. Lucey (2023) signals that the ***era of twenty-first-century medicine*** has begun, which has highlighted and brought to light that pressing need, the need also signalled by the World Health Organization years ago to *plan healthcare* for the next generation who will face a shortage of skilled workforce and functional or sustainable healthcare models.

Li (2023) recognises the types of innovative interventions in MH: "*web-based, mobile-based, and external device-based interventions*". The role and influence of these technologies as a disruptive force should not be diminished among those who provide these services. Not only should its positive impact on workplace well-being be emphasised, but both aspects should be understood, as Sun et al. (2022) said.

As early as 2009, Giveon et al. signalled the emergence of a new type of patient, "***e-patients***", who use the Internet and have transformed the patient-doctor relationship by empowering patients with readily available information. Sometimes, this information is misunderstood by "***e-patients***" and may instead increase therapeutic risks (Giveon et al., 2009) if they believe they can understand and follow therapeutic indications from unreliable or uncontrolled Internet sources. Deepfake phenomena can exist in any field. Perhaps even this appetite for the Internet should be exploited and used for self-care by implementing projects that provide qualified and especially safe information with beneficial effects on MH and improving the quality of life of people using this approach.

Park and Kwon (2018) also review the various concepts and patterns that have emerged through the use of the Internet for health purposes, such as "*electronic health literacy, health information-seeking behaviour, electronic health promotion, electronic health intervention, electronic mental health, help-seeking (online help-seeking), e-patient*".

The ubiquity of digitisation and the accessibility it implies is undeniable. Its involvement in mental health services, through digital innovations such as neural dialogue systems or emotional intelligence, can bring new possibilities by facilitating the flow of information and new concerns and limits (Balikuddembe & Reinhardt, 2020). The results of their possible use will be quantified by experimentation by most of the planet's inhabitants.

## 5.2 Mobile Apps for Mental Health Therapy

Parker et al. (2018) conveyed two essential messages, namely that poor or fragile mental health is pervasive and that individuals can easily manage their mental health issues with apps, which was later demonstrated.

Litke et al. (2023) pointed out that the use of mobile technology has increased and made therapies much more available than a decade ago, especially for young people who are the heaviest users of digital devices, but there is little scientifically sound data on the resulting effects.

Eder et al. (2023) present The European Psychiatric Association Expert Group's idea on the need for rules and laws to protect medical data that are available electronically, with the conservation of medical human capital resources. Apps are considered vital elements to ***fill the gaps*** in mental disorder care, and the results of their use in therapy for various MH problems belong to the future (Guracho et al., 2023), possibly not very distant due to the increase in such services.

According to Litke et al. (2023), health apps target a wide range of mental health challenges in young people (e.g. depression, self-injury, autism spectrum disorder, anxiety and obsessive-compulsive disorder), but one problem reported is that the apps do not appear to be *culturally appropriate.* In interpreting, the same messages may generate different expected outcomes in a different cultural paradigm. At some point, with the evolution of digitisation, generations growing up and being educated under the same technological auspices may interpret and react similarly to stimuli from the Internet or apps, according to the expected outcomes of the developers of these technologies.

The problem of the world's ageing population is another challenge in providing resources for mental healthcare in adults over 65 years of age, in whom the processes of degeneration of the neurological system constitute a significant challenge. For them, there is a need for an update of sound scientific knowledge demonstrating the effectiveness of therapeutic digital applications. Chou et al. (2023) identified elderly interaction with chatbots as an element that can improve well-being and reduce stress or depressive symptoms. However, there remains the element of openness to a new therapeutic approach and communication unfamiliar to today's seniors but possibly familiar to those over a decade older who are more familiar with these devices.

Lin et al. (2023) recommended 18 ***conversational agents—CA*** used with apps to treat problems such as stress, anxiety and depression. These address more than one mental disorder through cognitive behavioural therapies that aim to improve symptom self-management and mental disorders with fewer therapeutic elements and resources.

Tani et al. (2024) have classified 11 categories of technologies "*for mental health interventions: heart rate estimation, exercise or physical activity, sleep estimation, contactless heart rate/pulse wave estimation, voice and emotion analysis, self-care/ cognitive behavioural therapy/mindfulness, dietary management, psychological safety, communication robots, avatar/metaverse devices, and brain wave devices*".

And while diagnosis by "***digital phenotyping***" is important according to them many other technologies may become more widespread (Tani et al., 2024).

Ethical standards for application and use also need to be expanded (Cheng et al., 2023) as users grow in number. These apps should provide users with security (of data and in case of therapeutic emergencies) and quality information by using quality scientific data on which artificial intelligence will respond effectively to mental health (Lin et al., 2023).

Many apps can support users who want to change unhealthy behaviours (for example, Bold et al., 2023, studied those related to smoking cessation). There is much digital information available, in accessible formats, presenting knowledge and ways to adopt a healthy lifestyle, available to all by paying for an Internet subscription or by paying the additional cost of pages that require creating individual accounts. However, what is important to point out is that few apps offer support based on solid scientific evidence. There is no way of verifying much of the information, so the danger and challenge for a ***"home user"*** arises.

Dhirani et al. (2023) bring into the debate ethical standards that are being developed, and the lack or gaps in this area may lead to ethical violations and possible ethical dilemmas related to new technologies. Reluctance due to fear of fraud or fear of a device that works differently than the previous ones can be observed, especially in the elderly and may be one of the issues that decrease the degree of trust more than the qualitative value of a therapeutic application.

***A European report on digitisation*** proposes that the approach to risks related to the use of digital devices should be based on ethical principles and the same principles of respect for human rights, concepts that should not be missing from all regulations of online activities.

If these principles have been validated in human activities that do not take place using the Internet, it would be natural that the same ethical rules and principles that preserve the safety of the individual should also apply to activities in the virtual environment. Extensive regulations are needed to modulate use and ensure greater safety without the emergence of mental health problems. Recent discussions from various sources and social or political bodies mention concerns about Internet and digital technology use because some negative effects associated with use have begun to be more visible in younger users' behaviours (especially). These behaviours no longer fit within the statistical boundaries prior to the expansion of use. Different statistical values have emerged and are likely to continue to emerge.

## 6 Discussions

The world without the Internet was not a world without any form of addictions, but rather one with different behaviours and different social interactions, slower. Communication was more face-to-face (direct communication, gatherings of people or events in more localized social circles), or the dissemination of information was done at a slower pace (through books, physical correspondence or mail, newspapers, landline telephone, radio, television), the media was dominated by radio or television, entertainment was primarily based on the physical component and presence. Planning and patience were also behaviours necessary for the accessibility to entertainment, the gratification was somehow delayed, and the constant stimulation does not exist. Social circles were often confined to the neighbourhoods' places, with stronger bonds between community members.

Geographic proximity was the key factor that limited access in many aspects, polarizing knowledge and information, any type of expression or socialisation etc. Pre-Internet world had its own set of challenges and problems, but it was fundamentally different from the digital world we inhabit and definitively fundamentally different from the one the younger of us will experience.

The complexity of the shift with both positive and negative changes it's essential to be considering in the complexity of opportunities and challenges.

## 6.1 Rules for Avoiding the Dangers of Using the Internet

When someone buys a device, they do not take medical courses explaining the dangers. They are sometimes just the subject of online media stories that do not reach the user. Alternatively, they arrive via a source like a sibling, friend, or parent.

Teachers should also be, together with parents, the main advocates for safety awareness because they require students to use devices extensively. Teachers have the opportunity to keep the balance in favour of the benefits and advantages of DT in the long term because they can regulate or control the time of use to solve homework, for example, and create healthy usage behaviours.

One possible solution may be to use parental control features on devices and apps or secure devices to limit access to age-inappropriate content or monitor online activity and interactions with other virtual users to protect against malware and viruses.

In terms of education and knowledge, users (especially children) need to be educated and warned about the risks to their health present online. They need to become aware of the importance of privacy and personal information, how they can be affected by those who have this information illicitly, identify phishing attempts, or the credibility of the information.

It is particularly important to set time limits and to use methods to quantify the time spent using a digital device, apps and online activities so that they do not only have adverse effects (addiction, altered behaviour, physical disorders, etc.).

Open communication between parents, teachers, and children and promoting an environment where they can discuss their online experiences and concerns can be another way to improve DT usage.

Ensure an adequate legal framework to combat cyberbullying by educating (especially children) with healthy moral and social values that apply online, just as in the real world.

Psychoeducation on screen time in adolescents should be a focus of interventions in adult mental well-being (Wiguna et al., 2023).

### 6.2 Ways to Maintain Good Mental Health

Good or positive mental health can be maintained by actively involving the individual in the following types of activities: socialising, sports, recreational activities, hobbies, volunteering or helping others, getting more in touch with nature, sunbathing, activities that provide mental relaxation or mindfulness such as yoga, meditation, breathing exercises, psychotherapy, and many others. Combining these activities can be an even better solution, for example, systematically and regularly practising a sport or movement activity as a team outdoors.

Also recommended to support good mental health is the development of resilience-building methods to help people cope with stress or unforeseen situations, real or virtual, by identifying solutions, support, or help appropriate to the crisis.

Maintaining optimal physical condition through nutrition, a healthy sleep schedule, proper hydration, avoiding traumatic risks, and avoiding risk factors for the development of non-communicable diseases is in line with the saying "mens sana in corpore sano", which has been the basis of medical therapies and has brought into question the importance of the new holistic concept.

In some cases, maintaining mental health involves the individual level by knowing and being aware of the risk, permanently giving up or avoiding for certain periods those activities that induce high stress, or giving up those relationships that affect mental health and decrease well-being.

All these modalities can be used as prevention or (adjuvant) therapy for ***mental disorders associated or not with Internet use, digitalisation and digital devices*** by any of us as individuals responsible for our long-term health.

## 7 Conclusions

Indeed, new methods, devices, and action directions will be applied in the mental health field. Some modern technology facilities may even surpass traditional practices in low-resource settings regarding outcomes, diagnosis, therapies, and treatment, either with patients who prefer digital distance or cannot be physically present.

Electronic health records can provide specialists with large amounts of highly accessible data, reducing bureaucracy and intervention time, especially in emergencies.

The controversy and the question that arises from this is whether people are prepared, can intuit or already know all the changes that this DT has produced and those that it may produce, especially those in the MH area. Nevertheless, tracking, understanding, and studying the relationship between Internet use and mental health through medical studies would help generalise and generate data on especially beneficial behaviours in this area.

The novelty of DT-generated behaviours requires new definitions and legal frameworks (cyber aggression, for example) and the adaptation of the bodies that regulate these changes for security to create mechanisms that provide security for users.

New health problems need standardised therapies for standardised diagnoses.

Everyone can become a victim of using Internet and DT like any other device or human activity. For some of us, the negative effects are more extensive than others, sometimes due to age or other personal factors. The most significant risk is among new users (especially those of vulnerable ages) if they are unaware of the risks and the focus is only on the beneficial effects or outcomes of using a device. The emergence of psychiatric disorders is directly linked to excessive use of DT, at inappropriate ages for inappropriate purposes, with new risks of which the user has not been made aware.

Is safety just a personal user issue, or is it a societal one? Who can keep the balance?

Usually, when a new digital device is introduced, people only talk about its technical features, compare it with other versions, and point out new improvements and what it does better than its predecessor. Risks and impacts on behaviour are not topics that are addressed very often.

The balance held only in favour of the beneficial effects and advantages of DT is very unstable, as many research studies on human behaviour point out, especially those of the last 3–4 years with many subjects.

Complexity of positive effects should be prevalent for good long-term DT user MH. Many other questions and controversies will possibly be answered as the time of human coexistence in the digital age quantifies the changes in human MH as has happened with other factors that we now know for sure are detrimental to health.

"Physical therapy of the mind" may be a benefit of controlled digitisation for the elderly, and the Internet may be the new "rebus", the completion of which brings the expected benefits to the MH of these users.

Perhaps even from childhood, learning and adapting the brain to this new digital environment would be advantageous.

Awareness of risks and negative effects, disadvantages, and publicity among the population remain the main way to counteract these negative effects on mental health. The prophylactic component through awareness apps to be installed on new devices for users can be a good strategy, as observed and confirmed in the fight against other conditions such as infectious diseases or addictive behaviours.

Maintaining and discovering the positive effects of digital technology on beneficial (time) values as an integral part of our lives in the current era is an ongoing process; digitisation will be increasingly present.

Knowledge transfer and management of the challenges of this process, such as the uncertainty of Internet and DT use policies, underlines the need for proactive risk awareness to protect oneself from these challenges.

Mental health was primarily focused on traditional forms of mental illness, digital technology help professionals to address the new issues form Internet addictions, public awareness and research of the potentials risks should grow and strategies for prevention and treating through multifaceted approach are desirable.

The recognised and quantify impact can lead to effective interventions, policy and regulation can be helpful for the vulnerable and control technology production, tools and platforms that promote mental wellness promoting a healthier balance between digital and physical activities can be a solvability intervention.

But the future of digital tools and mental health is not already determined, we need proactive and steady interventions to shape and promote the digital landscape that we desire with less potential harms for our mental health.

## References


Aimone, J. B., & Parekh, O. (2023). The brain's unique take on algorithms. *Nature Communications, 14*(1), 4910. https://doi.org/10.1038/s41467-023-40535-z

Alenoghena, C. O., Ohize, H. O., Adejo, A. O., Onumanyi, A. J., Ohihoin, E. E., Balarabe, A. I., et al. (2023). Telemedicine: A survey of telecommunication technologies, developments, and challenges. *Journal of Sensor and Actuator Networks, 12*(2), 20.

Arain, M., Haque, M., Johal, L., Mathur, P., Nel, W., Rais, A., Sandhu, R., & Sharma, S. (2013). Maturation of the adolescent brain. *Neuropsychiatric Disease and Treatment, 9*, 449–461. https://doi.org/10.2147/NDT.S39776

Balikuddembe, J. K., & Reinhardt, J. D. (2020). Can digitisation of health care help low-resourced countries provide better community-based rehabilitation services? *Physical Therapy, 100*(2), 217–224.

Beyari, H. (2023). The relationship between social media and the increase in mental health problems. *International Journal of Environmental Research and Public Health, 20*(3), 2383. https://doi.org/10.3390/ijerph20032383

Bold, K. W., Garrison, K. A., DeLucia, A., Horvath, M., Nguyen, M., Camacho, E., & Torous, J. (2023). Smartphone apps for smoking cessation: Systematic framework for app review and analysis. *Journal of Medical Internet Research, 25*, e45183. https://doi.org/10.2196/45183

Bouna-Pyrrou, P., Mühle, C., Kornhuber, J., & Lenz, B. (2015). Internet gaming disorder, social network disorder and laterality: Handedness relates to pathological use of social networks. *Journal of neural transmission (Vienna, Austria: 1996), 122*(8), 1187–1196. https://doi.org/10.1007/s00702-014-1361-5

Cai, Z., Mao, P., Wang, Z., et al. (2023). Associations between problematic internet use and mental health outcomes of students: A meta-analytic review. *Adolescent Res Rev, 8*, 45–62. https://doi.org/10.1007/s40894-022-00201-9

Cheng, S. W., Chang, C. W., Chang, W. J., Wang, H. W., Liang, C. S., Kishimoto, T., Chang, J. P., Kuo, J. S., & Su, K. P. (2023). The now and future of ChatGPT and GPT in psychiatry. *Psychiatry and Clinical Neurosciences, 77*(11), 592–596. https://doi.org/10.1111/pcn.13588

Chou, Y. H., Lin, C., Lee, S. H., Chang Chien, Y. W., & Cheng, L. C. (2023). Potential Mobile health applications for improving the mental health of the elderly: A systematic review. *Clinical Interventions in Aging, 18*, 1523–1534. https://doi.org/10.2147/CIA.S410396

Costardi, C. G., D'agostini, A. C. C., Pan, P. M., & Bressan, R. A. (2023). Digital mental health interventions for school teachers–a narrative review. *Early Intervention in Psychiatry, 17*(8), 749–758. https://doi.org/10.1111/eip.13431

Dhirani, L. L., Mukhtiar, N., Chowdhry, B. S., & Newe, T. (2023). Ethical dilemmas and privacy issues in emerging technologies: A review. *Sensors (Basel, Switzerland), 23*(3), 1151. https://doi.org/10.3390/s23031151

Eder, J., Dom, G., Gorwood, P., Kärkkäinen, H., Decraene, A., Kumpf, U., Beezhold, J., Samochowiec, J., Kurimay, T., Gaebel, W., De Picker, L., & Falkai, P. (2023). Improving mental health care in depression: A call for action. *European Psychiatry: The Journal of the Association of European Psychiatrists, 66*(1), e65. https://doi.org/10.1192/j.eurpsy.2023.2434

Giveon, S., Yaphe, J., Hekselman, I., Mahamid, S., & Hermoni, D. (2009). The e-patient: A survey of israeli primary care physicians' responses to patients' use of online information during the consultation. *The Israel Medical Association journal: IMAJ, 11*(9), 537–541.

Goodday, S. M., Karlin, D. R., & Friend, S. H. (2023). The digital redesign of mental health: Leveraging connected digital technologies for agency-driven patient-focused care. *The British Journal of Psychiatry: the Journal of Mental Science, 222*(2), 51–53. https://doi.org/10.1192/bjp.2022.152

Guracho, Y. D., Thomas, S. J., & Win, K. T. (2023). Smartphone application use patterns for mental health disorders: A systematic literature review and meta-analysis. *International Journal of Medical Informatics, 179*, 105217. https://doi.org/10.1016/j.ijmedinf.2023.105217

Harris, B., Regan, T., Schueler, J., & Fields, S. A. (2020). Problematic Mobile phone and smartphone use scales: A systematic review. *Frontiers in Psychology, 11*, 672. https://doi.org/10.3389/fpsyg.2020.00672

Hatos, A. (2019). The impact of digitalisation on educational achievement: A literature review from a sociological perspective. *Calitatea vieții, 30*(1), 3–16.

Huesmann, L. R. (2007). The impact of electronic media violence: Scientific theory and research. *The Journal of adolescent health: official publication of the Society for Adolescent Medicine, 41*(6 Suppl 1), S6–S13. https://doi.org/10.1016/j.jadohealth.2007.09.005

Ishfaq, N., & Mengxing, H. (2022). Consumer usage behavior of internet-based services (IBS) in Pakistan during COVID-19 crisis from the perspective of technology acceptance model. *Environmental Science and Pollution Research International, 29*(57), 85632–85647. https://doi.org/10.1007/s11356-021-15868-1

Jelenchick, L. A., Eickhoff, J., Christakis, D. A., Brown, R. L., Zhang, C., Benson, M., & Moreno, M. A. (2014). The problematic and risky internet use screening scale (PRIUSS) for adolescents and Young adults: Scale development and refinement. *Computers in Human Behavior, 35*. https://doi.org/10.1016/j.chb.2014.01.035

Jenkins, R., Baingana, F., Ahmad, R., McDaid, D., & Atun, R. (2011). Mental health and the global agenda: Core conceptual issues. *Mental Health in Family Medicine, 8*(2), 69–82.

Johar, S. S., Suhaimy, K. A. M., Ismail, F., Ibrahim, R. Z. A. R., & Saputra, J. (2023). Digitisation of mental health therapy in the era and post-pandemic COVID-19. In *Handbook of research on dissecting and dismantling occupational stress in modern organizations* (pp. 55–69). IGI Global.

Kalman, J. L., Burkhardt, G., Samochowiec, J., Gebhard, C., Dom, G., John, M., Kilic, O., Kurimay, T., Lien, L., Schouler-Ocak, M., Vidal, D. P., Wiser, J., Gaebel, W., Volpe, U., & Falkai, P. (2023). Digitalising mental health care: Practical recommendations from the European psychiatric association. *European Psychiatry: The Journal of the Association of European Psychiatrists, 67*(1), e4. https://doi.org/10.1192/j.eurpsy.2023.2466

Lecce, F., Smith, C. R., & Burbach, F. R. (2023). Digital mental health interventions for people with epilepsy: A systematic review. *Seizure, 107*, 91–103. https://doi.org/10.1016/j.seizure.2023.03.016

Leonardi, P. M., & Treem, J. W. (2020). Behavioral visibility: A new paradigm for organisation studies in the age of digitisation, digitalisation, and datafication. *Organization Studies, 41*(12), 1601–1625.

Li, J. (2023). Digital technologies for mental health improvements in the COVID-19 pandemic: A scoping review. *BMC Public Health, 23*(1), 413. https://doi.org/10.1186/s12889-023-15302-w

Lin, X., Martinengo, L., Jabir, A. I., Ho, A. H. Y., Car, J., Atun, R., & Tudor Car, L. (2023). Scope, characteristics, behavior change techniques, and quality of conversational agents for mental health and Well-being: Systematic assessment of apps. *Journal of Medical Internet Research, 25*, e45984. https://doi.org/10.2196/45984

Litke, S. G., Resnikoff, A., Anil, A., Montgomery, M., Matta, R., Huh-Yoo, J., & Daly, B. P. (2023). Mobile Technologies for Supporting Mental Health in youths: Scoping review of effectiveness, limitations, and inclusivity. *JMIR mental health, 10*, e46949. https://doi.org/10.2196/46949

Lockwood Estrin, G., & Bhavnani, S. (2019). Brain Development. *Structure*. https://doi.org/10.1016/B978-0-12-809324-5.23776-0

Lucey, J. V. (2023). A public health perspective on mental health: Lessons for population health in the 21st century. *Irish Journal of Psychological Medicine, 40*(1), 2–5. https://doi.org/10.1017/ipm.2021.44

Marsch, L. A. (2021). Digital health data-driven approaches to understand human behavior. *Neuropsychopharmacology: official publication of the American College of Neuropsychopharmacology, 46*(1), 191–196. https://doi.org/10.1038/s41386-020-0761-5

Matt, C., Trenz, M., Cheung, C. M., & Turel, O. (2019). The digitisation of the individual: Conceptual foundations and opportunities for research. *Electronic Markets, 29*, 315–322.

Mcguckin, C. (2015). *The coping with cyberbullying questionnaire (CWCBQ): The development of a new measure*.

Meerkerk, G. J., Van Den Eijnden, R. J., Vermulst, A. A., & Garretsen, H. F. (2009). The compulsive internet use scale (CIUS): Some psychometric properties. *Cyberpsychology & Behavior: The Impact of the Internet, Multimedia and Virtual Reality on Behavior and Society, 12*(1), 1–6. https://doi.org/10.1089/cpb.2008.0181

Merlo, L. J., Stone, A. M., & Bibbey, A. (2013). Measuring problematic Mobile phone use: Development and preliminary psychometric properties of the PUMP scale. *Journal of Addiction, 2013*, 912807. https://doi.org/10.1155/2013/912807

Meyer, H. C., & Lee, F. S. (2023). Translating developmental neuroscience to understand risk for psychiatric disorders. *The American Journal of Psychiatry, 180*(8), 540–547. https://doi.org/10.1176/appi.ajp.19010091

Milenkova, V., & Manov, B. (2020). Digitisation of society and challenges to children s behavior online. *Anthropological Researches and Studies, 1*(10), 112–121.

Moitra, M., Owens, S., Hailemariam, M., Wilson, K. S., Mensa-Kwao, A., Gonese, G., Kamamia, C. K., White, B., Young, D. M., & Collins, P. Y. (2023). Global mental health: Where we are and where we are going. *Current Psychiatry Reports, 25*(7), 301–311. https://doi.org/10.1007/s11920-023-01426-8

Momen, N. C., Plana-Ripoll, O., Agerbo, E., Christensen, M. K., Iburg, K. M., Laursen, T. M., Mortensen, P. B., Pedersen, C. B., Prior, A., Weye, N., & McGrath, J. J. (2022). Mortality associated with mental disorders and comorbid general medical conditions. *JAMA Psychiatry, 79*(5), 444–453. https://doi.org/10.1001/jamapsychiatry.2022.0347

Mootz, J. J., Chantre, C., Sikkema, K., Greene, M. C., Lovero, K. L., Gouveia, L., et al. (2024). Leveraging a digitised mental wellness (DIGImw) program to provide mental health Care for Internally Displaced People. *Psychiatric Services, 75*(1), 98–101.

Morea, A., & Calvete, E. (2022). Understanding the perpetuation of cyberbullying victimization in adolescents: The role of executive functions. *Research on child and adolescent psychopathology, 50*(10), 1299–1311. https://doi.org/10.1007/s10802-022-00926-0

Morrison, L., & Frank, C. J. (2023). Social determinants of mental and behavioral health. *Primary Care, 50*(4), 679–688. https://doi.org/10.1016/j.pop.2023.04.003

Nikitenko, V. (2019). The impact of digitalisation on value orientations changes in the modern digital society. *Humanities Studies, 2*(79), 80–94. https://doi.org/10.31014/aior.1992.06.02.514

Pan, Q., Lan, M., Tan, C. Y., Tao, S., Liang, Q., & Law, N. (2024). Protective factors contributing to adolescents' multifaceted digital resilience for their Well-being: A socio-ecological perspective. *Computers in Human Behavior, 108164*, 108164.

Pan, Y., Teng, Z., Gu, M., Chen, C., & Shek, D. (2024). Editorial: Internet use and psychological Well-being among children and adolescents. *Frontiers in Psychiatry, 14*, 1349082. https://doi.org/10.3389/fpsyt.2023.1349082

Park, E., & Kwon, M. (2018). Health-related internet use by children and adolescents: Systematic review. *Journal of Medical Internet Research, 20*(4), e120. https://doi.org/10.2196/jmir.7731

Parker, L., Bero, L., Gillies, D., Raven, M., Mintzes, B., Jureidini, J., & Grundy, Q. (2018). Mental health messages in prominent mental health apps. *Annals of Family Medicine, 16*(4), 338–342. https://doi.org/10.1370/afm.2260

Peker, A., Özhan, M. B., & Eroğlu, Y. (2015). Development of the Scale on Coping with Cyber Bullying towards Adolescents<p>Ergenlere Yönelik Siber Zorbalıkla Başa Çıkma Ölçeği'nin geliştirilmesi. *Journal of Human Sciences, 12*(1), 569–581. Retrieved from https://www.j--humansciences.com/ojs/index.php/IJHS/article/view/3156

Ramalho, R., Volpe, U., & Gaebel, W. (2023). WPA digitalization in mental health and care: Empirical report on action plan and related activities. *Journal of Technology in Behavioral Science, 9*, 1–5.

Schmidt, M., Frank, L., & Gimpel, H. (2021). How adolescents cope with technostress: A mixed-methods approach. *International Journal of Electronic Commerce, 25*(2), 154–180. https://doi.org/10.1080/10864415.2021.1887696

Serrano-Santoyo, A., & Cabrera-Flores, M. R. (2018). *Exploring the implications of digitisation in the post-digital era. E-health 2018 ICT, society and human beings 2018* (p. 235).

Shikwambi, J. N., & Angula, N. (2021). The impact digitalization has on students' behavior on a high-level institution as the University of Namibia. *International Journal, 1*(1), 6–14.

Shilko, R., Egorov, S., & Zinchenko, Y. (2021). Digitalisation of education and mental health. *European Psychiatry, 64*(S1), S349–S350.

Stern, E., Micoulaud Franchi, J. A., Dumas, G., Moreira, J., Mouchabac, S., Maruani, J., Philip, P., Lejoyeux, M., & Geoffroy, P. A. (2023). How can digital mental health enhance psychiatry? *The Neuroscientist: a review journal bringing neurobiology, neurology and psychiatry, 29*(6), 681–693. https://doi.org/10.1177/10738584221098603

Sun, J., Shen, H., Ibn-Ul-Hassan, S., Riaz, A., & Domil, A. E. (2022). The association between digitalisation and mental health: The mediating role of Well-being at work. *Frontiers in Psychiatry, 13*, 934357. https://doi.org/10.3389/fpsyt.2022.934357

Tani, N., Fujihara, H., Ishii, K., Kamakura, Y., Tsunemi, M., Yamaguchi, C., Eguchi, H., Imamura, K., Kanamori, S., Kojimahara, N., & Ebara, T. (2024). What digital health technology types are used in mental health prevention and intervention? Review of systematic reviews for systematisation of technologies. *Journal of Occupational Health, 66*(1), uiad003. https://doi.org/10.1093/joccuh/uiad003

Teepe, G. W., Glase, E. M., & Reips, U. D. (2023). Increasing digitalisation is associated with anxiety and depression: A Google Ngram analysis. *PLoS One, 18*(4), e0284091. https://doi.org/10.1371/journal.pone.0284091

Tooley, U. A., Bassett, D. S., & Mackey, A. P. (2021). Environmental influences on the pace of brain development. *Nature Reviews. Neuroscience, 22*(6), 372–384. https://doi.org/10.1038/s41583-021-00457-5

Turel, O., Cheung, C., Matt, C., & Trenz, M. (2018). *The digitization of the individual (DOTI)*. Information Systems Journal. Special Issue Editorial.

Vevera, A., & Albescu, R. (2019). Hărţuirea cibernetică. *Revista Română de Informatică şi Automatică., 29*, 10.33436/v29i2y201905.

Walker, E. R., McGee, R. E., & Druss, B. G. (2015). Mortality in mental disorders and global disease burden implications: A systematic review and meta-analysis. *JAMA Psychiatry, 72*(4), 334–341. https://doi.org/10.1001/jamapsychiatry.2014.2502

Wiguna, T., Minayati, K., Kaligis, F., Teh, S. D., Sourander, A., Dirjayanto, V. J., Krishnandita, M., Meriem, N., & Gilbert, S. (2023). The influence of screen time on behaviour and emotional problems among adolescents: A comparison study of the pre-, peak, and post-peak periods of COVID-19. *Heliyon, 10*(1), e23325. https://doi.org/10.1016/j.heliyon.2023.e23325

Xie, X., Cheng, H., & Chen, Z. (2023). Anxiety predicts internet addiction, which predicts depression among male college students: A cross-lagged comparison by sex. *Frontiers in Psychology, 13*, 1102066. https://doi.org/10.3389/fpsyg.2022.1102066

Yellowlees, P. (2022). Climate change impacts on mental health will Lead to increased digitization of mental health care. *Current Psychiatry Reports, 24*(11), 723–730.

Zheng, H., Ma, W., Li, J., & Botero, J. (2023). Relationship between internet use and negative affect. *Applied Research in Quality of Life, 18*(4), 1693–1713. https://doi.org/10.1007/s11482-023-10158-z

**Mihaela Zăvăleanu** is Associate Prof. MD, PhD, in University of Craiova (UCV) since 2017, and with 21 years of experience in teaching in the Faculty of Physical Education and Sport, Department of Physical Therapy and Sports Medicine. Since 2010, she is PRM Specialist and has worked in different treatment facilities for the disabled and adults and children with musculoskeletal disorders from Craiova.She is a member of the editorial board of the *Journal of Sport and Kinetic Movement* and performed evaluation activities for scientific events organized by UCV. She has written over 20 articles (regarding physical therapy, disability, rehabilitation), published in Journals or Conference Proceedings, and 4 books like first author materials for the training of disability persons (with neuromuscular diseases or Down syndrome). She has been a member in national and international research projects, also in different European Erasmus+ projects that extend her experience in Down syndrome and intellectual disabilities area. She has expertise and her interests are in rehabilitation, physiotherapy and electrotherapy, health-care, care of individuals at all stages of the lifespan, functional abilities and health, in prophylaxis of diseases and health and active and well ageing promotion programmes.

**Amalia-Raluca Stepan,** PhD, Lecturer at the University of Craiova since 2013, psychotherapist and autonomous psychological counsellor, trained in Unification Experiential Psychotherapy, P.E.U., clinical psychologist. She has written over 30 articles (on psychology, disability, sports psychology, rehabilitation, and intervention), published in Journals or conference proceedings, and training materials for people working with people with disabilities (e.g. adults with Down syndrome). She has been involved as a member of more than 10 national and international projects in the field of intellectual disabilities and has expertise in individual and group personal development; child, adult, and family psychotherapy and counselling; and vocational and career counselling. She has associated expertise in brief psychotherapy, narrative psychotherapy, and EMDR (Eye Movement Desensitization and Reprocessing). Her interests also include educational psychology, school and vocational counselling, work and organizational psychology, and transport psychology. She is a member of SPER (Society of Experiential Psychotherapy) and COPSI (College of Psychologists of Romania).

**Prof.hab. Germina Cosma,** PhD, has been in the University of Craiova since 2003; she is currently the Director of the Department of Theory and Methodology of Motor Activities within the Faculty of Physical Education and Sport. She is the author of more than 150 articles (on physical education, dance for people with disabilities, physical training), published in specialized journals or conference proceedings, and 6 books. She has been a member of more than 10 national and international European projects in the field of disability and sport. She has been a member of some basic bodies in the structure of the University as well as national and international structure: Faculty Council of FEFS Craiova 2016–present; Senate of the University of Craiova; and General Secretary of the Romanian Sports Science Council. She is editor in chief of the *Journal of Sport and Kinetic Movement* and Ambassador for International Organization for Health, Sport and Kinesiology in Romania (IOHSK).

**Mihaela Tinca Udriştioiu** works as a lecturer for the Department of Physics, University of Craiova, Romania. She is a Fulbright Alumna at Georgia Southern University, where she developed and implemented a project focused on STEM education, interactive physics teaching methods, and continuous assessment at the University of Craiova, Romania. For her, the standardised assessment is a reference system for quality assurance in education. She was a member of or coordinated the organizing team of the European Researchers' Night at the University of Craiova for five years. This event brought science closer to the general public each year and gathered in the hall of this university between 2500 and 4500 children, parents, and grandparents within a few hours. She coordinated three volunteering projects focused on developing a network of air quality monitoring sensors in the Oltenia region (southwestern part of Romania). In the framework of these projects, her team organizd public awareness campaigns in mass media about the importance of screening and prevention in detecting some health issues and the importance of a clean environment for health. She promotes physics with her colleagues through experiments in the network of high schools in Oltenia. She participates in summer schools for children from disadvantaged backgrounds to convince them to consolidate their education. Her professional experience includes four EU projects as a project manager and more than ten as an expert.

# Our Obtuse Obsolescence: A Medium Theory Analysis of Format Consumption

**Christopher Cohen**

**Abstract** From the dawn of fire to the invention of the wheel, our agricultural revolution, and the latter industrial revolution, humankind's unprecedented lust for advancement has changed the world beyond recognition. Whilst largely understood that societal advancement is a trade to the determinant of ecology and nature. In reality, what tends to be largely misunderstood by the majority of mainstream consumers, is the deeper, more eclipsed and far more complex inner mechanisms regarding the effects of consumption, and how far away humanity truly is from sustainability.

All findings in this chapter will be examined through the lens of critical realism and will incorporate a series of mixed methodological findings where questions will be raised and data presented. Through these means, a more elaborate tapestry is to be formed and by the chapter's end a greater understanding of social format consumption, through the lens of music mediums is to be unveiled. Moreover, no medium or technology is exempt from equal blame; you see it is our human condition that compels us with the desire for *new*. Nevertheless, dialogue in this chapter will primarily center around the music sector, predominantly that of data, suitability and the desuetude of contemporary technologies.

All things change as all things must, and advancement is of unparalleled importance, and essential for our growth as a species, this can be seen from the humble piano sheet roll to the mobile listening devices that litter the pockets of billions. Yet with all our understanding, how is it that new mediums, mediums that created billions in tons of $CO_2$, cannot be used with or interface with newer digital services? In what way did society accept that our far older technologies such as the VHS or vinyl can still complete their baseline function? Whereas devices little over 10 years old cannot. Even the Edison Tube or Reel-to-Reel video can deliver their data with the right perquisites. What changed?

This leaves questions; Is this the plan? Part of the scheme of innovation? Or just a stark warning in the footnote of humankind's urge for discovery, whilst leaving the

C. Cohen (✉)
London, UK


M. Todorovic et al. (eds.), *Sustainability and Social Responsibility of the Media and in the Media*, CSR, Sustainability, Ethics & Governance,
https://doi.org/10.1007/978-3-032-00086-6_3

pile of antediluvian technological waste in the review view mirror, hoping that future generations could clean up the clutter when the time is right. Or perhaps worse, when the end is nigh. You see it is paradoxical in nature, it is humankind's desire for the qualities of faster, quicker, easier, simpler, and more intelligent, coupled with a misunderstanding of everlasting abundance, which facilitates the great misunderstanding of our time.

**Keywords** Music · Technology · Format · Society · Environment · Nostalgia · Data · Vinyl · Tape CD · iPod · Digital music · Streaming · Mobile technology · Computing · Carbon footprint · Consumption · Medium theory · Deliverable theory · Culture · History · Planned obsolesce · Apple · Spotify · Napster · Deezer · Longevity · Lifespan

## 1 An Introduction: A Prelude

When discussing formats in this chapter, all formats or format technologies will have a basis in the field of music and music technology, science, culture, or history. It is essential to draw attention to how "format" is described ultimately as "the medium". This idea of the medium should be highlighted as a Medium Theory which has been researched extensively for over 70 years by writers such as Innis, 1949; McLuhan, 1964; Meyrowritz, 1994; and Nyre, 2008.

The word "medium" derives its origin from the Latin word "middle"; with this understanding, the Medium Theory could be broken down further into middle theory, where a format by definition is the medium or the middle, with music being on one side of this spectrum, the organic receiver on the other, and the format as the "middle" carrying information from one party to the next.

Seen above in Table 1, is how we derive that technological music formats are the medium connective points between humankind (organics) and music (audio frequencies or data types). It is therefore logical that the topic of discussion that follows should be about society, music format culture, and music format consumption.

This chapter is also accompanied by some essential data sets being:

- A historical timeline of music medium and data formats.
- A tabulated format chart.
- Created quantitative data sets from survey responses.

**Table 1** A simplified process of medium theory, with data on the left, technology in the middle and a receiver on the right

| Data type | Medium (format technology) | Receiver/Listener |
|---|---|---|
| Analog | LP 78 record | |
| Music data → | Medium technology → | Ear (human) |
| Digital | iPod | |

These data sets have been created to help further inform the diverse statistical analysis required to help piece together a far more succinct representation of the field as a whole.

The data sets in this chapter were created as part of a mixed methodological approach where historical and survey data sets were combined for the best accuracy possible at the time of their creation in year 2022.

This chapter aims to discuss data findings throughout this project to better reflect results regarding sociological format consumption. We follow five pillars highlighted as, Data, Shareability, Longevity, Environment, and Nostalgia. This breakdown will further illustrate information found prior.

It is clear that due to technological developments, humankind has created incredible advances in music technologies. The proliferation of consumptive audio devices alongside their data transition rates in the last half-century shows that significant progress has been made in sociological music format consumption.

## *1.1 Data Collection*

New research data sets for this document were created by way of surveys, with the population of the aforementioned surveys consisting of participants of different age groups, ethnicities, genders, and nationalities.

Initially, subscribers to chriscohenmedia.com were targeted as a part of an email mailout campaign to garner participants' attention, after which surveys were promoted on social media channels to acquire further attention. There was no boundary to who could be involved with the study. Prizes such as free music, software, event tickets, or more were employed to incentivize participants.

Further measures were taken to ensure the data collected was unique, clear, concise, and created in a way that could be analyzed.

By the end of Leey Survey Three, over 550 participants had participated. Oversight was necessary to ensure that the data collected was sufficient before analysis and that no one had submitted a set of answers twice from the same email address, as prizes would only be sent to an email address once. This is flawed as a user could create more than one email address. However, this is unlikely due to the number of participants, the time it takes to respond to a survey and the payoff an individual would receive for doing so.

Before any survey was sent to the public domain, a pilot study was conducted; based on this, amendments were made to questions and grammar.

A limiting factor to the data generated in research may be geographic. Whereby as all data was created online, respondents could participate from any part of the world, which is helpful as it allows the study to be of people from all nations but suffers due to its inability to be as focused on one region. A second limiting factor may be age. Although respondents were asked their age to help validate responses, individuals in a society of older or younger generations who may not use technology in such a manner may not have been able to participate. This leaves a gap in the generational study. As this chapter explores older technologies, such insight from older generations may have been helpful if the collection by way of interviews had been possible.

This chapter may also be limited to the available technology as of 2022, just like the many that have come before. With music formats not even 200 years old, various unknown developments are still to take place for a market that is relatively still in its infancy.

## 1.2 Collected Data Outcomes

Of the questions asked in Leey Survey One, various prominent data sets arose. This survey concentrated on asking questions in relation to format consumption habits and whether music format users had brought a physical format or had chosen to stream music instead as their format choice of music consumption.

As seen in Fig. 1, only 28% of music format users as part of this study had chosen to buy a CD (compact disk) since 2020. It was expected that results would be relatively similar to that of vinyl, with the CD being highly popular even in the years 2010–2015. However, by asking the question in this manner to participants, a clearer pattern emerges when considering user age as a part of Leey Survey One findings.

Questions asked in Leey Survey Two asked whether or not which streaming service respondents use. Afterwards, respondents were questioned about what format they were more likely to use when in the car. Survey questioning was concluded when asking about whether or not participants still use CD players at home. By asking this series of questions and combining the answers with previous survey answers, we create a broader image of general CD format use and consumption versus streaming habits in the public domain as of 2022.

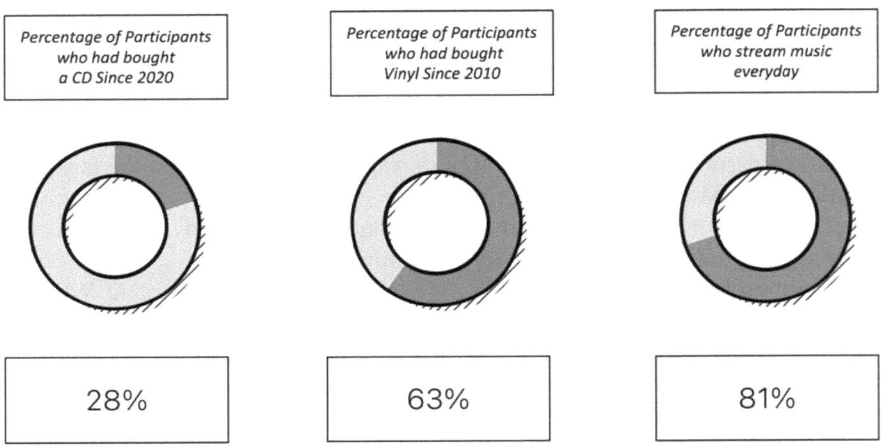


**Fig. 1** Percentages of purchase and streaming habits from Leey Survey One

Data in Fig. 2 shows that 21 respondents (15.56%) choose not to use a streaming service. This leaves a total of 84.44% of 114 respondents admitting to using a streaming service of some kind.

The highest percentile of our sample belongs to Spotify, with a total figure of 47.41% of responded answers (or 64 respondents). It should be of note that this percentile figure is over three times that of respondents who said they chose not to use a streaming service of any kind.

Respondent choices as seen in Fig. 3 show that over 70% of participants confirmed that they chose to now stream music in their vehicle over using a physical format. Shows that a majority of respondents in the sample base use streaming services as their music format of choice in their vehicle.

Figure 4, concerns CD consumption at home, a substantial 83.7% of users combined chose an answer that pertained to not using the CD format at home in any capacity as of 2022. More than half of the respondents used to consume CDs at home. However, as data suggests, this number has diminished significantly, showing that a greater number of respondents do not even own a CD player as of 2022.

Questions asked in Leey Survey Three pertained to participants' feelings towards their current listening device, with what is missing from their listening device? Furthermore, what leads an individual to purchase a listening device? Participants

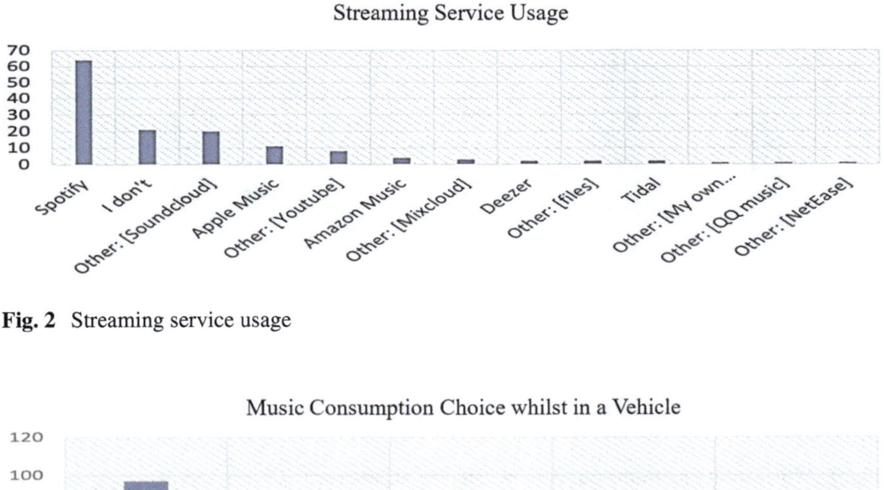


**Fig. 2** Streaming service usage

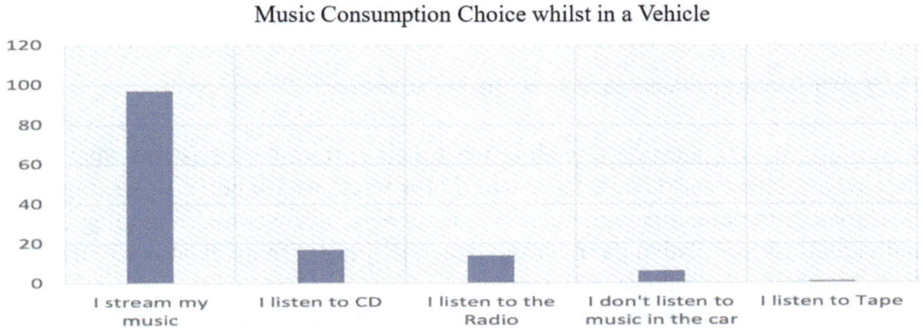


**Fig. 3** Music consumption choice whilst in a vehicle

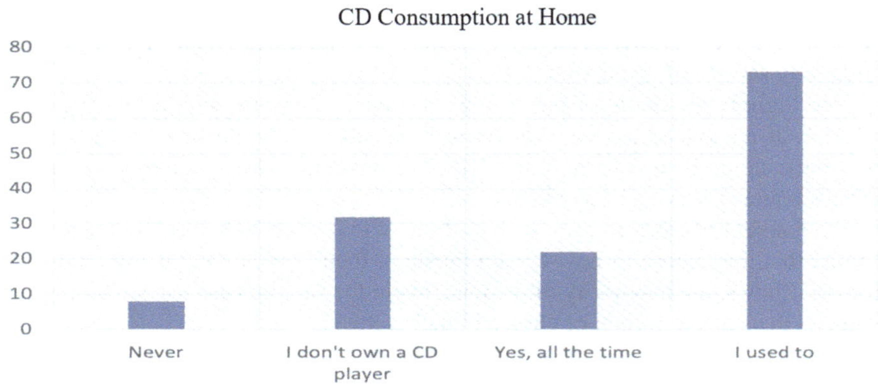


**Fig. 4** CD consumption at home

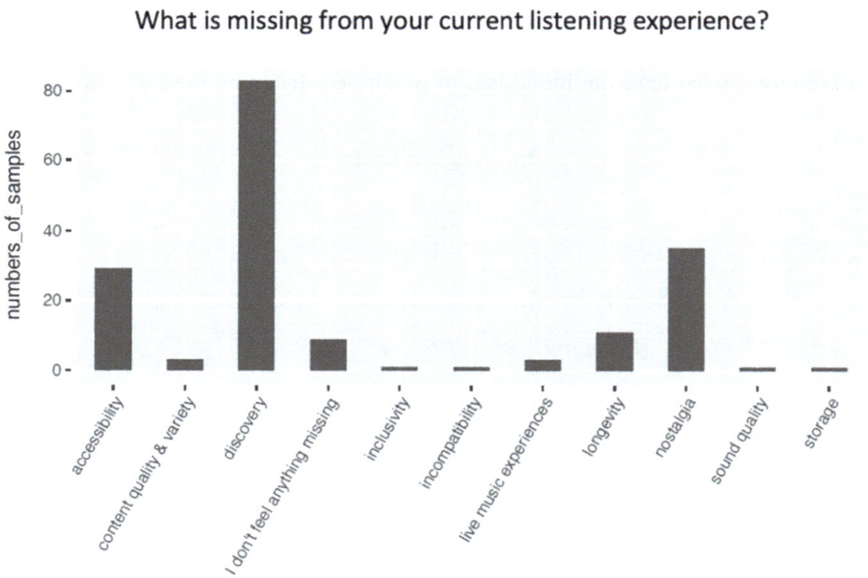


**Fig. 5** What is missing from your current listening experience?

were also asked a question regarding what their current device is and what rating they would give their device, from 1 to 5 (1 = worst to 5 = best). By asking these questions in combination with the previous surveys, a deeper understanding of why individuals like or dislike their current generation of listening experience in regard to technological format use can emerge.

Data in Fig. 5 shows that three critical areas are important and are missing from current respondents listening experiences, these being discovery (36.7%); nostalgia (33.9%); and accessibility (14.7%).

Whereby the section discovery suggests that respondents indicate that there are so many choices but find it hard to discover new music due to algorithmic music services. Nostalgia suggests that respondents miss their physical collection. A distinctive personal experience and connection that physical formats have to offer, despite the ease of digital music streaming services. The third highest point of contention with what users miss from their music listening experience today is accessibility, where respondents admit to frustration that they cannot listen to music when there is no Internet availability.

Table 2 shows how a handful of selected respondents answered concerning questions given in Leey Survey Three.

**Table 2** Leey Survey Three user responses

| User device as answered | Rating of device (1= worst - 5=Best) | What leads you to buy new music technology products? | What do you feel is missing from your current listening device? | What is vital to your listening experience? |
|---|---|---|---|---|
| iPhone | 3 | I see reviews online | Constant variation / get bored of the same song | Good quality headphones and connection |
| Android | 3 | I see what friends have | Phone with no goddam headphone jack | A long battery life. |
| iPhone or MacBook | 4 | New experiences (spatial audio) | Storage for more music | Good quality headphones and connection |
| iPhone | 4 | I see reviews online | The algorithm for finding new music on Spotify is beginning to get a bit repetitive and I feel like it's stopping me from branching out to new genres. I am now back to using radio stations to find new stuff. | Good quality headphones and connection |
| Android | 4 | New products | I need a new turntable for my old school vinyl, not much else. | Internet availability |
| iPhone | 4 | I need functionally | Equipment due to cost | No ads; Playlists |
| My mobile phone and iMac using streaming or YouTube | 4 | I see what friends have | I enjoyed the days of albums that had a concept. | Good quality headphones and connection |
| Spotify via iPhone App | 4 | New experiences (spatial audio) | I hate wireless devices. Always having issues and disconnecting. | Speakers |

Engaging assessments appear when polarities occur, where one respondent states "Playlists" are vital for their listening experience, and another highlights the "algorithm" being a problem for their listening experience.

Another such division arises around the use of hardware, with some respondents stating that "good quality headphones" are vital. In contrast, one respondent showed displeasure at the fact that the older technology of the "headphone jack" (AUX cable) had been removed from their current listening device. While important to some consumers, Apple explained the removal of the "headphone jack" or AUX cable as "courage" or, as reported by Lovejoy (2019), "The reason to move on: courage. The courage to move on and do something new that betters all of us." This should be of note as in Leey Survey One, part of the survey, participants were asked whether they owned Bluetooth headphones (Leey Survey One questions appendix), with over 60% of participants admitting they do own Bluetooth headphones (Fig. 10). Was the great wireless leap forward "courage"? Or something else?

Additional Survey Information.

Additionally, questions asked by participants relating to Leey Surveys One, Two, and Three may be found in the appendix, along with their respective data sheets.

## 2 A Historical Timeline of Music Medium and Data Formats

As part of Table 3, music formats are considered, as are data speed proliferations from the dawn of the first computer to our modern-day, high-speed Internet-connected devices. Doing so allows a far richer understanding of how format

**Table 3** A historical timeline of music medium and data formats

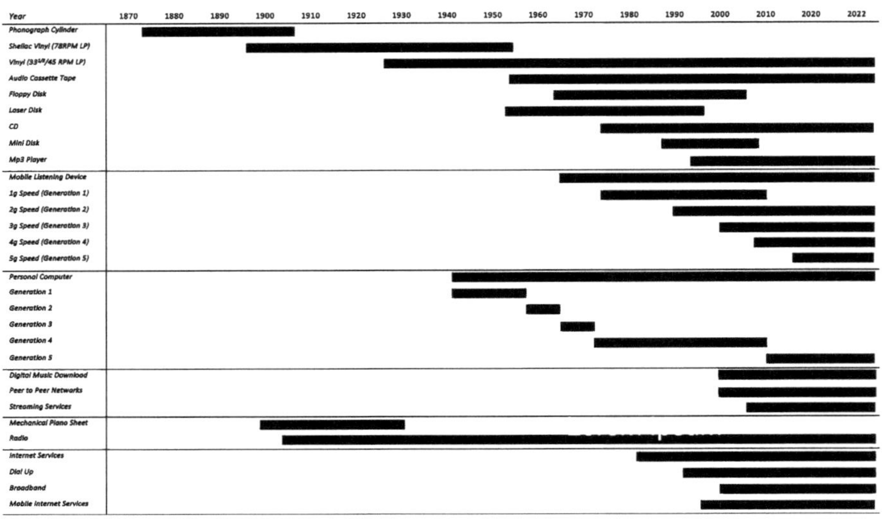

changes and developments can be crystallized, allowing for a more comprehensive understanding of why formats are consumed.

In the book "*Sound Media, From Live Journalism to Music Recording*" by Lars Nyre (2008, p.18–19), a similar timeline was created, and as such, the timeline present here as part of this chapter owes its inspiration to Nyre's work. As such, the timeline on the page following has been created to facilitate better other data points that were not yet available in 2008.

## *2.1 A Tabulated Format Chart*

Here, data from survey findings have been merged with the previous Table 3 (A Historical Timeline of Music Medium and Data Formats) in order to create a comprehensive tabulated format chart (Table 4). The data in the tabulated format chart goes a step further, combining other factors that have previously not been discussed, such as:

- $CO_2$ per unit creation
- Manufacturing by-products of unit production
- The material choice of a specific product
- Ease of shareability
- The data storage capability of a specific format

Moreover, after merging Tables 3 and 4 with data generated as part of Leey Surveys One, Two, and Three, far greater reasoning can be formed when understanding human usage of music and media format mediums as a choice of consumption.

Due to the considerable amount of data and references within the table created, explanations, validations and supporting evidence of figures can be found in the appendix of this chapter.

# 3 Time Ticking Backwards or Forwards?

## *3.1 Nostalgia*

Often it is when conversing about older formats that individuals get excited about their own memories' past, their days once gone, their immediate thoughts, beckoning back to greater times, the feelings of what was, the age-old feeling of nostalgia (particularly those who experienced them first-hand).

None are truly immune from the tendrils of what was, of before, and with somewhat rose-tinted glasses, it is humanity, with its addiction to reliving youth, or maybe our fear of death, that keeps us constantly and continually harkening back to

**Table 4** Tabulated format chart

| | Analogue Audio Formats | | | | Digital Audio Formats | | | | Digital Audio Services | |
|---|---|---|---|---|---|---|---|---|---|---|
| | Phonograph Cylinder | Shellac Vinyl (78RPM LP) | Vinyl ($33^{1/3}$/45 RPM LP) | Audio Cassette Tape | CD | Mp3 Player[1] | Mobile Listening Devices[2] | Personal Computers (PC)[3] | Peer to Peer Networks[4] | Streaming[5] |
| ***Year Of Formal Release to public***[6] | 1888 | 1901 | 1948 | 1963 | 1982[7] | 1997[8] | 2000[9] | 1995[10] | 1999[11] | 2008[12] |
| ***Data Creation Time (The amount of time it takes to print an album)***[13] | Real-time, in real life recording* | Real-time, real life recording* | Real Time creation | Real Time creation | Device Dependent** | – | – | – | – | – |
| ***Data Transmission (The copy of data from one format to another)*** | – | – | – | – | Device Dependent** | Device Dependent** | Device Dependent** | Device Dependent** | Internet Speed Dependent*** | Internet Speed Dependent*** |
| ***Data Copy during manufacture by company possible?***[14] | No | Yes | Yes | Yes | Yes | Yes | Yes | Yes | – | – |
| ***Data Copy for sharing Possible by customer?***[15] | No | No | No | Yes, Real-Time Recording copy or slice copy | Yes, User Device Dependant** | Yes, User Device Dependant** | Yes, User Device Dependant** | Yes, User Device Dependant** | Yes, User Device Dependant** | No, however, users can share music** |

| | | | | | | | | | | |
|---|---|---|---|---|---|---|---|---|---|---|
| ***Amount Of Data That Can Be Held on To the Device (In Mins)***[16] | 2-4M | 10" - 6M (3M per side)<br>12" - 10M (5M per side) | 33[13] -<br>30-44M (1522M per side)<br>45 - 24-30M (12-15M per side) | 90M (45M per side) | 74M | Device Dependant** | Device Dependant** | Device Dependant** | Device Dependant** | Device Dependant** |
| ***Amount Of Data That Can Held on To the Device (In Songs)***[17] | 1 | 2/3 | 10-14.3/8-10 | 30 | 24.3 | ^ | ^ | ^ | ^ | ^ |
| ***Amount Of Data That Can Held on To the Device (In Albums)***[18] | – | – | – | 3 | 2.3 | ^ | ^ | ^ | ^ | ^ |
| ***Manufacturing Materials***[19] | Ceresin, Beeswax, Stearic Wax[20] | Shellac, Other materials (various)[21] | Plastic (PVC)[22] | Plastic, Metal oxides (various)[23] | Plastic, Metals (Various)[24] | Silicone, Plastic, Glass, Metals (various) Elements (Various)[25] | Silicone, Plastic, Glass, Metals (various), Elements (Various)[26] | Silicone, Plastic, Glass, Metals (various), Elements (Various)[27] | N/A | N/A |

(continued)

**Table 4** (continued)

| | Analogue Audio Formats | | | | Digital Audio Formats | | | | Digital Audio Services | |
|---|---|---|---|---|---|---|---|---|---|---|
| ***Other Manufacturing By-products***[28] | Non-recyclable | Recyclable & Reusable | Difficult to recycle | Difficult to recycle (case can be recycled), can mould | Very Difficult to recycle (case can be recycled) | Mostly recyclable uses high-tech components | Mostly recyclable uses high-tech components | Mostly recyclable uses high-tech components | – | – |
| ***Co2(Pg/Pkg) Generated Per Product Creation***[29] | N/A | N/A | 2.2Kg[30] | 2.8Kg[31] | 172g[32] | 45Kg Per Typical MP3 Player[33] | 69K.gs Per Typical Mobile Device[34] | 185Kg Per Typical PC (Laptop)[35] | N/A | 55g (Per hour)[36] |
| ***Benefits of Format Consumption*** | Revolutionary, First music format, highly prestigious | More durable than previous generation | More durable than previous generation | Compact, easy to share music, enables record from other sources | First digital audio format, superior sound quality to previous generations | A library of non-internet dependant music | Any song at the touch of a button* | Can store a large library of owned music and internet based music | Unlimited access to music | Unlimited access to music |
| ***Ease of Sharing with Others*** | Difficult due to cylinder damage | More widely adopted, disks superior in durability | Easier to lend due to higher durability of format | Additional hardware required for recreation | Additional hardware required for recreation | Some MP3 players allow for the copy of files | Some Mobile Devices players allow for the copy of files | Can download and share music files with ease | Built on principles of shared media | Music can be shared; receiver may need to pay* |

| | | | | | | | | | | |
|---|---|---|---|---|---|---|---|---|---|---|
| ***Challenges of Formal Consumption*** | Would wear out very quickly thus requiring new and costly cylinders | Prone to break or wear, thus requiring a new disk after several plays | Format easily damageable by scratching disk, vinyl disk can warp in wrong storage environment | Tapes had to be rewound, audio tape prone to breakage, low fidelity audio | CD may be damaged easily if portable requires many CDs, can skip, or be scratched | Not connected to the internet, Battery life may be short, may become quickly obsolete | Battery life may be short, may become quickly obsolete | Inconvenient for portability, may become quickly obsolete | Requires an internet-connected device, some platforms were illegal to use | Can require an internet-connected device, can require a paid subscription |
| ***Nostalgic Values*** | - | Historical significance, gramophone stylization | Large artwork covers, may sometimes contain posters or other insets, sound quality | Mix tapes of personal music, shared with others | Jewel Case, artwork, the listening of a specific album | Music library collection ownership, device specific | Users may feel connection to previous mobile devices | Users may feel connection to previous personal computers | Users may miss old services or online communities | Users may have preferred previous software versions |

the past, to feel the warm embrace of "things" (in this case technology) which are fresh, novel or new.

It is suggested that formats in the age bracket of 50–80, such as vinyl, tape, or even CD, should be highly nostalgic for consumers. This may be for several reasons, such as the physicality of the jewel case, the artwork of a vinyl sleeve or other novel or new technology use cases. Another factor of this, however, as said by Jones (2010, p.xiii) is "customers are still willing to spend money if they can get an 'experience' that's not available online. You can't get free in-store gigs online…an atmosphere that you want to be part of with genuine human interaction online. You can with a record shop."

But what of formats gone, where does the music go? What happens to our art? Nostalgia seems to all but evaporate, when the technologies subset of users its "clique" all but disappears as after a format reaches past the age of 80; this may be due to a number of factors, such as the original user base may have all but deceased as the average lifespan of a human is 73 years old (The World Bank, 2022).

This leaves a format of over 80 years a subset of novel enthusiasts, who sit in dark rooms, analyzing the way in which the format existed, and although, this being said, nostalgia should not be confused with the exploration of historical formats.

Where individuals may feel a sense of nostalgia for the 33 or 45RPM vinyl, the same cannot be said for the 78 or the phonograph cylinder.

Personal findings of interest from previously generated data (Figs. 1, 2, and 3), asked questions to individuals who had purchased CDs or vinyl. This was queried due to both formats having very similar offerings when removing the data side of their musical aspects, such as requiring an additional object to play the format or other offerings in terms of art or physicality.

The situation suggests that nostalgia for purchasing vinyl over CD may be due to the analogue signal of vinyl playing with a psychoacoustic listening response curve that CD does not have, known as the RIAA (Recording Industry Association of America) curve (Hull, 2014). Or possibly that CD is seen subconsciously as only a vessel for digital audio files, the same digital audio files which can be delivered at a lesser cost and in a quicker fashion from streaming services rather than owning a CD and CD player for deliverable consumption purposes. Conversely, Bowe (2022) argues that the quality of CD audio is higher than that of steaming or even vinyl, stating: "The best compressed digital music files are referred to as lossless because they don't lose information." However, the idea the lossless audio quality of a CD is better than that of streaming an HQ (high quality) 320Kbps song is put into question, with many arguing that consumers cannot hear the difference between uncompressed audio files and MP3s (Fisher & Ganz, 2015; Broom, 2022).

Furthermore, what cannot be understated is how media has changed as it has passed from the physical, tangible realm into the digital intangible realm. Moreover, due to the ever-sweeping digital transformation that has taken place in the last half-century, and whilst progressive gains have been made, a loss of other significant technological and cultural elements of consumer music format technologies have also been lost to time.

Kramer (2022) summarizes when concerning how music is listened to by a consumer by beautifully affirming, "Listeners select not only where they start and stop listening to a piece of music but also in what order they hear its sections. As recordings have progressed from tapes and records to compact discs, random access by listeners has become easier"; however, the randomness of being lost in the flow of a vinyl or tape, the continual listening of an album via CD or even that of the randomness of the iPod shuffle, may have been lost forever.

#### 3.1.1 What Society Has Gained from Advances in Music Technologies?

It has been noted that due to the proliferation of mobile data devices, music has become more accessible now than ever. To explain, smartphones have replaced previous technologies such as cameras or MP3/4 players (Reitemeier, 2020; Bhalla, 2021).

However, it is this advancement in mobile listening that has allowed for unprecedented music consumption on a scale never seen before.

Elements of technological use in human society have also been lost due to advancements in music technologies. As Straw (2009) discussed prior, the tape is now the cemetery for mixtapes, as it was this part of the culture that was lost when CD became the mainstream format of choice.

And CD, too, with its packaging; however, even if still manufactured, is now on the same mountain that tapes lie. Straw (2009) notes how the "jewel box" (p.86) offered an experience where a consumer could enjoy a booklet (insert) and a CD containing music in a tidy package.

Where did that experience go? Was it lost in the shuffle?

Or was it the shuffle that was lost? Ultimately consumer habits shifted, becoming more time and need-bound; the desire for shuffle or randomness in music was replaced by the want of any music available instantaneously. This discussion on advancement in regard to the disuse of shuffle is explored by Powers (2014), who concludes that it was curated playlists by the likes of Apple and Spotify combined with "algorithmically curated musical recommendation" (p.261).

As such we are presented with a question on nostalgia? A question on our future. When all physicality is gone, and the algorithms of Spotify, Apple Music and their many counterparts that dictate our taste endure, what will remain to be nostalgic about music? Will it be the association of the time of listening rather than the devices we listened with, rather than the art, the collection or the ceremony that was?

## 4 Let's Share Data!

### *4.1 Data & Shareability*

A crucial component of any format is that of data. As this chapter highlights music formats, the discussion of data should be thought about in several specific ways. This is important due to the different kinds of data that exist, as music formats may be either analogue or digital, the data in which both produce, although different, serve the same purpose.

When looking at the progression of music formats over time, it cannot be understated how far music formats have developed. Table 4 shows a well-defined representation of linear progression regarding music format data that has been continual since the foundation of the phonograph cylinder in 1888. This linear progression is due to a number of factors, such as technological, societal, or environmental shifts in the music industry and the world as a whole.

Due to this, an understanding of how music data has exponentially grown can be formed—from a single unit of a phonograph cylinder that is only able to hold a single song in 4 minutes; 90 minutes available on a cassette tape; the iPod with its capability of having 4000 minutes of capacity—famously said by Steve Jobs 1000 songs (Wilkins, 2021); or to the near limitless storage of modern devices.

However, other aspects of data proliferation that can be seen in modern life, can be seen in Table 3. Here, two factors have been critical to the development of data proliferation, these being (1) the advancements in computing power of the personal computer from the mid-1940s to today and (2) the transformation of Internet services over time, from stationary to mobile.

Newer technologies such as digital downloading and peer-to-peer networks such as Napster, in which the technology had a new and novel concept that would allow individuals globally to connect through music.

The idea of shareability or digital music ownership is all but lost when considering digital streaming services, as music is locked behind paywalls or advertisements. Whilst true that a user may be able to share music far more efficiently or by quicker means, problems arise when sharing data with another individual who may use another streaming platform provider.

With this being said, there are both positives and negatives regarding analogue or digital tangible audio formats and shareability. For instance, by the 1960s, the end user of an Audio Cassette Tape may have painstakingly created what was previously known as a "mixtape" to give to friends, family, or lovers on a special occasion. In the creation of a "mixtape" is found the action of copying audio data from the radio or another tape with as much precision as possible to give the gift of their favorite music to another (whether legal or otherwise).

In reality, this was the first time that individuals outside of the manufacturing process could easily share their love of music with one another by way of copying audio data, as then both parties (from another to a friend or otherwise) could keep the music they desired to listen to.

This idea of shareability was only amplified in the 1990s with the advent of CD technology becoming the mainstream format of choice, and by the end of the decade, the power of computers was making its way into more homes than ever. For better or worse with this computing boom, technologies like the CD-RW or the rewritable CD would allow for sharing audio files on a scale never seen before (Fig. 6).

With the rise of the Internet as a popular method of communication in the 1990s file sharing networks known as P2P, or peer-to-peer networks, would be born. The most famous P2P networks, Napster (in 1999) and LimeWire (in 2000), would not only change the music industry forever but also encapsulate humanity's love of sharing music, later epitomizing human sociological culture in terms of media sharing.

At the behest of the profit for musicians and international music conglomerates alike, the likelihood of shareability in terms of music ownership peaked with the introduction of peer-to-peer networks. The ubiquity of digital audio file sharing is all but lost when a user is stuck in the walled gardens of Apple's Apple Music, Spotify Music streaming platform or the like (Fig. 7).

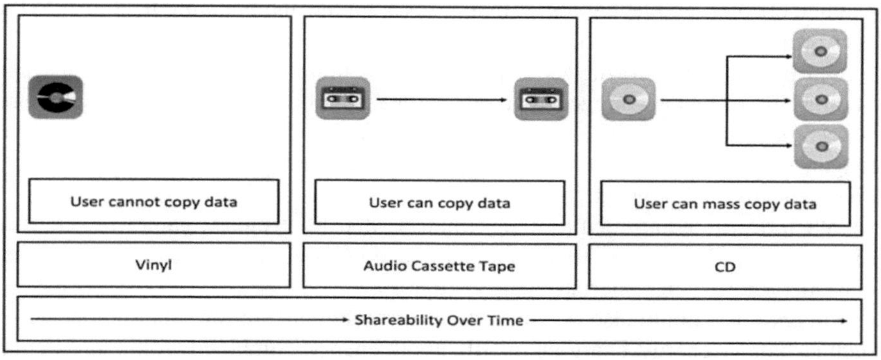


**Fig. 6** A simplified diagram showing user ease of audio shareability behavior over time

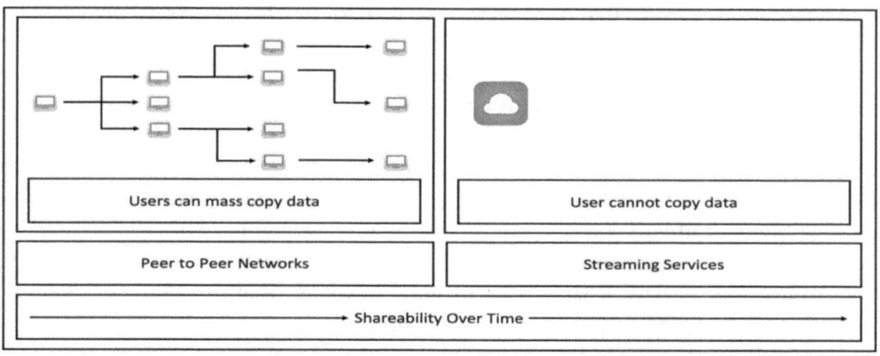


**Fig. 7** A simplified diagram showing user ease of audio shareability concerning digital files

# 5 Long Live the Environment?

## *5.1 Longevity Environment*

Another factor closely studied as one of the five pillars concerning research created was that of longevity; when regarding music formats, the longevity or lifespan of a format is of parallel importance to the environmental damage a format may create.

For example, when the lifespan of an older mobile device that would work perfectly fine in the past ceases to function due to planned obsolescence or is outpaced by the growth of technologies, one of two things may happen—a consumer may feel either compelled to purchase a new listening device for a variety of reasons or be forced into having no other choice than to buy a new listening device. It is the planned obsolesce of technologies that poses a unique and unparalleled threat to the environment.

Our human wants for newer, faster, better, or more convenient is evident when findings in this chapter suggest that, on average, a mobile listening device's lifespan is 2 years. However, the lifespan of a mobile listening device or a smartphone is far longer, with Stegner (2021) suggesting the typical software lifespan of an Apple iPhone device is 5–7 years.

When downloading media of any kind, be that film, music, games or other, as Mayers et al. (2015) note, there is an associated $CO_2$ emission, indicated as either "Product Energy" or "Internet Energy" (p.409).

Energy cost should also be considered for mobile music formats in other ways, such as battery energy, charging, or Bluetooth energy, combined with data calculations from users who stream their music or download music for later consumption—after which data points can coincide with previous analyses, creating a clearer image of this research topic. Weber et al. (2010) have created a tabulated data set that coincides well with that of Mayers et al. (2015), in which analyses physical and digital music distribution.

However, what about installable software such as Spotify or Deezer? Surely any mobile listening device with an Internet connection should be able to interface properly with simple music-listening software. However, this is not the case with Spotify or Deezer, where software support for Spotify ends at IOS13 (2019) (Fig. 11). Deezer, as of, 2022, software support is quoted as "Deezer app will no longer be compatible with older versions than iOS 14 by mid-November 2021," meaning that devices that no longer support IOS 14 (2020) (Fig. 11) or before will no longer be able to use the service. It should be of note, however, that the iPhone 6 can still interface with Apple's "Apple Music" streaming service as of 2022 (Fig. 11).

To put this into perspective, this means that the iPhone 6 and 6 Plus, released in 2014, could no longer receive or, even worse, interface with software 5 years after its release. Apple (2022) reveals that a single unit iPhone 6 created 95 Kg of $CO_2$.

Statista (2018) reports that in the release year 2014 169.22 million units of the iPhone were sold worldwide, therefore a figure of (Table 5):

With this understanding, further hypotheses surrounding music format consumption in regards to longevity can be made with roughly 200 billion compact disks being manufactured over the span of the format's popularity (BBC, 2019).

This is important to highlight as is it the devices in question, released in the years 2014 and prior, that can no longer interface with popular music software, giving a generous lifespan of the iPhone 6 anywhere from 5 to 8 years maximum. Wielen (2017), however, refutes that all older hardware is completely obsolete, "with over 63% of all iPhones still in use as of 2017." Fast forward to 2022, and a radically different picture is being painted with Law (2021), stating "older cellphones, including the iPhone 5, iPhone 5S and Samsung Galaxy S4, will become obsolete in 2022 as mobile providers make the switch to 5G networks." Although $CO_2$ per unit has been reduced year on year, this is only the tip of the obeisance iceberg.

Findings in research from Leey Survey Two as part of this study suggest that for the billions of tonnes of $CO_2$ created by the manufacture of CDs, only 16.3% of individuals still choose to use CDs as a listening format. And where the car was considered the final bastion of the CD's legacy, a similar pattern emerges where individuals would rather consume music via a streaming provider.

This is not without fault. However, Table 2 illustrates that physical formats have had an incredible lifespan for years of public adoption, with Audio Cassette Tape having a lifespan in the public domain of 70 years, CD 50 years and vinyl topping out at almost 100 years of longevity in the public sphere.

Therefore, the amount of $CO_2$ generated in just 1 year from the manufacture of the iPhone equals almost half that of all CDs ever produced (46.7%) (Table 6).

This is not without fault. However, Table 2 clarifies that physical formats have had an incredible lifespan for years of public adoption, with Audio Cassette Tape having a lifespan in the public domain of 70 years, CD 50 years and vinyl topping out at almost 100 years of longevity in the public sphere.

Moreover, comparative studies between CD (Blu-ray) versus digital download technologies with regards to disk manufacturing $CO_2$ emissions; however, Mayers et al. (2015) do briefly mention the music industry in their findings stating,

**Table 5** Calculation of $CO_2$ created due the manufacture of the iPhone 6 in 2014 verses vinyl or CD

| |
|---|
| 95 kg $CO_2$ x 169.22 million iPhone = 16,085,900,000 Kg $CO_2$ emissions |
| With $CO_2$ emissions equivalent to the manufacture of |
| 7,004,954,545,454.5 Vinyl or 89,598,255,813.9 compact disks |

**Table 6** Calculation of $CO_2$ of 1 year of iPhone manufacture versus the entire lifespan of CD manufacture

| |
|---|
| 200,000,000,000 compact disks at 34,400,000,000 Kg $CO_2$ |
| Versus |
| 95 kg $CO_2$ x 169.22 million iPhone = 15,410,900,000 Kg $CO_2$ emissions |

"downloading data over physical distribution should be considered desirable…Consumer focused studies within the music industry support this view, results of this study for gaming did not" (p.413).

Whilst humanity is making gains through music streaming services as a format of choice, consumers think little of how their devices have a larger scale impact on the environment when concerning the creation of carbon as a by-product of manufacturing.

Moreover, the findings in table cited can help create a comparison between vinyl versus a mobile listening device (iPhone or equivalent), or CD versus a mobile listening device (iPhone or equivalent), and can be made where a calculation value would be proposed, such as (Table 7):

The mathematics above, whilst considering a typical smartphone used currently, may vary slightly depending on the model. However, this is only the tip of the iceberg when considering how digital formats may affect the climate. As again in Table 3, it can be seen that a typical MP3 player creates 45 Kg of $CO_2$ output per device, or (Table 8):

However, this is where a mobile listening device such as an iPhone may differ significantly from a typical MP3 player of yesteryear. Where an MP3 player was a non-Internet connected device, and where a listener would acquire their music collection through any means, legal or otherwise, music was stored upon the said MP3 device. With the proliferation of data services allowing for the ubiquity of streaming that is of current times, not only does a user buy a device in which to listen to audio, whereby carbon is created, but also data streaming in itself creates carbon on both ends of music consumption as a service.

A revised model of the figures above should take not only the hardware into account but also the usage of data. It is suggested that listeners consume 18.4 hours of music weekly on average (Green & Gibson, 2021), so a new calculation should be made whereby (see Table 9):

**Table 7** Comparison of $CO_2$ emissions of one mobile listening device versus vinyl and CD

| |
|---|
| 1 x typical mobile listening device at 69 Kg $CO_2$ = 31.3 33RPM vinyl at 2.2 Kg $CO_2$ |
| Or |
| 1 x typical mobile listening device at 69 Kg $CO_2$ = 401.1 compact disks at 172 g $CO_2$ |

**Table 8** Comparison of $CO_2$ emissions of one MP3 player versus vinyl and CD

| |
|---|
| 1 x MP3 player at 45 Kg $CO_2$ = 20.4 33RPM vinyl at 2.2 Kg $CO_2$ |
| Or |
| 1 x typical mobile listening device at 69 Kg $CO_2$ = 261.6 compact disks at 172 g $CO_2$ |

**Table 9** Calculation of an average user's listening habits over 1 year

| |
|---|
| **18.4Hr per week ÷ 7 (days) x 365 (1 year) = 959.4Hr per year** |

Assuming this is music that is only streamed data, and the user has not downloaded music to their mobile device, a total of 52.7 Kg per year of $CO_2$ has been created, or as is presented in Table 10:

Ng (2019) states that "American smartphone owners used their phones for 22.7 months on average before upgrading. By 2018, that number had increased to 24.7...Users in Great Britain logged the longest average of 27.7 months in 2018...China was relatively shorter at 20.2 months in 2016 and 21 months for both 2017 and 2018."

For the year 2018, a medium average can be extrapolated from the three nations prior, where users, on average, kept their mobile devices for 24.4 months.

So, what $CO_2$ is typically generated per user in the case of modern music digital formats per year? A calculation of listening habits and format consumption over one year in relation to the creation of $CO_2$ is presented in Table 11.

There is however another side to this argument, where yes, a music format consumer may create the aforementioned amount of carbon per year due to the purchase of a new device and data streaming.

A consumer who purchases the physical format CD creates not only a by-product of $CO_2$ but also a largely unrecyclable tangible asset that ends up littering the shelves of second-hand shops. Alternatively, worse on the pile of ever-growing consumer electronics, discarded, never to be used again.

RIAA (2022), has a website that lists a database where sales of popular formats are shown over the past half-century. It can be seen that due to declining sales of CD technology sales, fewer CDs are being manufactured. By doing so, America alone has gone from using 15,080,000 Kg (942.5 million units) of various earth elements for the production of CDs in 2000 to just 505,600 Kg (31.6 million units) of various earth elements in 2021, a decrease of 14,574,400 Kg or 96.66% potential e-waste material, or:

**Table 10** Calculation of user listening habits over 1 year in relation to the creation of $CO_2$

| |
|---|
| **959.4 listening hours x 55 g $CO_2$Ph = 52.7 Kg $CO_2$ per year** |

**Table 11** The creation of $CO_2$ in one year in relation to listening habits and format consumption

| |
|---|
| 1 x typical mobile listening device at 69 Kg $CO_2$ ÷ 24.4 months x 1 year (12 months) = 33.9 Kg $CO_2$ per year average |
| + |
| 52.7 Kg $CO_2$ per year average (data streaming) |
| = |
| 86.6 Kg $CO_2$ created per year per consumer |

***162,410,000 Kg $CO_2$ per year in 2000 to 5,439,200 Kg Per year $CO_2$ per year as of 2021***

For some perspective, that is roughly the same amount of e-waste saved in kilograms (Kg) as the entire weight of the Shard or the equivalent carbon ($CO_2$) output of 36,843,402.5 cars on the road for the whole of an entire year (EPA, 2022).

What cannot be understated is how media (in the case of research within this document, music) has changed as it has passed from the physical, tangible realm into the digital intangible realm. However, due to the digital transformation that has taken place in the last half-century, whilst progressive gains have been made, a loss of other significant technological and cultural elements of consumer music format technologies has also been lost to time.

## 6 Who Wants to Live Forever?

### *6.1 Our Obtuse Obsolesce*

When considering modern audio formats, there seems to be a great schism that occurs when consumers consider digital products versus analogue counterparts. Moreover, if the same law of consumption and obsolescence that takes precedent today was in effect on all music formats, the likes of vinyl and tape would be but an echo of the past in the age of digital consumerism that is this digital era.

It should be of great concern to all in how on micro and macro levels, the use, replacement and obsolesce of our technologies is handled. For it is widely understood that humanity is causing global warming through means such as manufacturing. What is largely unknown is how even though innovation and progress may be of benefit for all, the lifespan or longevity of technologies used as of current pale in comparison to those of the past. Where a vinyl record from 1945 can interface with a record player created in the 2020s, many newer, more carbon-creating devices cannot play music or even interface with music services, even though these devices are only one-tenth the age of the vinyl.

This chapter proposed an analysis of how format usage and progression occur in society—focusing on five key points relevant to linearity and contextual historical development regarding format progression.

Much of the research in this chapter pays homage to the medium theorists and format theorists and follows the same sect of knowledge in pursuit of creating the reasoning regarding humanity's fascination with media, technology, history, and futurism.

A common theme in Medium Theory and medium theorists is that of how humans interact with media of all types and how media has an effect on individuals. For this, it is concluded that on one end of the medium spectrum, a receiver exists; for this, we will refer to the receiver as the human ear, in the middle is the device in question, and on the opposing end is what one could regard as a deliverable.

It is the idea that one side of the medium spectrum is a deliverable that has put into motion many questions regarding sociological format consumption within this research. For context, Medium Theory could be further segmented into three parts (Table 12).

What makes the idea of deliverables attractive is that deliverables never change. Society, technology, and human perception may change, evolve, and upgrade. But the deliverable? That always stays the same (Table 13).

This reasoning may also work for deliverables in the form of media formats, such as visual formats. For instance (Table 14):

For this, an addition to Medium Theory is called for, which I call Deliverable Theory, where the medium may change. However, the media type delivered by a medium and received by an organic (such as a human ear) is the same.

If it is medium theory that is the model in which a human exists on one end of the spectrum and the output on the other, such as presented in Fig. 8:

Then Deliverable Theory proposes that the medium is to be on one side of the spectrum, with a receiver on the other, where the media type delivered is unchanged (Fig. 9).

What is clear from research in this chapter is, that no matter how old the technological medium may be (for instance an iPod, CD or vinyl), the deliverable (in this case, audio regardless of its nature of analogue or digital), will always stay the

**Table 12** A simplified process of media being delivered from a media type to a receiver in this case a human

| Media Medium | Deliverable Media | Receiver (Human) |
|---|---|---|
| → | → | ◎ |

**Table 13** Deliverable theory model in relation to music media mediums as deliverables to a receiver

| | | |
|---|---|---|
| Media Medium (CD) | | |
| Media Medium (Vinyl) | | |
| Media Medium (Tape) | Deliverable Media (Music) | Receiver (E.g., Human ear) |
| Media Medium | | |
| (Mp3 Player) | | |
| Media Medium (Streaming) | | |
| → | → | ◎ |

**Table 14** Deliverable theory model in relation to visual media mediums as deliverables to a receiver

| | | |
|---|---|---|
| Media Medium (VHS) | | |
| Media Medium (DVD) | | |
| Media Medium (Blu-ray) | Deliverable (Media Film) | Receiver (E.g., Human eyes) |
| Media Medium (Streaming) | | |
| → | → | ◎ |

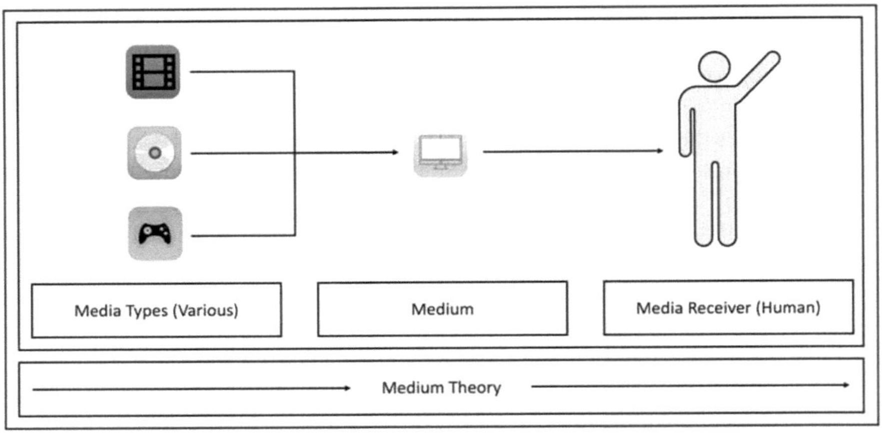


**Fig. 8** A simplified diagram explaining the process of medium theory

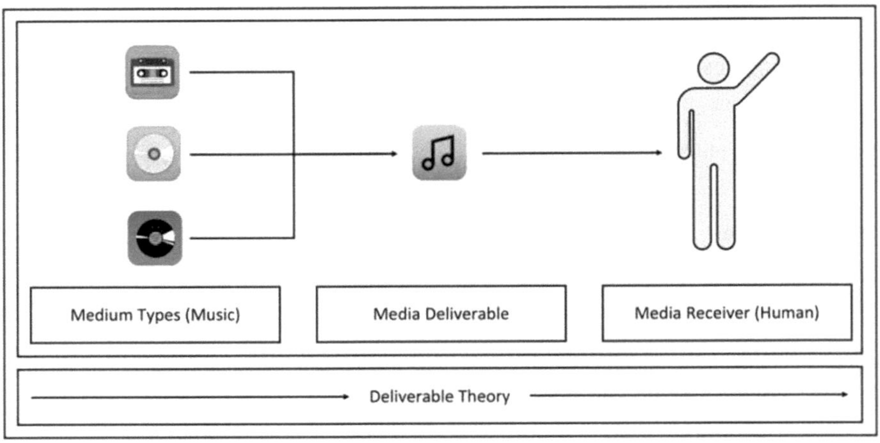


**Fig. 9** A simplified diagram explaining the process of deliverable theory

remain for the receiver (again, in this case, the listener). Therefore, with this understanding, a compelling argument can be made whereby the majority of mainstream listeners may believe that they are becoming far greener since they are purchasing fewer physical material products.

A crucial part of the conversation surrounding media formats and technology obsolescence lies in educating the public about the environmental consequences of their choices. While many people may feel that using digital music streaming services is the "green" option due to reduced physical product manufacturing, many may not realize the immense carbon footprint of data centers and the manufacturing cycles of mobile devices, which are updated much more frequently than the formats they replace. Raising awareness about these hidden environmental costs can empower consumers to make more informed choices. For instance, promoting the longevity of devices like vinyl turntables, CDs, or MP3 players could help reduce e-waste, which is one of the fastest-growing environmental concerns globally. Furthermore, advocating for a more sustainable approach to technology (such as repairability, durability, and upgradability) can shift the focus away from perpetual obsolescence toward more conscious consumption.

The purchase of a mobile listing device heavily outweighs the carbon output of purchasing various other formats that may only be used in a best-case scenario for only a handful of years. You see perhaps it may be humankind's desire for faster, newer or more convenience that compels the future, or maybe, just maybe it is *Our Obtuse Obsolescence.*

## Appendix

### *Supporting Evidence and Explanations for Table 4*

1. This means all MP3 players, but only MP3 players that are non-Internet connected devices.
2. This means all mobile devices, smartphone or other that has the ability to play MP3 or different music file types.
3. This means all computer devices that have the ability to play MP3 as a file type regardless of the operating system.
4. All Internet-connected networks with the ability to share files in a manner that is from one to another over an Internet connection.
5. Any service that allows for either the listening or downloading of MP3 or other audio file types as a service.
6. Format year of release implicitly implies from the first time the first device became available, no matter its level of success.
7. The CD was first released for public sale in 1982 (BBC, 2019; Phillips, 2022).
8. MPMAN was released in 1997, the first MP3 released in history for public use.
9. The release of the Samsung SPH-M100 was the first mobile telephone that could also be utilized as an MP3 player.
10. Winplay3, released in 1995 by the Fraunhofer institution as an add-on for the then-popular Windows 1995, was the first time that a listener could hear MP3 files on their computer.
11. The release of Napster and how this would ultimately change music forever with its peer-to-peer file transference systems.
12. The release of Spotify as the world's biggest streaming platform and the new change of music into the digital Internet age.
13. The amount of time it takes to print an album.
14. Can a company copy data or mass produce data easily?
15. Can a consumer copy data or mass produce data easily?
16. The amount of data that can be held on a format in minutes (device dependant*).

17. The amount of audio data that can be held upon a device in terms of songs, with the average consensus of a song being roughly 3.5 minutes (McKinney, 2015; Longdon, 2018; Wiebe, 2021).
18. The amount of audio data that can be held upon a device in terms of albums, with the average album being roughly 8–12 songs leaving a medium average of 10 songs (Marie, 2021).
19. Manufacturing materials of selected formats.
20. The phonograph cylinder was made of various materials, primarily bee's wax or other waxes such as ceresin or stearic wax (LOC, 2022).
21. Shellac disk manufacturing was made primarily from the resin of the female Shellac Bug (Yale, 2022).
22. PVC plastic is the main composition for $33^{1/3}$/45RPM vinyl disks ().
23. A tape can consist of various metals, metal oxides, magnets, and plastics (Britannica, 2022).
24. The compact disk (CD) comprises various metals such as aluminium, silver or gold and plastic (Madehow, 2022).
25. MP3 players manufacturing materials are extensive and contain various elements such as glass, metals, plastic and more (Perez, 2012).
26. Mobile listening devices (Mobile Phones) have an extensive list of materials used in their construction, from glass, metals, plastic, rare earth elements such as lithium and more (Vol, 2020).
27. Personal computers are made of various materials, some of which used in their construction are glass, metals, plastic, and other rare-earth elements such as gold (Wheeler, 2018).
28. During the manufacture of consumable products, by-products of the manufacturing process will be created, such as hazardous waste or other wastage.
29. During the manufacture of consumable products, the creation of $CO_2$ or Carbon Dioxide may be created as a by-product.
30. The average amount of $CO_2$ emitted per unit of vinyl manufacture is 2.2 kg (Perison- Hagger, 2021).
31. The average amount of $CO_2$ emitted per unit of Audio Cassette Tape manufacture is 2.8 kg (Perison-Hagger, 2021).
32. The average amount of $CO_2$ emitted per unit of compact disk (CD) manufacture is 172 gs (Perison-Hagger, 2021).
33. 45 Kg, the $CO_2$ for a typical MP3 player, in this case, the iPod Classic (released in 2009) (Apple, 2009).

    (a) 69 Kg, the $CO_2$ for a typical mobile listening device, in this case, the iPhone 13 Pro (released in 2021) (Apple, 2021).
    (b) 185 Kg, the $CO_2$ for a typical mobile listening device, in this case, the Macbook Pro 13 (released in 2020) (Apple, 2020).
    (c) The $CO_2$ is created via listening to one hour of non-downloaded music via a streaming service (Peirson-Hagger & Swindells, 2021).

        - Real-time recording means that a recording must be copied in real-time, therefore 1 minute of recording will take 1 minute exactly.

**—Device-dependent means that the format in question may have one or more iterations.

This creates a variable number that could not be quantified as part of this study.

– Data not present.

^—Data value the same as data above.

N/A—Data not currently available or established.

## *Additional Figures 10 and 11*

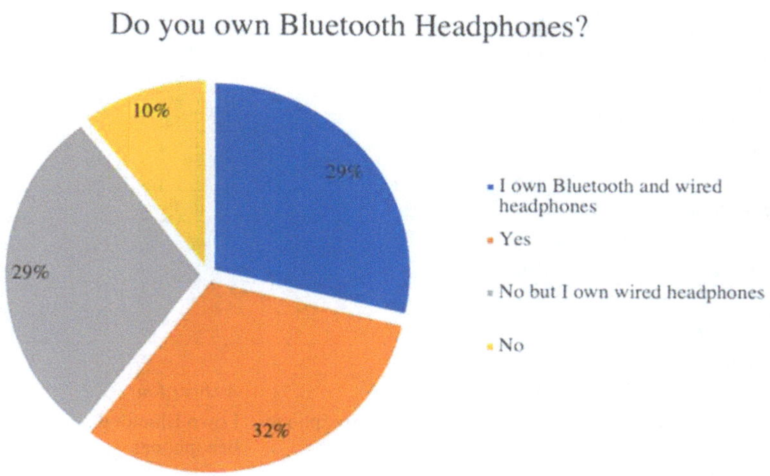


**Fig. 10** Appendix 1— Leey Survey One Question Five—Do you own Bluetooth headphones?

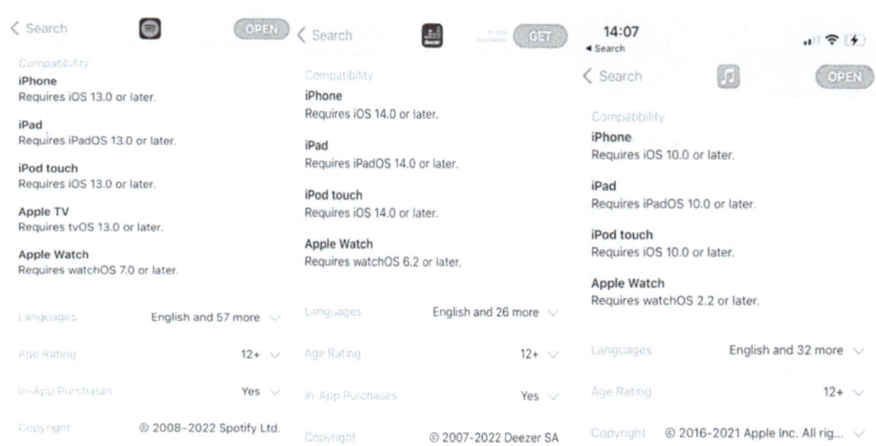


**Fig. 11** Appendix 2—Spotify, Deezer, Apple Music iOS compatibility in 2022

## *Survey One Questions*

Questions asked of participants in Leey Survey One, dated between the 26th of April 2022 and the tenth of May 2022 and does not contain questions pertaining to prizes offered to participants.

| | | | | |
|---|---|---|---|---|
| ***Question One*** | | | | |
| What is your email address? | | | | |
| *Answer* | | | | |
| Participant entered answer | | | | |
| ***Question two*** | | | | |
| What is your age? | | | | |
| *Answer one* | *Answer two* | *Answer three* | *Answer four* | *Answer five* |
| 18–24 | 25–31 | 32–45 | 46–55 | 55+ |
| ***Question three*** | | | | |
| Have you bought a CD (compact disk) since 2020? | | | | |
| *Answer one* | | *Answer two* | | |
| No | | Yes | | |
| ***Question four*** | | | | |
| Have you bought a vinyl since 2010? | | | | |
| *Answer one* | | *Answer two* | | |
| No | | Yes | | |
| ***Question five*** | | | | |
| Do you own Bluetooth headphones? | | | | |
| *Answer one* | *Answer two* | *Answer three* | *Answer four* | |
| No | Yes | No but I own wired headphones | I own Bluetooth and wired headphones | |
| ***Question six*** | | | | |
| How often do you stream music? | | | | |
| *Answer one* | *Answer two* | *Answer three* | *Answer four* | |
| Never | Once a month | Once a week | All the time | |

## *Survey Two Questions*

Questions asked of participants in Leey Survey One, dated between the tenth of June 2022 and the 24th of June 2022 and does not contain questions pertaining to prizes offered to participants.

| *Question One* |
|---|
| What is your email address? |
| *Answer* |
| Participant entered answer |

| *Question two* | | | | |
|---|---|---|---|---|
| What is your age? | | | | |
| *Answer one* | *Answer two* | *Answer three* | *Answer four* | *Answer five* |
| 18–24 | 25–31 | 32–45 | 46–55 | 55+ |

| *Question three* | | | | | | |
|---|---|---|---|---|---|---|
| What music streaming service do you use? | | | | | | |
| *Answer one* | *Answer two* | *Answer three* | *Answer four* | *Answer five* | *Answer six* | *Answer seven* |
| Amazon music | Apple music | Deezer | I don't | Spotify | Tidal | Other (participant entered Answer) |

| *Question four* | | | | |
|---|---|---|---|---|
| How do you listen to music in a car? | | | | |
| *Answer one* | *Answer two* | *Answer three* | *Answer four* | *Answer five* |
| I don't listen to music in the car | I listen to CD | I listen to radio | I listen to tape | I stream music in the car |

| *Question five* | | | |
|---|---|---|---|
| Do you use a CD player at home? | | | |
| *Answer one* | *Answer two* | *Answer three* | *Answer four* |
| I don't own a CD Player at home | I used to | Yes, all the time | Never |

## *Survey Three Questions*

Questions asked of participants in Leey Survey One, dated between the 13th of August 2022 and the 27th of August 2022 and does not contain questions pertaining to prizes offered to participants.

| *Question One* |
|---|
| What is your email address? |
| *Answer* |
| Participant entered answer |
| ***Question two*** |

| *Question One* | | | | |
|---|---|---|---|---|
| What is your age? | | | | |
| *Answer one* | *Answer two* | *Answer three* | *Answer four* | *Answer five* |
| 18–24 | 25–31 | 32–45 | 46–55 | 55+ |
| ***Question three*** | | | | |
| What leads you to buy new music technology products? | | | | |
| *Answer one* | *Answer two* | *Answer three* | *Answer four* | *Answer five* |
| I see reviews online | I see what friends have | New experience | Portability (size of a device) | Other (participant entered answer) |
| ***Question four*** | | | | |
| What do you feel is currently missing from your current music listening experience? | | | | |
| *Answer one* | *Answer two* | *Answer three* | *Answer four* | *Answer five* |
| I miss my | My battery | Sometimes I get | There is so | Other |
| Physical | Always dies | No internet | Much choice, | (participant |
| Collection | | Service, so I | But it's hard to | Entered answer) |
| | | Cannot listen to | Find new music | |
| | | music when I | | |
| | | Want to | | |

| ***Question five*** | | | | | |
|---|---|---|---|---|---|
| What is your current music listening device? | Rate your current music listening device (1 worst—5 best) | | | | |
| *Answer one* | *Answer one* | *Answer two* | *Answer three* | *Answer four* | *Answer five* |
| Participant entered answer | 1 | 2 | 3 | 4 | 5 |

## References


Innis, H. A. (1949). The bias of communication. *Canadian Journal of Economics and Political Science/Revue canadienne de economiques et science politique, 15*(4), 457–476.

Jones, G. (2010). *Last shop standing: Whatever happened to record shops?* Omnibus Press.

Mayers, K., Koomey, J., Hall, R., Bauer, M., France, C., & Webb, A. (2015). The carbon footprint of games distribution. *Journal of Industrial Ecology, 19*(3), 402–415. https://doi.org/10.1111/jiec.12181

McLuhan, M. (1964). *Understanding media, the extensions of man*. Massachusetts Institute of Technology.

Meyrowritz, J. (1994). Medium theory. *Communication Theory Today*, 50–77.

Nyre, L. (2008). *Sound media: From live journalism to music recording*. https://doi.org/10.4324/9780203869055.

Powers, D. (2014). Lost in the shuffle: Technology, history, and the idea of musical randomness. *Critical Studies in Media Communication, 31*(3), 244–264. https://doi.org/10.1080/15295036.2013.870347

Straw, W. (2009). The music CD and its ends. *Design and Culture, 1*(1), 79–91. https://doi.org/10.2752/175470709787375751

Weber, C. L., Koomey, J. G., & Matthews, H. S. (2010). The energy and climate change implications of different music delivery methods. *Journal of Industrial Ecology, 14*(5), 754–769. https://doi.org/10.1111/j.1530-9290.2010.00269.x

## Websites

Apple. (2020). *Product environmental report, 13-inch MacBook Pro.* Accessed Sep 20, 2022, from https://www.apple.com/environment/pdf/products/notebooks/13inch_MacBookPro_PER_Nov2020.pdf

Apple. (2021). *Product environmental report iPhone 13 Pro.* Accessed Sep 20, 2022, from https://www.apple.com/environment/pdf/products/iphone/iPhone_13_Pro_PER_Sept2021.pdf

BBC. (2019). *History of the CD: 40 years of the compact disk.* Accessed Sep 7, 2022, from https://www.bbc.co.uk/newsround/47441962

Bhalla, A. (2021). *6 gadgets that were killed by smartphones that we never thought would leave our lives ever.* Accessed Jun 7, 2022, from https://www.mensxp.com/technology/news/76425-6-gadgets-that-were-killed-by-smartphones-that-we-never-thought-would-leave-our-lives-ever.html

Bowe, T. (2022). *Yes, you should still be buying CDs, Here's Why.* Accessed Sep 11, 2022, from https://www.gearpatrol.com/tech/audio/a731474/reasons-to-buy-cds/

Britannica. (2022). *Casette.* Accessed Aug 29, 2022, from https://www.britannica.com/technology/cassette

Broom, J. (2022). *Apple music boss admits most can't tell the difference between compressed vs lossless audio.* Accessed Aug 9, 2022, from https://www.mrafterparty.com/apple-music-boss-admits-most-cant-tell-the-difference-between-compressed-vs-lossless-audio

Deezer. (2022). *Update to IoS 15.* Accessed Aug 29, 2022, from https://support.deezer.com/hc/en-gb/articles/4400035203357-Update-to-iOS-15

EPA. (2022). *Greenhouse gas emissions from a typical passenger vehicle.* Accessed Aug 29, 2022, from https://www.epa.gov/greenvehicles/greenhouse-gas-emissions-typical-passenger-vehicle

Fisher, T., & Ganz, J. (2015). *How well can you hear audio quality?* Accessed Sep 15, 2022, from https://www.npr.org/sections/therecord/2015/06/02/411473508/how-well-can-you-hear- audio-quality

Green, A., & Gibson, M. (2021). *Hours spent listening to music on the rise globally—Report.* Accessed Aug 29, 2022, from https://www.walesonline.co.UK/news/UK-news/hours-spent-listening-music-rise-21922829

Hull, S. (2014). What Is The RIAA Curve? Accessed Sep 21, 2022, from https://sessionville.com/articles/what-is-the-riaa-curve

Kramer, J. D., (2022). *The impact of technology on the musical experience.* Accessed Sep 10, 2022, from https://www.music.org/index.php?option=com_content&view=article&id=2675:the-impact-of-technology-on-the-musical-experience&catid=220&Itemid=3665

Law, J. (2021). *Here's why your old cellphone may be obsolete in 2022.* Accessed Aug 29, 2022, from https://www.boston25news.com/news/local/heres-why-your-old-cellphone-may-be-obsolete-202

LOC. (2022). *History of the cylinder phonograph.* Accessed Sep 11, 2022, from https://www.loc.gov/collections/edison-company-motion-pictures-and-sound-recordings/articles-and-essays/history-of-edison-sound-recordings/history-of-the-cylinder-phonograph

Longdon, V. (2018). *Why are pop songs 3 minutes long?* Accessed Aug 10, 2022, from https://www.classicfm.com/discover-music/why-are-pop-songs-3-minutes/

Lovejoy, B., (2019). *Poll: Looking back now, did Apple exhibit 'courage' in removing the headphone jack from iPhones?* Accessed Sep 15, 2022, from https://9to5mac.com/2019/04/08/courage

Madehow. (2022). *Compact disk.* Accessed Aug 29, 2022, from https://www.madehow.com/Volume-1/Compact-Disc.html

Mckinney, K. (2015). *A hit song is usually 3 to 5 minutes long. Here's why.* Accessed Aug 10, 2022, from https://www.vox.com/2014/8/18/6003271/why-are-songs-3-minutes-long

Marie, Y. (2021). *How many songs should be on an album?* Accessed Sep 11, 2022, from https://yonamariemusic.com/yona/blog/46/how-many-songs-should-be-on-an-album

Ng, A., (2019). *Smartphone users are waiting longer before upgrading—here's why.* Accessed Aug 29, 2022, from https://www.cnbc.com/2019/05/17/smartphone-users-are-waiting-longer-before- upgrading-heres-why

Peirson-Hagger, E., & Swindells, K. (2021). *Is Spotify bad for the environment?.* Accessed Sep 20, 2022, from https://www.newstatesman.com/environment/2021/11/how-environmentally-damaging-is- music-streaming

Perez, C., (2012). *Mp3 player manufacturing process.* Accessed Sep 20, 2022, from https://prezi.com/lpp- q5sfcotu/mp3-player-manufacturing-process/

Phillips. (2022). *The history of the CD–The CD family.* Accessed Sep 4, 2022, from https://www.philips.com/a-w/research/technologies/cd/cd-family

Reitemeier, M., (2020). *The sad truth: Smartphones killed the MP3 player.* Accessed Jun 7, 2022, from https://phenixxgaming.com/2020/03/12/the-sad-truth-smartphones-killed-the-mp3-player/

RIAA. (2022). *U.S. sales database.* Accessed June 10, 2022, from https://www.riaa.com/u-s-sales-database/

Statista. (2018). *Unit sales of the Apple iPhone worldwide from 2007 to 2018.* Accessed Aug 29, 2022, from https://www.statista.com/statistics/276306/global-apple-iphone-sales-since-fiscal-year-2007

Stegner, B., (2021). *How long do iPhones last? An analysis.* Accessed Aug 29, 2022, from https://www.makeuseof.com/how-long-do-iphones-last

The World Bank. (2022). *Life expectancy at birth, total (years).* Accessed Aug 29, 2022, from https://data.worldbank.org/indicator/SP.DYN.LE00.IN

Vol, D. (2020). *What materials are used to make SmartPhones?* Accessed Sep 20, 2022, from https://curiosityguide.org/technology/what-materials-are-used-to-make-smartphones/

Wheeler A. (2018). *What raw materials are used to make hardware in computing devices?* Accessed Sep 20, 2022, from https://www.engineering.com/story/what-raw-materials-are-used-to-make-hardware-in-computing-devices

Wielen, B., (2017). *Celebrating 10 Years of iPhones: 63% of all iPhones Ever Sold Are Still in Use—728 million.* Accessed Aug 29, 2022, from https://newzoo.com/insights/articles/63-percent-of-all-iphones-ever-sold-still-in-use

Wiebe, D. A. (2021). *How long should a song be, both for radio and other?* Accessed Aug 9, 2022, from https://www.musicindustryhowto.com/how-long-should-a-song-be-both-for-radio-and- other/

Wilkins, S. (2021). *"1,000 songs in your pocket".* Accessed Aug 15, 2022, from https://influence.digital/1000-songs-in-your-pocket/

Yale. (2022). *The history of 78 RPM recordings.* Accessed Aug 29, 2022, from https://web.library.yale.edu/cataloging/music/historyof78rpms

**Christopher Cohen** (MA/BA/MCMI) is a distinguished media expert with a deep passion for music, technology, and the intricate relationship between humans and their socio-bio-mechanical extensions. His academic journey began with notable success at the University of Brighton, where he was honoured with the "Outstanding Achievements Award 2016", and a brief fascination with Medium Theory that broadened his intellectual horizons.

After university, Christopher moved to China, where he managed a highly successful art company in Shenzhen. Here, his team created music that resonated with over 130 million people globally, including a single event that reached an audience of over 10 million. This experience ignited his passion for the intersection of art, design, management, technology, and social interactivity, laying the foundation for his deep involvement with Web3 technologies.

Returning to the UK, Christopher pursued a master's degree at the University of West London, earning a distinction with an impressive average grade of 76%. His dedication to research and data analysis during this period was pivotal, culminating in his current work toward his first publication in media sustainability. His academic excellence and global presentations on Web3, music, and technology have solidified his reputation as a thought leader in these fields.

Christopher's research focuses on the complex relationships between humans and their technological counterparts. His unique perspective is shaped by his experiences living in some of the world's most futuristic cities, enhancing his understanding of what he calls our computing counterparts or "socio-bio-mechanical extensions". This research is vital in pioneering new understandings of human–technology interactions and their future trajectories.

Looking ahead, Christopher aspires to pursue a PhD in media and communications. His ultimate goal is to lecture, research, and mentor future generations, inspired by the mentors who shaped his journey. He is fluent in three languages and has travelled to 32 countries. Christopher's global experiences and a deep appreciation for diverse cultures enrich his worldview.

Christopher's mission is to share his story and inspire his students to pursue their dreams, demonstrating that education is the key to overcoming adversity and achieving greatness. His journey from a challenging childhood to becoming a global media expert exemplifies the transformative power of education and determination.

# Sustainability Standards in Film Production

**Carmela Gulluscio, Cecilia Chirieleison, and Teresa Turzo**

**Abstract** Reflecting a global trend, this chapter examines the integration of sustainability standards in the film industry, comparing standards from different countries and highlighting their significance. The main objective is to provide a comprehensive analysis of the primary sustainability standards in the film industry from an international perspective. This includes evaluating each standard's contribution to sustainability and comparing their respective advantages and challenges.

In Europe, the commitment to sustainable filmmaking is clearly demonstrated by several initiatives. The UK's Albert Certification, provided by the British Academy of Film and Television Arts (BAFTA), is a relevant certification standard focused on the reduction of carbon footprints and encouraging eco-friendly production methods. In France, the EcoProd initiative, from a film-environment consortium, addresses environmental impacts minimization. Italy Green Film rating system, developed by national industry bodies, offers a process for evaluating films based on their sustainability. North America has also made significant advances in promoting sustainable practices in film production. The PGA (Producers Guild of America) Green Initiative is a principle-based standard that includes resources like best practices, a carbon calculator, and a vendor guide. Similarly, Canada's Ontario Green Screen initiative provides resources such as climate and sustainable production training, a carbon calculator, and case studies, demonstrating a robust commitment to reducing environmental impacts and fostering a sustainable culture in film industry.

C. Gulluscio (✉)
Department of Law and Economics, University of Rome UnitelmaSapienza, Rome, Italy
e-mail: carmela.gulluscio@unitelmasapienza.it

C. Chirieleison
Department of Political Science, University of Perugia, Perugia, Italy
e-mail: cecilia.chirieleison@unipg.it

T. Turzo
Department of Law and Economics, University of Rome UnitelmaSapienza, Rome, Italy
e-mail: teresa.turzo@unitelmasapienza.it



M. Todorovic et al. (eds.), *Sustainability and Social Responsibility of the Media and in the Media*, CSR, Sustainability, Ethics & Governance,
https://doi.org/10.1007/978-3-032-00086-6_4

This chapter reviews the primary sustainability standards in the film industry and performs a critical comparison of them, exploring their effectiveness and challenges. It concludes by highlighting the role of these standards in shaping a more sustainable future for the film industry, thereby contributing to an evolving conversation on sustainability and SDGs (Sustainable Development Goals).

**Keywords** Accountability · Accountability tools · Albert certification · Best practices · Carbon footprint · Certification standards · Certification-based standards · Corporate sustainability · EcoProd initiative · EcoProd label · Environmental management · Environmental policy · Environmental stewardship · Film industry · Green film rating system · Green production · Industry-specific standards · Media sustainability · Ontario green screen initiative · PGA green initiative · Principle-based standards · Resource efficiency · SDGs · Sustainable development · Sustainable development goals · Sustainable filmmaking · Sustainability · Sustainability-related strategies · Sustainability standards · Waste management

## 1 Introduction

In the last decades, sustainability gained momentum, driven by factors such as increased stakeholder awareness of environmental, social, and ethical issues, as well as the release of the Sustainable Development Goals (SDGs) by the United Nations. SDGs, particularly the Goal 13 ("Climate Action") and the Goal 12 ("Responsible Consumption and Production"), highlight the necessity for all industries to implement sustainability strategies (Pizzi et al., 2020).

Historically viewed as a low-impact industry with regard to sustainability related issues, it has recently become evident that film industry is a relevant contributor to environmental problems due to its high energy consumption, substantial waste production, and significant $CO_2$ emissions (Bozak, 2011; Maxwell & Miller, 2012). In this regard, a report released by British Academy of Film and Television Arts (BAFTA) albert consortium, the British Film Institute (BFI) and Arup (2020) indicates that producing a big-budget movie, i.e., a movie costing over $70 million, can produce about 2840 tons of $CO_2$, the equivalent of 11 trips to the moon and the amount of $CO_2$ that 3709 acres of forest would absorb in a year (BAFTA albert consortium et al., 2020).

Some examples can be emblematic to clarify the impact of film production in terms of sustainability. For instance, a large part of the famous Fox Studios' movie entitled "Titanic" was filmed in a water tank in the Mexican village of Popotla, Baja California. During filming, significant environmental damage occurred, such as the chlorination of seawater, which destroyed local sea urchin populations and reduced overall fish levels by 33%. The statistics indicate that the financial resources spent on the production of this film could have provided safe drinking water for 600,000 people for an year (Maxwell & Miller, 2012). Similarly, during the recording of the movie "The Beach" in the Phi Phi Islands of the National Park of Thailand, the production crew bulldozed natural landscapes, rearranged native plants, and planted coconut palms to match the studio's vision of a tropical paradise. These actions led to the collapse of sand dunes during the next monsoon season, deleting natural erosion defenses (Maxwell & Miller, 2012). The beach was subsequently closed due to the overwhelming number of tourists that the film attracted. In 2022, Fox was ordered to compensate the Thai government with ten million baht, roughly 273,000 dollars, for the environmental damage caused during filming and the subsequent overtourism (The Guardian, 2022).

As people have become more aware of these issues, companies from the film industry have faced with a urgent need for sustainability strategies and sustainability-focused accountability strategies (Campbell et al., 2003; Elalfy et al., 2021; Gulluscio, 2023; Kääpä & Vaughan, 2022a; Victory, 2015). This acknowledgment has led to the development and implementation of various sustainability standards in order to reduce the environmental impact of film productions (Elalfy et al., 2021). These standards serve as tools of indirect accountability, which ensure that companies follow sustainable practices by making their efforts clear and measurable (Turzo et al., 2024).

From a theoretical standpoint, two theories can help to frame the reasons behind the adoption of sustainability standards by a film producer. In detail, under the lens of legitimacy theory, the adoption of a sustainability standard can help the film producer to enhance transparency and obtain the "social license to operate" by following practices that meet the expectations of stakeholders (Gehman et al., 2017; Gulluscio, 2023; Suchman, 1995). In the meanwhile, according to the fundamental assumptions of the signaling theory, the adoption of a sustainability standard may represent an explicit signal of the deep commitment of the film producer to sustainability, thereby promoting the creation of a positive perception among stakeholders regarding the product and attracting investors and audiences who care about the environmental protection and other sustainability-related issues (Connelly et al., 2011).

Several sustainability standards there exist to promote sustainable practices within the specific field of film industry. In the UK, the Albert Certification, managed by the BAFTA, focuses on reducing carbon emissions and encouraging sustainable production methods. Similarly, the EcoProd initiative in France, developed by a consortium of environmental entities, aims to reduce the environmental impacts of film productions. In Italy, the Green Film Rating System, developed by national industry bodies, require a structured process for the evaluation and the enhancement of the sustainability levels of films. In the USA, the Green Initiative of Producers Guild of America (PGA) provides resources and guidelines to foster sustainability-related practices in the film and television industry. Finally, the Ontario Green Screen initiative of Canada offers resources and training aimed at reducing the impacts of film production on the natural environment and promote sustainability in the culture of film industry.

The chapter is organized as follows. After the introduction, the second section presents a detailed analysis of the main sustainability standards of the film industry, including the Albert Certification in the UK, the EcoProd Initiative in France, the

Green Film Rating System in Italy, the PGA Green Initiative in the USA, and the Ontario Green Screen initiative in Canada. The third section contains a comparative analysis of the mentioned sustainability standards. The last section outlines the conclusions and the limitations of the standards included in the analysis.

## 2 The Main Sustainability Standards in the Film Industry

### *2.1 Albert Certification (UK)*

The Albert Certification is a sustainability certification standard released by the BAFTA in collaboration with the BFI and Arup. BAFTA is an independent charity aimed at support, develop, and promote the art forms of the moving image, such as film, television, and videogames. It is widely known for its annual awards ceremonies that award the excellences of these creative fields. Within BAFTA, the BAFTA albert consortium is specifically dedicated to advancing sustainability in the film and TV industry. The initiative aims to reduce the environmental impact of these industries by offering practical tools, comprehensive training, and robust certification standards for sustainable production practices. A relevant action is represent by the introduction of the Albert Certification in 2011, with the goal of reduce the environmental impact of UK film and television production (BAFTA albert consortium et al., 2020).

Albert Certification is mandatory for all new commissions and recommissions of TV broadcast content, including television, children's, education, and non-news related sports programs. For digital contents such as YouTube videos, certification is encouraged but not mandatory, although an Albert carbon footprint is required. Exemptions apply to very short, low-budget, or acquired content, as well as programs made entirely from archive footage or post-production only, and co-productions where the BBC represents a minority funder.

The certification process involves three main pillar: (1) the completion of an Albert carbon footprint; (2) the development of a carbon action plan; and, (3) the offset of any residual emissions that cannot be eliminated (Kääpä & Vaughan, 2022a).

The carbon footprint form, which is the main part of the first pillar, requires data input on production's travel, accommodation, energy use, staffing, and post-production activities. This form is created using the Albert carbon calculator (available at https://calc.wearealbert.org) and should be completed as early as possible during pre-production phase. Once the carbon footprint form is completed, it must be checked and approved by the production company's nominated Albert reviewer and then submitted to an external auditor for a standard assurance review. After the approval, the reviewer will receive an email confirming the final carbon footprint, which must be forwarded to the BBC delivery contact within 6 weeks of the final episode delivery.

The second pillar, which is the elaboration of a carbon action plan, assists film producers to identify how they can reduce emissions descending from their products. The carbon action plan should be discussed, finalized, and submitted to Albert before the beginning of the filming process.

Offsetting is the core of the third pillar and require film producers to compensate for the carbon emissions that cannot be eliminated through investments in environmental projects aimed at the reduction of carbon emissions.

The Albert Certification process includes regular audits conducted by an independent auditor to ensure the data submitted is accurate and reliable. During the audit process, the external auditors perform a deep review of the carbon footprint and carbon action plan submitted by the production company. They conduct a detailed assurance review, checking the data against industry standards and benchmarks. The auditors verify if the production has correctly computed its carbon footprint and taken appropriate steps to reduce emissions. If everything results correct, the auditors approve the data and provide a feedback, confirming that the production meets the requirements of the Albert Certification standard (Kääpä & Vaughan, 2022b). Once certified, productions can use the Albert logo in their program endboard, in order to signal their effective commitment to sustainable practices and acquire legitimacy (BAFTA albert consortium et al., 2020).

## *2.2 EcoProd Initiative (France)*

EcoProd is the primary initiative in the promotion of sustainable film production in France. The main aim is the reduction of the environmental footprint in the film and TV industry. EcoProd is a nonprofit association founded in 2009 and established by Audiens, the CST, Film Paris Region, and major broadcasters like Canal+, France Télévisions, and TF1 (EcoProd, 2024).

EcoProd serves a diverse group of stakeholders in the film industry, including production companies, broadcasters, service providers, unions, film schools, and public institutions. With over 370 companies involved, EcoProd offers a collaborative platform to promote green production practices both in France and internationally (EcoProd, 2024). As an independent resource center, it provides various free tools, such as the Carbon'Clap carbon calculator certified by the CNC, green production guides, and specialized training for industry professionals and students (Ruiz Guzmán & de Monterrey, 2022).

In 2022, EcoProd introduced the EcoProd Label in partnership with Association Française de Normalisation (AFNOR) Certification organization. The last is the leading French organization specialized in certifying management systems, products, and services across various industries. With respect to EcoProd Label, ANFOR Certification organization is responsible for the conduction of the audits and the verification that film productions meet the established sustainability criteria before the obtainment of certification.

The EcoProd Label is a voluntary certification and can be applied to a film, advertisement, or TV program produced in a sustainable manner. The certification process involves several steps, starting with the application to AFNOR Certification. To qualify for the EcoProd Label, a production must score at least 65% based on a comprehensive framework listing 80 actions designed to reduce environmental impact, including eight mandatory criteria related to: (1) carbon footprint calculation; (2) energy efficiency; (3) waste management; (4) sustainable transportation; (5) eco-friendly materials; (6) water conservation; (7) biodiversity protection; and, (8) green communication.

The key element of the certification process is the use of Carbon'Clap (available at https://www.ecoprod.com/en/carbon-clap.html), a carbon calculator that helps producers to measure their environmental footprint and identify the main sources of $CO_2$ emissions. Producers have to elaborate a detailed action plan to reduce emissions, which is submitted before the beginning of the filming process. The action plan details the steps the film production will take to reduce its environmental impact, such as using energy-efficient equipment, reducing waste, and optimizing transportation and logistics to lower carbon emissions.

During the filming process, each of the member of the production team should collect evidence related to sustainability actions taken and upload them to a shared "green evidence folder," facilitating easy documentation procedure. Such a collaborative approach ensures that all members of the production team are aware of and committed to the sustainability goals. The collected evidence can include receipts for eco-friendly materials, logs of energy usage, records of waste disposal, and documentation of sustainable practices implemented on set.

AFNOR Certification organization conducts audits to verify the evidence provided. These audits, performed by means of on-site visit, online, or through document reviews, ensure the credibility of the certification through a deep assessment of compliance with the sustainability requirements proposed by the EcoProd standard. Auditors check for consistency and completeness in the documentation and can conduct interviews with the production team to gather additional information.

Following the audit, the adherence of the production to the EcoProd Label criteria is evaluated. If all requirements are met, the production is awarded the EcoProd Label and can use the certification logo in its end-boards. The final evaluation report is reviewed and validated by EcoProd, which then issues the labeling certificate specifying the level of certification achieved. The use of the EcoProd logo indicates that the production has successfully implemented sustainable practices and met the rigorous requirements set by the standard.

Since its introduction, about 50 productions have applied the EcoProd Label framework, in order to attest their attempt to minimize the environmental impact of their production. Feedback from these productions has helped refine the certification requirements and audit process. EcoProd continuously updates its standards through ongoing research and collaborations with institutions like Agence de la Transition Écologique (ADEME)—which is the French agency for the ecological transition—and Institut de Recherche sur le Cinéma et l'Audiovisuel (IRCAV)—the most important French institute for the research on media industry, engaged in the promotion of sustainability and innovation in the same field—, ensuring the framework remains relevant and effective.

## *2.3 Green Film Rating System (Italy)*

The Green Film Rating System is an initiative designed to enhance environmental stewardship within the Italian film industry. Created in 2019 by the Trentino Film Commission in collaboration with various film industry stakeholders, including environmental consultants and film professionals, this certification standard emerged from the growing awareness of the significant environmental impact associated with film production and the necessity for a standardized approach to mitigate these impacts (D'Urso, 2022).

The Green Film Rating System focuses on individual film and television projects, offering a voluntary certification for implementing sustainable practices. Producers and production companies can adopt this standard to align their projects with sustainability principles, meeting the growing demand from stakeholders for eco-friendly practices in the film industry (Green Film, 2022).

The certification process starts with producers downloading and reviewing the Green Film guidelines, which outline the steps needed to minimize environmental impact during production. Producers then complete a checklist within the guidelines, specifying which sustainability criteria they will commit to meeting. These criteria include energy saving, transport and accommodation optimization, eco-friendly catering, sustainable material selection, waste management, and effective communication of sustainable practices (Green Film, 2022).

After completing the checklist, producers must ensure compliance with the selected criteria throughout the production phase. An independent verifying body, which could be an environmental certification company or a public entity with expertise in environmental matters, audits the production to verify adherence to the criteria.

To obtain the Green Film certification, the production undergoes a comprehensive verification process involving both document verification and on-set inspections. Before filming begins, the production must submit a sustainability plan, including the appointment and curriculum vitae of a Green Film Manager with at least 2 years of experience in environmental consulting. The sustainability plan outlines the chosen criteria and details the actions to ensure the production is environmentally sustainable. This plan is then sent to the certifying body, any appointed party, the verifying body, and all production crew and staff members.

An essential part of the process is a kick-off meeting led by the Green Film Manager, involving all departments to discuss the sustainability plan and ensure everyone understands their role in its implementation.

The verification procedure includes the review of several documents, including the sustainability plan, the nomination of the Green Film Manager, and communication records sent to the crew. On-set verification checks the actual application of the plan, including the use of vehicles, fuel consumption, and the number of trips made during the production. In this regard, the Standard underlines the importance of transport optimization. A Mobility Manager must be appointed to create a "Transport optimization plan," which describes the number and type of vehicles used, the number of trips, fuel consumption, and the number of occupants per vehicle. The Transport optimization plan must be adhered to and updated throughout the production to ensure a minimal environmental impact.

The certification is awarded based on a scoring system where each criterion earns points. To obtain the certification, a production must achieve a minimum score of 20 points, ensuring compliance with all mandatory prerequisites. Productions that hold the Eco-Management and Audit Scheme (EMAS) registered benefit from a reduced minimum score requirement by five points.

The Green Film Rating System also includes provisions for continuous improvement and feedback. Productions that receive certification are encouraged to maintain and enhance their sustainable practices in future projects.

## 2.4 PGA Green Initiative (USA)

The PGA addresses sustainable filmmaking through the establishment of the Green Production Guide, a comprehensive standard designed to reduce the environmental impact of film and TV productions (Kääpä & Vaughan, 2022a). The Green Production Guide, created in a collaboration with Green Media Solutions LLC and Earth Angel Sustainable Production Services LLC, was launched in 2014. This initiative aims to encourage producers to implement sustainable practices across all phases of the film production (Victory, 2015). The guide is targeted at independent producers and production teams, providing detailed strategies for reducing carbon emissions and promoting sustainability in their projects.

The Green Production Guide is a principle-based standard, as it is a comprehensive resource that provides guidelines and best practices for each production phase, covering areas such as the production office, locations, water use, transportation, set design, props, catering, lighting, camera work, sound, special effects, wardrobe, hair, makeup, and post-production. The aim is to reduce consumption, waste, and pollution, promoting a more sustainable production environment (Green Media Solutions, 2014).

The process described within the Green Production Guide includes several critical steps and requirements. Firstly, producers are encouraged to embrace sustainability in the pre-production phase, involving both filmmakers and funders in the conversation. They should set achievable sustainability goals, hire an "Eco Supervisor," and hold meetings to communicate these goals to department heads and the whole crew.

To facilitate sustainability, producers are urged to use the PGA's carbon calculator (available at https://greenproductionguide.com/tools/#pgc-w64289dc9aad37-0-1) to track their production's carbon footprint and identify areas for improvement. The carbon calculator helps measure emissions from various activities, such as travel, energy use, and waste management. The results guide the development of a sustainability plan, which includes strategies for reducing emissions and implementing green practices.

A relevant element of the Green Production Guide is the emphasis on regular audits and monitoring. Although the guide does not include a formal certification process, it does involve both internal reviews and external audits by accredited auditors to ensure compliance with the requirements of the standard. These audits check for the accuracy of data submitted, assess the implementation of green practices, and ensure the integrity of the process. The involvement of external auditors allows producers to maintain high levels of transparency and enhance their level of accountability, through on-site inspections and the review of the documentation to validate the sustainability efforts.

Key requirements of the Green Production Guide include energy-saving measures, such as using fluorescent or LED lighting and opting for renewable energy sources when possible. The guide also emphasizes the relevance of sustainable set design and construction, encouraging the use of recycled materials and designing sets for easy disassembly and reuse. Catering services are advised to use locally sourced, organic, and fair-trade food products, with an emphasis on composting and waste reduction.

Throughout the filming process, the Eco Supervisor monitors progress, provides guidance, and ensures compliance with predefined sustainability practices and the requirements of the standard. Regular updates and evaluations are conducted to keep the crew engaged and address any challenges that arise. At the end of the filming process, a comprehensive sustainability report is created, detailing both successes and areas for improvement.

## *2.5 Ontario Green Screen Initiative (Canada)*

The Ontario Green Screen Initiative represents a significant step towards the promotion of sustainability within the film industry. It was launched by Ontario Creates in collaboration with key industry stakeholders in 2020. Ontario Creates is a governmental agency in Ontario designed to support the creative industries of the province—film, TV, music, publishing, and interactive digital media.

The primary aim of the Ontario Green Screen Initiative is to provide comprehensive resources and training to reduce the ecological footprint of film and television productions in Ontario.

The initiative is targeted at all professionals within the Ontario film industry, from producers and directors to crew members and support staff. Its inclusive approach provides everyone involved in a film production with the knowledge and tools necessary to implement sustainable practices.

The Ontario Green Screen Initiative is a principle-based standard. Thus, similarly to the PGA initiative of the USA, the Canadian initiative provides a framework and guidance to promote the voluntary adoption of sustainable practices.

The training provided by the Ontario Green Screen Initiative is a core component of its mission. The training covers a broad range of topics related to climate and sustainable production, offering film workers the knowledge needed to understand the climate emergency and the specific actions they can take on set to make a difference. The education program is designed to foster a deep understanding of environmental stewardship and discover how sustainability can be integrated into everyday production activities.

The Ontario Production Guide (2020), an essential resource offered by the initiative, is the most comprehensive guide to goods and services related to film and television production in Ontario. It includes contact information and details for a wide variety of services such as labor, studios, post-production, 3D, VFX & animation, equipment, accommodations, transportation, casting, personal services, government, and financial services. The guide is designed to support productions at every stage, from preparation to wrap-up to post-production, ensuring that all aspects of a production can be managed sustainably (Ontario Creates, 2020).

The requirements outlined by the Ontario Green Screen Initiative emphasize the importance of planning and commitment to sustainability. Productions are encouraged to develop and implement a sustainability plan that includes specific goals and strategies for reducing their environmental impact. This includes measures to minimize waste, reduce energy consumption, and promote the use of sustainable materials and practices throughout the production process.

While the initiative does not mandate formal certification, it does advocate for a structured approach to sustainability. Productions are encouraged to document their sustainability efforts and share their progress with stakeholders. The documentation process serves as a form of indirect accountability, ensuring that producers are transparent about their environmental impact and the steps they are taking to mitigate it.

Furthermore, the Ontario Green Screen Initiative underlines the importance of regular audits and monitoring. Productions are encouraged to conduct internal reviews and assessments to ensure compliance with their sustainability plans. These audits, when conducted, help maintain the integrity of the sustainability efforts by verifying that the practices outlined in the sustainability plan are being implemented effectively. External auditors may also be involved, providing an additional layer of transparency and assurance.

## 3 Comparison of the Standards

A detailed comparison of the sustainability standards analyzed in the previous sections reveals several interesting insights. For example, the most recent certification is the Ontario Green Screen Initiative, launched in 2020, while the oldest is the

EcoProd Initiative, established in 2009. The Albert Certification, EcoProd Initiative, and Green Film Rating System offer formal certification processes, whereas the PGA Green Initiative and Ontario Green Screen Initiative are principle-based standards that emphasize voluntary adoption and internal reviews. Each standard's scope varies, with the Albert Certification focusing on TV broadcast content, EcoProd and Green Film targeting film and TV productions, and the PGA and Ontario initiatives encompassing broader film and TV production activities. The mandatory versus voluntary nature of these standards also differs; Albert Certification is mandatory for certain TV content, while all the others are voluntary.

Table 1 provides a summary of the comparative analysis of the key aspects of the sustainability standards in the film industry.

From the analysis of the main sustainability standards in the film industry, several common points and differences emerge. Each standard addresses the need for

**Table 1** Comparative analysis of sustainability standards in film industry

| Standard / Standard feature | Albert Certification (UK) | EcoProd Initiative (France) | Green Film Rating System (Italy) | PGA Green Initiative (USA) | Ontario Green Screen Initiative (Canada) |
|---|---|---|---|---|---|
| **Issuance year** | 2011 | 2009 | 2019 | 2014 | 2020 |
| **Standard nature** | Certification-based standard | Certification-based standard | Certification-based standard | Principle-based standard | Principle-based standard |
| **Scope** | Film and TV broadcast content | Film and TV productions | Film and TV productions | Film and TV productions | Film and TV productions |
| **Primary geographic scope** | UK | France | Italy | USA | Ontario province |
| **Mandatory/ voluntary** | Mandatory for TV contents | Voluntary | Voluntary | Voluntary | Voluntary |
| **Key requirements** | • Carbon footprint, action plan<br>• Offsetting | • Carbon footprint<br>• Action plan<br>• 8 mandatory criteria | • Energy saving transport catering<br>• Materials<br>• Waste<br>• Communication | • Energy saving<br>• Set design<br>• Catering | • Sustainability plan<br>• Waste<br>• Energy<br>• Materials |
| **Audit** | External audits | External audits by AFNOR certification organization | Document verification and on-set inspections | Internal and external audits | Internal and external audits |
| **Auditor** | External auditors | AFNOR certification organization | Trentino Film commission | Accredited auditors | External auditors |

Source: authors' own elaboration

the implementation of sustainability strategies and sustainability-related accountability strategies in film industry, with a different focus in terms of approaches, requirements, and auditing.

In analyzing the scope, primary geographic scope, mandatory/voluntary nature, and key requirements of these standards, there are notable similarities and differences. The Albert Certification mandates completing an Albert carbon footprint, developing a carbon action plan, and offsetting residual emissions. It is primarily applied in the UK, with exemptions for very short, low-budget, or acquired content, and certain co-productions (BAFTA albert consortium et al., 2020). The EcoProd Label, a voluntary certification, is based on a framework of 80 actions, including eight mandatory criteria such as carbon footprint calculation, energy efficiency, waste management, sustainable transportation, eco-friendly materials, water conservation, biodiversity protection, and green communication. Primarily used in France, productions must score at least 65% to qualify (EcoProd, 2024).

The Green Film Rating System, applied in Italy, outlines criteria in six key areas: energy saving, transport and accommodation optimization, eco-friendly catering, sustainable material selection, waste management, and communication of sustainable practices. A production must achieve a minimum score of 20 points for certification (Green Film, 2022). Conversely, the PGA Green Initiative, primarily in the USA, emphasizes regular audits and monitoring but does not provide formal certification. It includes guidelines for energy-saving measures, sustainable set design, and eco-friendly catering (Green Media Solutions, 2014). The Ontario Green Screen Initiative, specific to the province of Ontario, Canada, promotes sustainability planning and commitment, encouraging the development and implementation of sustainability plans. While it does not mandate formal certification, it advocates for thorough documentation and audits to ensure compliance (Ontario Creates, 2020).

Critically analyzing these standards highlights their strengths and limitations. The Albert Certification, mandatory for TV broadcast content in the UK, ensures high compliance and uniformity across productions with its comprehensive approach encompassing carbon footprint calculation, action planning, and offsetting (BAFTA albert consortium et al., 2020). However, its mandatory aspect might pose challenges for smaller or independent productions due to associated costs and administrative burdens. In contrast, the EcoProd Label's framework, with its 80 actions and eight mandatory criteria, offers a detailed roadmap for significant environmental improvements (EcoProd, 2024). Yet, its voluntary nature depends on producers' willingness to adopt it, potentially limiting its widespread impact.

The Green Film Rating System's flexible scoring approach, requiring a minimum of 20 points, allows productions to select criteria most relevant to their context, making it adaptable to different production needs (Green Film, 2022). However, its reliance on self-reporting and the variability in interpreting criteria can lead to inconsistent implementation. The PGA Green Initiative's focus on audits and monitoring without formal certification offers a robust framework for continuous improvement (Green Media Solutions, 2014). While its principle-based nature encourages adaptability and innovation, the lack of formal certification might reduce perceived legitimacy and incentives for compliance.

Finally, the Ontario Green Screen Initiative emphasizes training and resource provision, which helps build capacity within the film industry in Ontario (Ontario Creates, 2020). Encouraging sustainability plans fosters a proactive approach to environmental protection, but similar to the PGA Green Initiative, its principle-based standard may struggle with widespread adoption due to the lack of formal certification's signaling power.

With specific reference to the primary geographic scope, it is interesting to note that, although the standards analyzed have the potential for worldwide application, they are designed to be adopted within their specific geographical contexts. Interestingly, all the standards are intended for national-level application, except for the Ontario Green Screen Initiative, which is suitable for application at the provincial level.

The comparative analysis highlights the different approaches taken by each standard regarding their mandatory or voluntary nature in promoting sustainability within the film industry. Albert Certification and the EcoProd Initiative offer formal certification processes with mandatory criteria and regular audits, providing a high level of assurance and accountability (BAFTA albert consortium et al., 2020; EcoProd, 2024). The Green Film Rating System also provides formal certification, emphasizing a comprehensive verification process (Green Film, 2022). The PGA Green Initiative and Ontario Green Screen Initiative, even if they do not offer a formal certification, promote principles aimed at encouraging voluntary adoption of sustainable practices through training, resources, and audits (Green Media Solutions, 2014; Ontario Creates, 2020).

Among the certification-based standards, the features of the certification processes are noteworthy. The Albert Certification involves three main steps: completing the carbon footprint, developing a carbon action plan, and offsetting emissions (BAFTA albert consortium et al., 2020). The carbon footprint form is created using the Albert carbon calculator and is audited by an external reviewer. The action plan is submitted before filming begins, and emissions are offset through environmental projects. The EcoProd Label certification process starts with the application to AFNOR Certification. Producers must use the Carbon'Clap calculator to measure their environmental footprint and develop an action plan. Evidence of sustainable practices is collected and verified through audits by AFNOR Certification. Productions meeting the criteria are awarded the EcoProd Label (EcoProd, 2024). The Green Film Rating System requires the submission of a sustainability plan, led by a Green Film Manager. The plan is reviewed by a verifying body through document verification and on-set inspections. A scoring system determines whether the production meets the certification threshold (Green Film, 2022). PGA Green Initiative and the Ontario Green Screen Initiative are principle-based standard, thus they do not contemplate a certification process (Green Media Solutions, 2014; Ontario Creates, 2020).

## 4 Conclusions

The comparison among the main sustainability standards in the film industry reveal that each standard has developed unique strategies to promote sustainability-related practices in the film industry. The Albert Certification stands out as the only mandatory certification, specifically for certain types of productions like TV broadcast content. This mandatory nature ensures consistent application of sustainable practices, setting a high standard for compliance. The EcoProd Initiative is notable for its strong emphasis on external audits conducted by AFNOR Certification. This reliance on third-party verification adds a layer of credibility and thoroughness to the certification process. The rigorous audits ensure that productions not only commit to sustainability but also provide substantial evidence of their efforts. The Green Film Rating System focuses heavily on mobility, requiring productions to optimize transportation and accommodation to minimize environmental impact. This includes appointing a Mobility Manager responsible for creating a detailed Transport Optimization Plan, covering everything from the number of vehicles used to fuel consumption and occupancy rates. This focus on mobility addresses one of the significant contributors to greenhouse gas emissions in film production. The PGA Green Initiative is unique in its attention to catering, advocating for the use of organic, locally sourced, and fair-trade food products. By promoting these practices, the initiative highlights the importance of sustainable food consumption within the production process, addressing both food miles and supporting local economies. The Ontario Green Screen Initiative is distinct in its provincial focus, tailored to the specific environmental challenges and regulatory frameworks of Ontario. This localized approach allows for solutions that are contextually relevant and impactful, supported by comprehensive resources and training.

However, these standards are not free of limitations.

One significant limitation is represented by the nonmandatory nature of audits for some standards, which can weaken the reliability of their sustainability claims and undermine their signaling power. Without the prevision of mandatory audits, there exists a relevant risk of inconsistent application and potential discrepancies in how sustainability strategies are implemented and reported.

Moreover, the creation of a national standard for sustainability in the film industry is challenging, and establishing an international standard is even more complex. The diversity of regulatory settings, cultural features, and industry practices across countries makes the development of a global sustainability standard not an easy task. The lack of uniformity among national environments make very difficult to perform a benchmark and a comparison of the sustainability efforts across different world areas, potentially hindering overall progress toward global SDGs.

To overcome these challenges, the establishment and the development of an ongoing collaboration and dialogue among industry stakeholders are fundamental. The identification of more inclusive and flexible standards with the potential of adoption by productions of all sizes and budgets is only possible through sharing best practices and the harmonization of standards' criteria and requirements.

Furthermore, the enhancement of transparency and accountability levels of the companies operating in film industry is more likely whether robust verification processes and third-party audits are implemented, as they can help mitigate the risk of greenwashing and reinforce the reliability of ensure that sustainability-related disclosure.

Another significant limitation of the analyzed standards is represented by their unique focus on the environmental dimension of sustainability. Although the reduction of carbon footprints and the wide reduction of the environmental impact of film productions are relevant, the sustainability sphere includes a broader spectrum of issues related to the social and economic dimension of sustainability. For instance, meaningful topics are those related to fair labor practices, diversity and inclusion, community engagement, and economic viability. The exclusive focus on the environmental dimension can lead companies operating in the film industry to an inappropriate approach to sustainability. For example, a film productions may achieve higher environmental performance, but still face criticism for inadequate labor conditions or a lack of diversity in casting and crew. For these reasons, a more comprehensive approach to the different dimensions of sustainability is required.

Every standard included in the analysis offers unique contributions and faces specific sustainability challenges. The adoption and maintenance of the analyzed sustainability standards by companies operating in the film industry represent a meaningful step toward the reduction of the environmental impact of film productions. Through their implementation and continuous improvement, the film industry can make its contribution to advancing Sustainable Development Goals (SDGs), achieve a sustainable development, as well as align with broader objectives of ecological stewardship, social responsibility, and ethical values.

## References


Albert, BFI, & Arup. (2020). *A screen new Deal: A route map to sustainable Film production*. The British Film Institute. https://wearealbert.org/wp-content/uploads/2021/03/Screen-New-Deal-Report-1.pdf

Bozak, N. (2011). *The cinematic footprint lights, camera, natural resources*. Routledge University Press.

Campbell, D., Craven, B., & Shrives, P. (2003). Voluntary social reporting in three FTSE sectors: A comment on perception and legitimacy. *Accounting, Auditing & Accountability Journal, 16*(4), 558–581. https://doi.org/10.1108/09513570310492308

Connelly, B. L., Certo, S. T., Ireland, R. D., & Reutzel, C. R. (2011). Signaling theory: A review and assessment. *Journal of Management, 37*(1), 39–67. https://doi.org/10.1177/0149206310388419

D'Urso, F. (2022). Green policies and new green protocols for the audiovisual industry promoted in Italy by state and regions. *Journal of Media, Performing Arts and Cultural Studies, 2*, 164–179. https://doi.org/10.26350/001200_000155

EcoProd. (2024). *Synthèse de l'étude d'impact de l'éco-production audiovisuelle*. EcoProd.

Elalfy, A., Weber, O., & Geobey, S. (2021). The sustainable development goals (SDGs): A rising tide lifts all boats? Global reporting implications in a post SDGs world. *Journal of Applied Accounting Research, 22*(3), 557–575. https://doi.org/10.1108/JAAR-06-2020-0116

Gehman, J., Lefsrud, L. M., & Fast, S. (2017). Social license to operate: Legitimacy by another name? *Canadian Public Administration, 60*(2), 293–317. https://doi.org/10.1111/capa.12218

Green Film. (2022). *Green Film rating system for sustainable film production*. https://www.green.film/wp-content/uploads/2023/02/Green-Film-Rating-System-2022.pdf

Green Media Solutions. (2014). *PGA unified best practices guide*. https://greenproductionguide.com/wp-content/uploads/2014/04/PGA-Green-Unified-Best-Practices-Guide-REV.pdf.

Gulluscio, C. (2023). Legitimacy theory. In S. O. Idowu, R. Schmidpeter, N. Capaldi, L. Zu, M. Del Baldo, & R. Abreu (Eds.), *Encyclopedia of sustainable management* (pp. 2209–2215). Springer International Publishing. https://doi.org/10.1007/978-3-031-25984-5_175

Kääpä, P., & Vaughan, H. (Eds.). (2022a). *Film and television production in the age of climate crisis: Towards a greener screen*. Springer International Publishing. https://doi.org/10.1007/978-3-030-98120-4

Kääpä, P., & Vaughan, H. (2022b). From content to context (and Back again): New industrial strategies for environmental sustainability in the media. In M. Hjort & T. Nannicelli (Eds.), *A companion to motion pictures and public value* (1st ed., pp. 308–326). Wiley. https://doi.org/10.1002/9781119677154.ch14

Maxwell, R., & Miller, T. (2012). *Greening the media*. Oxford University Press.

Ontario Creates. (2020). *Ontario production guide*. https://digitallibrary.ontariocreates.ca/DigitalLibrary/ProductionGuide/ProductionGuide.aspx?from=menu

Pizzi, S., Caputo, A., Corvino, A., & Venturelli, A. (2020). Management research and the UN sustainable development goals (SDGs): A bibliometric investigation and systematic review. *Journal of Cleaner Production, 276*, 124033. https://doi.org/10.1016/j.jclepro.2020.124033

Ruiz Guzmán, N. D., & de Monterrey, T. (2022). Ecological cinema with best practices and reducing environmental impact. *Avanca Cinema Journal*, 833–839. https://doi.org/10.37390/avancacinema.2022.a463

Suchman, M. C. (1995). Managing legitimacy: Strategic and institutional approaches. *Academy of Management Review, 20*, 571–610.

The Guardian. (2022). *Thai court orders repair of the beach location 22 years after filming | Thailand | the Guardian*. The Guardian.

Turzo, T., Montrone, A., & Chirieleison, C. (2024). Social accountability 8000: A quarter century review. *Journal of Cleaner Production, 441*, 140960. https://doi.org/10.1016/j.jclepro.2024.140960

Victory, J. (2015). Green shoots: Environmental sustainability and contemporary Film production. *Studies in Arts and Humanities, 1*(1), 54–68. https://doi.org/10.18193/sah.v1i1.6

**Carmela Gulluscio** completed her PhD at the Faculty of Economics of the Roma Tre University, Italy. She is now Full Professor of Accounting at the University of Rome UnitelmaSapienza. Her academic journey includes some visiting periods at the LMU University of Munich, Germany. Her research interests include corporate social responsibility and sustainability, public sector accounting, accounting history, and accounting education. She is the author of several books, book chapters, and numerous articles. Her studies are featured in international journals, including Business Strategy and the Environment and Sustainability.

**Cecilia Chirieleison** completed her PhD at the Faculty of Economics of the University of Pisa, Italy. She is now Associate Professor of Management in the Department of Political Science at the University of Perugia, Italy. She has been Visiting Professor at prestigious foreign universities, including the London School of Economics, University College of Dublin, and St. John's University in New York. Her research interests include corporate social responsibility and sustainability, tourism and event management, and economic impact analysis. She is the author of several books and numerous articles published in prestigious international academic journals including the *Journal of Sustainable Tourism*, *Journal of Cleaner Production*, and *Tourism Management Perspectives*.

**Teresa Turzo** is Assistant Professor of Accounting at the University of Rome UnitelmaSapienza. She previously served as Research Fellow at the University of Perugia, where she completed her PhD with a thesis on the relationship between non-financial reporting and religiosity. Her academic journey includes a visiting period at the University of Lincoln, UK. She currently teaches courses on Auditing and Analysis of Corporate Economic and Social Performance. Her research interests encompass non-financial reporting, sustainability standards, religiosity, and informal institutions. Her studies are featured in international journals, including *Meditari Accountancy Research* and the *Journal of Cleaner Production*.

# Artificial Intelligence and its Implications for Media Sustainability in Romania

**Georgiana Camelia Stănescu**

**Abstract** Artificial intelligence tools' widespread emergence has revolutionized how media content is produced, sold, and consumed. From chatbots that can write a news story in seconds if they have the necessary data to tools that generate images or photos based on the text provided, they have all changed procedures and how newsrooms work. With these phenomena has come a very dangerous one: the risk of fake news. Several websites and apps generate fake news on demand, especially when it comes to topics of interest. In this context, the media play an overwhelmingly important role. There is a need for credible media outlets to provide quality, verified content. However, artificial intelligence is increasingly weakening media institutions. To maintain their profits, some companies in the industry are starting to replace journalists with artificial intelligence tools. For example, a famous Romanian trust already has an AI-generated news anchor for a sports website. So, from avatars to tools for automating other tasks in newsrooms, they are all putting journalists' jobs at risk and, thus, the economic and social sustainability of the whole industry. This article analyses how artificial intelligence affects media organizations in Romania through qualitative methods. In this context, the authorities should take measures to protect the press in particular, but also the population in general, because the reduction of specialized media trained in this field may endanger democracy itself due to fake news. The article could be helpful for journalists, researchers, and national decision-makers.

**Keywords** Artificial intelligence · Sustainability, Mass media · Automation · Algorithms · Personalization · Innovation · Technological development · Efficiency · Digital transformation · Big data · Machine learning · Ethics · Journalism · Ai-generated content · Social impact · Fake news · Distribution · Audience · Future of media · Deep fakes · Data privacy · Media ethics · Disinformation · Virtual reality · Augmented reality · User experience · Democracy · Media industry · Environmental sustainability

---

G. C. Stănescu (✉)
Faculty of Letters, University of Craiova, Craiova, Romania
e-mail: georgiana.stanescu@edu.ucv.ro



M. Todorovic et al. (eds.), *Sustainability and Social Responsibility of the Media and in the Media*, CSR, Sustainability, Ethics & Governance,
https://doi.org/10.1007/978-3-032-00086-6_5

# 1 Introduction

This research focuses on examining how artificial intelligence (AI) affects the sustainability of the media sector in Romania, with specific attention to its economic, social, and environmental dimensions. Artificial intelligence has had a meteoric rise in the last 2 years, and the media industry has been profoundly affected by the transformations of the new digital environment (Diakopoulos, 2019; Hermann, 2021; Serrano et al., 2015, Voinea, 2025). The rapid deployment of new technologies in newsrooms and the rapid advancement of AI-driven software, coupled with the adaptation of organizations to meet the preferences of media consumers, have brought significant changes to the media industry (Broussard et al., 2019) but also challenges and risks for the sustainability of the entire sector (Larsson & Heintz, 2020).

New tools based on artificial intelligence allow media companies to replace journalists and automate repetitive processes (Underwood, 2019; Lewis et al., 2019a, 2019b). Thus, simple activities such as transcription, translation, proofreading, and even generating articles or editing images or photos can be done by applications backed by AI-based technologies.

However, as AI-based applications begin to replace journalists' jobs, one of the sustainability concerns remains the economics of the field. This may result in inaccurate portrayals and the marginalization of particular social groups in the media (Chen & Chekam, 2021).

Another concern related to the integration of AI applications in journalism that affects the long-term sustainability of the media is the risk of fake news being disseminated widely. We saw the effects of the spread of fake news after the outbreak of the two wars in Israel and Ukraine, when, in addition to conventional fighting, we witnessed a real informational conflict.

On the other hand, AI-based algorithms can adapt to new consumer preferences that have been evolving rapidly in the digital era (Parratt-Fernández et al., 2021). Users want access to personalized content that they can have instantly without wasting time searching for the information they need. Nowadays, media companies can provide personalized information based on readers' interests and behavior, thanks to algorithms capable of performing complex analysis (Chen & Chekam, 2021). This personalization enhances the user experience and contributes to loyalty and audience expansion for media companies.

Data security and privacy are also key issues for the sustainability of the media system, particularly in contexts where media organizations collect and utilize large volumes of personal data to deliver personalized content. This information could, therefore, be stolen and misused (Couldry & Turow, 2014).

The media industry can also play a crucial role in promoting environmental sustainability, primarily due to its significant influence on educating society (Hansen, 2018).

This study hypothesizes that the widespread development and adoption of artificial intelligence tools can affect media sustainability.

To achieve this, we will address the following key question: What are the economic, social, and environmental implications for media companies that adopt AI technologies? How can media organizations ensure long-term sustainability while integrating AI technologies?

Following the introductory section, Sect. 2 presents a brief literature review on the application of artificial intelligence in the media sector, accompanied by an analysis of AI's impact on media sustainability from economic, social, and environmental perspectives. The next section offers a detailed explanation of the methodology used in the study. Following this, the results obtained through qualitative analysis are presented, accompanied by relevant tables that illustrate the participants' responses. The article concludes with key findings and suggests potential directions for future research in this area.

## 2 Artificial Intelligence: Applications and Challenges for the Media Industry

Artificial intelligence is one of the most revolutionary technologies of the twenty-first century. It has enormous potential to bring about radical changes in numerous fields, from health and education to transportation and filmmaking, and the media industry is no exception. We are witnessing the rapid development of machine learning algorithms and increased data processing and utilization capabilities, all of which have enabled AI-based systems to perform complex tasks that previously required human intervention.

Although the concept of artificial intelligence has been increasingly used and encountered since 2022, when the company OpenAI launched the first chatbot-type artificial intelligence tool capable of providing coherent answers to real-time questions (Márquez, 2023), the term AI is not new at all. In 1950, Alan Turing began testing the ability of machines to imitate human intelligence. Thus, he proposed the Turing test to evaluate a machine's ability to imitate human intelligence. In 1956, John McCarthy proposed the use of machines to simulate human intelligence and introduced the concept of artificial intelligence (AI) at the Dartmouth Conference (McCarthy, 1998). Since then, AI has evolved and developed into an independent field of study. Gradually, in the following years, search algorithms made the first significant advances in AI. The true potential was fully realized with the emergence of machine learning techniques and deep neural networks, which experts refer to as deep learning (Choudhary et al., 2021).

Artificial intelligence finds its applicability in the media sphere almost immediately after the first Turing tests. Cox (2000) explains that journalists have been using computers to produce news since 1952, when a newly created model was used to predict the outcome of the presidential elections in the United States. Other specialists (McGregor, 2013) mention that applications based on artificial intelligence have been used in newsrooms across the United States even before the advent of Web 1. Thus, computer-assisted journalism can be traced back to 1960 and involves the use of technology and computer tools to collect, analyze, and present data within newsroom activities. Computer-assisted journalism has become increasingly important in the digital age, allowing professionals in this field to manage large volumes of information and uncover complex stories through data analysis. Once the data has been collected and analyzed, journalists use this information to write detailed articles and investigations.

Parratt-Fernández et al. provided a concrete example of Computer-Assisted Reporting (CAR), in which journalist Philip Meyer created a profile of the protesters in Detroit in 1967 using a survey based on computer-assisted research, and the article won the Pulitzer Prize. Moreover, this was officially the first method of using complex analysis systems based on computer intelligence (Léchenet, 2014). Later, Meyer (2002) defined this type of investigation as precision journalism. The next stage in the use of intelligent tools in journalism was marked by the emergence of data journalism (Coddington, 2015; Hermida & Young, 2019).

What truly began to bring about changes in the flow of mass media was the phenomenon that specialists refer to as algorithmic journalism or robotic journalism (Anderson, 2012; Coddington, 2015).

This type of journalism is backed by natural language generation technology (NLG). It is specifically about the automatic generation of texts from structured digital data. This technology "has been produced and marketed in the last decade by several companies, such as Arria, Applied Semantics, Automated Insights, and Narrative Science" (Thäsler-Kordonouri, S., & Barling, K., 2023). According to Reiter (2010), NLG technology first emerged in the 1950s but was only used to its full potential in the last decade. In newsrooms, such technologies are used to automate specific repetitive, routine tasks within news organizations (Dörr, 2015).

The true revolution in newsrooms emerged in 2022 with the advent of ChatGPT, an AI-based tool featuring a conversational interface that generates coherent responses for users (Gutiérrez-Caneda et al., 2023). ChatGPT was launched by the company OpenAI, and shortly after, tech giants announced that they had developed and were launching similar models. Google has taken the lead with Bard, a chatbot model also based on conversational artificial intelligence, while Microsoft has integrated a chat feature into its search engine. The latter has similar conversational functions and, in addition, allows the user to create images with the help of artificial intelligence. These tools are easily accessible to journalists and can be effectively used to carry out their professional tasks.

# 3 Editorials Taken over by Artificial Intelligence

## *3.1 Automatic Transcription of Texts and Real-Time Monitoring*

Thus, several applications are helpful in newsrooms, and one of them is related to automatic transcription. In the past, transcribing interviews was a laborious task that consumed much time. Journalists had to listen to audio or video recordings, frequently pause them to capture every word, and then review the transcription for corrections. This process could take hours for a single interview, which limited the time available for analysis and writing articles. AI-based speech recognition technologies have transformed this process. Now, intelligent software can automatically transcribe conversations in real-time or from pre-recorded files, providing excellent and accurate results in a much shorter time (Gondwe, 2023). These tools can identify different speakers and insert time markers, thus facilitating review and editing. Thus, artificial intelligence now enables the automatic monitoring of media streams, including news, TV shows, radio, and podcasts. AI software can analyze content in real-time, automatically transcribing and indexing broadcasts based on keywords or topics of interest.

## *3.2 TV Presenters of the Avatar Type*

The rapid development of artificial intelligence (AI) tools has also led to the creation of innovative methods for presenting news (Wu, H. et al., 2024; Baraniuk, 2018). For example, artificial intelligence generates avatars that serve as presenters in the audiovisual field (Xue et al., 2022; Wang & Zhu, 2022; Wu et al., 2024).

Avatar presenters can be defined as digital characters created through the integration of various advanced technologies, and they exist exclusively within the realm of display devices (Huang, Yuke, and Yu, 2023). These digital presenters have physical and personality traits similar to humans and closely imitate human gestures, expressions, and actions. Previous research has provided evidence regarding the benefits that a show can have in the context of using digital presenters.

For example, a presenter created through artificial intelligence does not make mistakes, always expresses themselves coherently, and does not require a salary or sick leave (Stănescu, 2023).

Furthermore, these can be customized for different linguistic and cultural markets, presenting news in multiple languages without the need for additional staff. Avatars can interact in new and engaging ways with the audience, using elements of augmented and virtual reality to enrich the visual experience of news consumers. A famous international press agency announced back in 2020 that it uses AI for this purpose, but a customized one. More precisely, after filming a presenter in a studio reciting keywords about a specific topic, the media company could generate an indefinite number of news articles on that subject using its image. Moreover, these reports are almost impossible to distinguish, and the agency's clients were impressed (Chandler, 2020).

However, there is a risk of depersonalization of news and the loss of the essential human element in journalism. Ethical questions arise regarding manipulating information and potentially creating fake news, considering AI's ability to generate content autonomously.

In Romania, several news websites have started using virtual presenters to ensure a continuous flow of live information, such as E-Mil, the first virtual sports presenter (Hațieganu, 2024). Although it represents an important step in the automation of content production, E-Mil poses a number of significant challenges. These include a lack of emotional expressiveness and human empathy, and difficulty in adapting the message to the social or cultural context.

## 3.3 Writing Texts with AI

Text generation through AI is based on advanced machine learning techniques and neural networks. Models like GPT (Generative Pre-trained Transformer), developed by OpenAI, can learn the structure and linguistic style from large volumes of textual data and produce coherent and contextual texts (Gutiérrez-Caneda et al., 2023). In journalism, these models generate articles on various topics and other types of content that involve repetitive or structural data. In modern newsrooms, AI produces standardized articles at a speed and accuracy that surpasses human capabilities. For example, news agencies use AI to generate financial reports or cover sporting events automatically. Additionally, AI allows for content personalization, providing readers with articles tailored to their specific interests, thereby increasing engagement and audience satisfaction (Henestrosa et al., 2022). Among the activities that such software can supplement in the editorial process are automatic grammar and spelling correction, suggestions for rephrasing for clarity and conciseness, and adapting the linguistic style to suit the target audience. Additionally, these tools can detect redundancies and provide stylistic alternatives, allowing editors to maintain consistency in the tone and structure of the text.

## 3.4 Photo and Video Editing with AI

AI-based photo and video editing platforms have innovated newsroom creative procedures, providing users with advanced tools that significantly simplify and accelerate visual production. These platforms use algorithms to enhance the quality of images and videos automatically. Thus, anyone in the editorial team can easily edit without needing advanced editing skills.

Among the features of these sites are automatic color adjustment, background removal, video stabilization, and the application of complex filters (Rooth, 2024). For example, applications like Adobe Photoshop and Premiere Pro integrate AI to make precise selections, remove unwanted objects from images, or automatically enhance lighting and clarity. In video, AI can automatically sync lip movements with voice, generating high-quality content with minimal effort from the user. These technologies not only save time but also democratize access to professional editing tools, allowing anyone to create quality content.

## 4 Media Sustainability

Media sustainability refers to the ability of media organizations to operate responsibly and viably in the long term, while maintaining their core journalistic functions and adapting to technological, economic, and social changes. This study bases its analysis of the concept of sustainability on the three pillars model proposed by the World Business Council for Sustainable Development (WBCSD), which defines sustainability as a balance between economic viability, social equity, and environmental protection (WBCSD, 2002). This conceptual framework is widely accepted and used in research in the field of sustainable development (Elkington, 1997; Purvis, Mao & Robinson, 2019), being adapted here to reflect the specifics of the media industry in the context of the transformations generated by artificial intelligence.

In the last 20 years, the entire media industry in Romania has undergone a profound transformation, evolving from the cumbersome transmission of images via relays to real-time broadcasts through digital technologies. Moreover, the entire digital revolution has also left its mark on the media industry, which has been radically transformed. Among the emerging technologies, artificial intelligence is making its way in, having begun to be integrated into newsrooms and impacting specific media processes such as producing, distributing, and consuming media content. Nevertheless, the rapid transition to AI-based digital media automatically brings significant challenges. Rapid changes in consumer preferences, the decline in traditional advertising revenues, and the spread of misinformation are just a few challenges facing media organizations. In this context, adopting sustainability strategies and advanced technologies becomes essential for ensuring the long-term viability of the media.

## 4.1 The Economic Sustainability of the Mass Media

Rapid technological progress has fundamentally changed the way information is produced, distributed, and consumed (Gondwe, 2023). These technologies have made access to information extremely easy, but they have also brought significant challenges to traditional business models in the media industry.

Advertising, which has long been the backbone of media funding, has largely moved online, where costs are lower and targeting is more precise (Cham et al., 2022). This has led to immense economic pressure on traditional media institutions, resulting in cost-cutting, layoffs, and even the closure of some publications. In this context, economic sustainability has become a significant concern for media organizations in the last years. The traditional business model of the media has primarily relied on advertising revenues. New technologies, including artificial intelligence, virtual reality, and augmented reality, have fundamentally changed businesses, promoting significant innovations in research and marketing practices. This evolution has acted as a driving force for marketing, fueling the emergence of new trends and patterns in digital marketing and data analytics (Grewal et al., 2020).

To ensure economic sustainability (Florea et al., 2021), media organizations have implemented a series of measures, such as introducing subscription models and membership programs. Major international publications successfully implemented these models, which attracted readers willing to pay for quality content (Pantic, 2021). Other publications have focused on developing effective digital advertising strategies and social media campaigns. Nevertheless, not all have managed to adapt, and many media institutions, especially local ones, are on the brink of bankruptcy.

Thus, the industry is facing significant challenges related to sustainability, especially regarding job retention. The widespread use of artificial intelligence in the media sector can profoundly impact the workforce in the field and generate cascading social effects. Generative artificial intelligence can replace many routine tasks and unskilled labor (Nah et al., 2023) in the media industry. As previously argued, activities related to text editing, proofreading, and generating standardized content can be taken over by artificial intelligence algorithms. These measures can lead, in the long term, to job losses for individuals engaged in such activities.

The replacement of human labor in the media sector with artificial intelligence could lead to an increase in unemployment, especially among those who do not possess advanced technological skills—and here we are talking about many older journalists. This transition may further widen the income gap between those who can upgrade their skills to work with advanced technologies and those who cannot.

On the other hand, with the emergence of artificial intelligence, competition among media organizations is becoming increasingly intense. Large media corporations with significant financial resources can afford to invest in cutting-edge technologies that automate the processes of writing, editing, and distributing content. These organizations can use AI to generate articles in real time, personalize content according to user preferences, and optimize Search Engine Optimisation strategies, thus remaining competitive and relevant in the global media market.

In contrast, small organizations, which often operate with limited resources, risk being left behind.

A media outlet with limited resources may need help to compete with large media conglomerates in terms of accuracy, timeliness, and content appeal, which can lead to a decrease in audience and, consequently, in revenue. In a digital era where media consumers expect quick and personalized information, the inability to adopt cutting-edge technologies can place local journalism at a long-term disadvantage.

Sustainability in mass media targets both the economic aspect and the social and environmental responsibility of organizations in the field. Media institutions must adopt practices that allow them to operate efficiently in the long term, maintaining the quality and integrity of journalism. Additionally, in the current context, reducing the negative impact on the environment is essential.

## *4.2 Social Sustainability of the Media*

On the other hand, the use of AI in the media also raises numerous social and ethical challenges. The mass media play an essential role in informing and educating the public, influencing its opinions and behaviors. Recent global issues such as the COVID-19 pandemic, the war in Ukraine, and the conflict in Israel have once again highlighted the dangers of misinformation and the importance of having accessible, well-trained journalists to provide accurate information and combat this phenomenon.

The advent of artificial intelligence has raised concerns about the rise of deep fake technology, which allows for the creation of highly convincing yet entirely fabricated audio and video content. This threatens to blur the lines between truth and fiction and poses serious risks to personal privacy, public safety, and the integrity of information.

One of the main concerns is related to algorithms' biases. Since AI relies on historical data to learn and make predictions, there is a risk that existing biases in the data may be perpetuated or even amplified (O'Neil, 2016). This can lead to unfair representations and the exclusion of certain social groups from the media landscape. For example, algorithms can favor specific sources or types of content at the expense of others, creating an unbalanced informational landscape.

Data confidentiality and security also represent significant challenges. As media organizations collect and analyze large amounts of personal data to provide personalized content, the risk arises that this data may be compromised or misused. The industry must develop and implement strict data protection policies to safeguard user privacy and maintain public trust. The recent scandals related to security breaches and the improper use of personal data have highlighted the importance of this issue.

### *4.3 Environmental and Sustainability Aspects*

The impact of AI on the sustainability of the media is not limited to economic and social aspects; it also has significant implications for the environment. The use of digital technologies, including AI, can help reduce the media industry's ecological footprint.

Algorithms can analyze and predict consumer trends, allowing companies to plan production and content distribution more efficiently, minimizing resource waste. Additionally, AI-powered cloud computing technologies can reduce energy consumption by optimizing the use of IT infrastructure, thereby contributing to a smaller carbon footprint. Moreover, by optimizing energy consumption, reducing waste, and promoting eco-friendly content, AI can help media organizations lessen their negative impact on the environment and play an active role in educating the public about environmental issues. Nevertheless, addressing the challenges related to AI's energy consumption, unequal access to technology, and ethical and privacy issues is essential. By implementing appropriate frameworks and promoting equitable access to technology, AI can play a crucial role in protecting the environment and promoting sustainability in the media industry.

Climate change and environmental degradation are significant challenges of the twenty-first century. In this context, the mass media industry plays a vital role in educating and raising public awareness about environmental issues. AI can be used to generate media content that promotes sustainability and environmental awareness automatically. AI-based software can create articles, blog posts, and other content that educate the public about environmental issues and sustainable solutions. Personalizing media content for different audience segments can maximize the impact of sustainability messages. AI can analyze user preferences and behaviors to deliver relevant and engaging eco-friendly content, thereby encouraging the adoption of sustainable practices among the audience (Yeh et al., 2021).

By adopting these strategies, the mass media industry can maximize AI's sustainability benefits and contribute to building a greener and more sustainable future.

## 5 Research Methodology

This research adopts a qualitative methodology to understand how artificial intelligence tools influence the sustainability of the Romanian media industry. Sustainability, as applied in this study, follows the framework proposed by the World Business Council for Sustainable Development (WBCSD), which comprises three main dimensions: economic viability, ecological balance, and social progress.

This research is based on the hypothesis that the development and widespread adoption of artificial intelligence tools significantly affects the sustainability of the media sector, particularly in terms of its economic, social, and environmental dimensions.

To explore this intersection between AI and sustainability, a series of semi-structured interviews was conducted with editors from local media outlets in Romania who actively use artificial intelligence tools in their workflows for the production and distribution of information. This targeted sampling ensured the relevance and applicability of the responses to the specific objectives of the research.

The interview questions were intentionally designed to assess both the perceived opportunities (e.g., efficiency, personalization, cost reduction) and challenges (e.g., job losses, misinformation) associated with AI. This methodological approach provided a rich contextual understanding of how AI contributes to—or challenges—the long-term sustainability of media organizations in Romania, while also offering insights into broader implications for journalism, ethics, and society (Table 1).

The data obtained was analyzed and sought to identify common patterns and relevant differences in participants' perceptions of the impact of AI on sustainability in the media.

The responses were anonymized, and the respondents were labeled as R1, R2, R3, R4, R5.

## 6 Results

The results collected from the interviews provided a clear picture of how AI is integrated into editorial processes and the perceptions related to its short-term and long-term impact.

From the perspective of changes in the production and distribution process of media content, most editors-in-chief explained that they have started to use AI to a limited extent without implementing radical changes.

R1 explained that they introduced artificial intelligence software into the editorial process. The journalists are currently using the premium version of Chat GPT. They use this tool for Search engine optimization, to gather background information on specific topics, and to synthesize statistical data. He mentioned, however, that he has noticed the risk of errors, which is why information is always passed through human filtering, even though it requires time and resources. However, they want to maintain the image they have earned in the media market as a very credible

**Table 1** Questions asked during interviews

| |
|---|
| 1. How has the process of producing and distributing news and programs in your newsroom changed following the implementation of artificial intelligence? |
| 2. What do you think are the potential implications of using artificial intelligence on economic sustainability within the media industry? |
| 3. How do you assess artificial intelligence's impact on the social sustainability of the media? |
| 4. To what extent do you believe that artificial intelligence can influence the ecological sustainability of the media industry? |
| 5. What do you consider to be the most significant risk associated with the integration of artificial intelligence in the media sector? |

journalistic source. A similar response was provided by R2, another local editor who mentioned that he tried to technologize and automate much of the content production process but encountered several drawbacks with the AI tools. He used limited resources to acquire some avatars but found that they needed to meet expectations. They smiled when presenting negative news and did not empathize with the viewers. So it was left only with human presenters. The other respondents use AI in similar editorial processes and explained that, in general, they have not necessarily given up certain functions in the editorial teams. For example, the translation process is overseen by a journalist who knows English, so there is no need for a translator to take a specific news item and interpret it (Table 2).

In conclusion, regarding the practical use of artificial intelligence in newsrooms, all respondents reported integrating AI-based tools, particularly for SEO optimization, photo-video content editing, and writing assistance. However, the degree of

**Table 2** Answers to the question "What changes have been implemented in your newsroom as a result of using AI?"

| | |
|---|---|
| R.1 | We use artificial intelligence tools for SEO optimization, but the results are not always as expected. Either they don't sound good journalistically, or they dramatize a lot. Similarly, we can synthesize data to generate automated news articles that involve a lot of statistical data. However, we have to keep in mind that artificial intelligence can still make mistakes, which can put us at risk of losing credibility. We can also background a story with a generative text system. However, we are still at the stage where our information is checked through the human filter because we are part of a credible media organization with many years of experience behind us. We cannot risk losing readers as a result of an AI error. |
| R.2 | At the moment, we only use AI for photo/video editing. We are a media platform with classic TV and site-online and distribute social media content. We haven't had very big changes. |
| R.3 | We use AI for SEO optimization in particular. AI helps us optimize titles, descriptions, and keywords, ensuring articles are more accessible and visible in online searches. In addition, we tried using avatars for presentation and live, but the result was not as expected. It should be noted that we used cheap technology, and the avatars purchased had several minuses. They smiled when presenting serious news and failed to synchronize their attitude with the presented topic. Temporarily, we gave up, but the avatars will probably be a workable option as technologies improve and costs come down. |
| R.4 | The changes are more on a subconscious level. There is also a comfort for employees who can handle specific topics from an AI perspective. Professionals in the field have years of experience behind them without the need for AI in their jobs. Now, this support is more of a validating one for them. If a topic is broached, AI is not asked in the first instance. Instead, reporters/editors act according to their own experience and professionalism, methodically constructing their news or story, and only then can AI tools be used as an extra check. It's not bad at all; it's just that, at the moment, most employees use it not as a source of inspiration but as a source of validation. This approach can also be put down to personal hubris, as professionals want to signal that human input is still essential in tackling a subject with moral/ethical connotations and at the level of human consciousness. |
| R.5 | We use artificial intelligence in newsrooms. We've tried several tools, but not all of them are useful. We even tried using generative text tools for news production but gave up. We found that they still had to go through the human filter, and we were just wasting time. Now, we use chatbots to SEO to find a better headline variant. We also edit photos with AI and sometimes generate images when we don't have them from our sources. |

use and technological purpose differ. For example, one editor (R3) tested AI avatars but gave up due to a lack of expressiveness and naturalness. R4 uses AI strictly as a validation tool, maintaining the emphasis on the role of human professionalism in journalism, while R2 uses AI in a limited way, mainly for multimedia production, without substantial changes in the editorial workflow. These variations reflect both the level of trust in the technology and the degree of strategic integration of AI into editorial work.

## 6.1 *Impact on Economic Sustainability*

Regarding economic sustainability, most respondents believe that AI will have little impact on solid editorial teams in the short term. However, in the long run, the improvement of AI tools could replace, for example, live performers or presenters. Everyone agreed that field reporters cannot be replaced at this time and that human journalists are needed to verify information from the scene of events. Additionally, AI is a risk in light of new competitors who might use the technology to cut costs and produce lower-quality content that is still very appealing to the audience. This could lead to decreased revenues for traditional organizations that invest in quality journalism. At the same time, the high costs of traditional media production are recognized as a challenge for long-term sustainability, requiring continuous adaptation to new economic and technological conditions. Some media organizations have tried new ways to generate revenue to ensure economic sustainability, but these have yet to have the expected effect. Most have focused their attention on social media to attract readers to their websites, basing their income on digital advertising. Another issue raised by one of the interviewees was related to the approach to the topics. He runs a highly reputable news site and explained that the classic model through which everyone provides content delivered by institutions is becoming increasingly challenging to operate. A news report received from the police will have approximately the same content on other sites if the journalist does not go on the ground and approach the subject in their way. This could be the solution for people who consume information to turn towards a different kind of content (Table 3).

In terms of the economic dimension, most respondents (R1–R4) believe that artificial intelligence does not have a major economic impact on established newsrooms, at least in the short term. However, there are concerns about competitive pressure from new media players that make intensive use of AI and reduce costs through automation. R5 has a more pessimistic view, anticipating a gradual replacement of human functions, while R4 highlights market adaptability and the role of demand as the main decision-making factor. These perspectives reflect differences in the perception of risks, but also in the economic adaptation strategies adopted by newsrooms.

**Table 3** Answers to the question "What do you think are the potential implications of using artificial intelligence on economic sustainability within the media industry?"

| | |
|---|---|
| R.1 | In the next 10 years, I don't think the impact will be very big. Solid newsrooms and media organizations that have built a credible image are not going to give up people for AI. I think journalists can use AI tools to make their work easier. I think the risk will be competition. There will be all kinds of new organizations that will overuse artificial intelligence, that will be low cost and will override a lot of ethical issues to get traffic, and that's something that we might feel because people will be attracted to the sensational, perhaps in favor of a subject treated seriously and with objectivity. It all depends very much on the cultural baggage of the audience. Also, the news and the traditional press have become very expensive and very difficult to sustain. For example, if you send a team to cover a story 100–200 kilometers away, it consumes resources. Many organizations prefer to search for images on social networks or maybe contact someone local to send some images/information. The quality of both documentation and images decreases, plus there is even a risk of serious errors. But without significant resources, quality journalism is not possible. We are already feeling it, and this phenomenon will likely grow over time. One option would be to subscribe for quality content, but in Romania, this will not work, at least as far as the press is concerned. People are not used to paying for content in the context that they find it for free, whether it is truncated or subjective. |
| R.2 | From an economic point of view, I don't think there will be major changes, at least in the next 10 years. Under no circumstances are we going to give up journalists because the subjects we cover require the physical presence of journalists. We mainly produce video and it's very difficult at the moment to replace journalists who go out in the field, do interesting stories with artificial intelligence. Oh, they can use tools that generate text, for a better idea, maybe in a moment when they feel tired, uninspired. In addition, we can use it to save time for routine, repetitive, time-consuming activities and leave them more time. |
| R.3 | AI has the potential to profoundly transform the media industry, offering opportunities for efficiency and innovation but also bringing significant challenges related to economic sustainability, ethics, and diversity. Adapting to these changes will be crucial for the future of the industry. Larger companies with the resources to deploy advanced AI technologies can gain a significant competitive advantage, which may lead to greater market concentration and difficulties for smaller organizations, especially local media like ours with limited resources. |
| R.4 | Economic sustainability is a sensitive issue, especially from the perspective of an employer interested in this direction. It may tip the balance in favor of initial investments, perhaps more costly, but with long-term applicability, if this is desirable. At the same time, without people, this craft is hardly up to quality standards. There are multiple variables that are taken into account when calculations are also made. Normally, the price/quality ratio in the equation of transferring information to the general public also has to take into account the specifics of the newsroom, the target audience, the market requirements, "what sells". The media has taken into account the quality of the audience and has reacted accordingly each time, perhaps not correctly, but in line with market requirements. So, let the market decide what is demanded and I believe that the media will also upgrade or not upgrade with AI tools, if they will economically respond to the demands. |
| R.5 | I think a lot of the newsroom functions will be replaced by AI. Yes, we will no longer have people in text editing, as this is easily done with AI, maybe not even in photo-video editing as the software for this is efficient. |

## 6.2 Impact on Social Sustainability

Most respondents mentioned that socially, AI is perceived as a tool with a significant impact on the equitable distribution of information and the quality of public discourse. If AI continues to be used to create and distribute manipulated or false content, it could lead to heightened social polarization and an erosion of trust in the media. The main risk identified by respondents regarding the use of AI in the media is related to the spread of misinformation and fake news. AI algorithms can facilitate the rapid creation and distribution of manipulated content, which can undermine public trust in the media and negatively impact the democratic health of society. Moreover, algorithmic biases and the lack of transparency in the use of AI can exacerbate issues related to informational equity and media pluralism. One of the respondents mentioned that, for example, if the paid content model were introduced at the local level, it would raise accessibility issues for the general public, which could exacerbate social inequalities (Table 4).

All participants expressed concerns about the social impact of AI integration in the media, particularly regarding the proliferation of misinformation, loss of public trust, and polarization of discourse. They emphasize the urgent need for regulation and ethical responsibility in the use of these technologies. For example, R1 draws attention to the risks associated with unequal access to information, especially in the context of the possible introduction of paid models. R4 proposes media literacy as a solution to limit negative effects. Overall, a common vision is emerging on the critical role that the media must play in supporting social equity and cohesion in the AI era.

## 6.3 Impact on Ecological Sustainability

Although not a central aspect of editorial activity, ecological sustainability has been addressed through concrete measures to reduce resource consumption. The interviewed media organizations have invested in energy-efficient equipment and are exploring solutions to generate energy from renewable sources, such as solar panels. Additionally, AI can potentially reduce dependence on physical materials by digitizing processes and optimizing the energy consumption of IT infrastructure (Table 5).

In terms of environmental sustainability, all respondents recognize the potential of artificial intelligence to help reduce environmental impact by digitizing content, eliminating physical materials and optimizing resources. Some participants (R1, R2, R5) highlight the educational role of the media in promoting environmentally friendly behaviors. R3 notes that the use of AI has led to a reduction in editorial team travel, which implies a decrease in carbon footprint. However, R5 warns about the high energy consumption associated with AI infrastructure, drawing attention to the responsibility of large technology companies.

**Table 4** Answers to the question "How do you assess artificial intelligence's impact on the social sustainability of the media?"

| | |
|---|---|
| R.1 | From a social point of view, the effects could be multiple. If the paid content option is chosen, then we are talking about inequity in society because those with limited resources will not have access to this information. On the other hand, the use of artificial intelligence by those who aim to make profits in any form or those who have all sorts of obscure interests can provide false content which can lead to a series of social knock-on effects. This phenomenon has been on the rise since 2016 and is growing every year. The control of advanced AI technologies by a few large corporations may lead to an excessive concentration of power in their hands, affecting the pluralism and diversity of media sources. |
| R.2 | The biggest problem related to AI and media is related to fake news. The risk of this phenomenon is incredible at the moment. We are facing a wave of fake news, and from a social point of view, the impact of AI is huge. False information can generate mass social reactions that can jeopardize the balance of society. The media has a decisive role to play here. That is why the sustainability of the media must be guaranteed, ensured by serious regulation, because it is the pillar of any stable democracy. I believe that a set of regulations similar to those in the television sphere, a universal code of media organizations could limit this phenomenon. |
| R.3 | In terms of social sustainability, the impact of AI on the media will be complex and multi-layered. On the one hand, the use of AI can facilitate access to information and democratize the content production process, allowing for quick and efficient creation of articles, analyses or reports. On the other hand, there is a risk of social polarization, especially if AI is predominantly used by entities that are more profit-driven than aiming to inform the public correctly. If AI is used to produce sensationalist content or misinformation, this can amplify social divisions and spread misinformation widely, undermining social cohesion and trust in media institutions.<br>In conclusion, the impact of AI on social sustainability in the media will largely depend on how these technologies are integrated and managed. A responsible and ethical approach is essential to ensure that AI contributes to a balanced and inclusive media landscape that reflects the diversity and complexity of society. |
| R.4 | Normally, a balance should be maintained between human input and that of AI tools, precisely in order to relate as effectively as possible to human values, but also to current requirements. The quality of the audience is also important because the verdict comes from there. The current period is one in which fake-news is quite present, and unfortunately there are many examples of this. It can happen that the target audience is reluctant to differentiate between real and fake, between AI and human, even after certain regulations in this area have intervened. On a social level, the effects of poor quality journalism can be reflected in the actions and opinions of people who are prone to hasty judgments as a result of poor education. Sound public education will lead to responsible journalism, even with AI as an aid. Like any tool, AI can be considered a double-edged sword. But in the right hands, it can also be the tool to help media in the future. |
| R.5 | I believe the biggest risk of using AI in the media will be the erosion of public trust in journalism and information. As AI becomes increasingly involved in the production and distribution of content, there is a real danger that information will be manipulated or distorted, either intentionally or due to errors. |

**Table 5** Answers to the question "To what extent do you believe that artificial intelligence can influence the ecological sustainability of the media industry?"

| | |
|---|---|
| R.1 | As a media organization, we try to address sustainable solutions. For example, we use resource-saving equipment and computers that use less electricity. We are looking to find our own sustainable solutions. We have invested in computers and other IT equipment that consume less electricity. These devices help us save resources and reduce carbon emissions associated with energy consumption. We believe that our role could be a significant one. We believe that our role is not only limited to adopting sustainable solutions internally but also to educating and raising awareness among the general public. Our reports and articles promote the importance of protecting the environment and provide information on how each individual can contribute to a sustainable future. In doing so, we aim to inspire other organizations and individuals to adopt environmentally friendly practices. |
| R.2 | From an environmental point of view, we, as a media institution, consume significant electricity resources. But we already have a project to install solar panels to produce environmentally friendly energy. In addition, we are equipping our newsroom with state-of-the-art equipment that consumes less electricity. So we are working hard and realize that we can only have a clean environment through responsible decisions. And we need to be an example to others and take action. We also try to set examples of good practice, including through the topics we cover in our news and broadcasts. |
| R.3 | In terms of environmental sustainability, the impact of AI on the media will be both significant and ambivalent. On the one hand, the use of artificial intelligence in the media can help reduce the carbon footprint. For example, automating content production and distribution processes can reduce the need for physical travel by teams of journalists, thereby reducing fuel consumption and greenhouse gas emissions. AI can also optimize the use of digital resources, such as servers and data centers, by efficiently managing energy consumption. |
| R.4 | The use of AI tools in the context of environmental sustainability is both economically and culturally related. On the one hand, investments in the field need to be correlated with the current requirements to comply with the level of electricity consumption and the requirements of necessity in terms of equipment. Carbon emissions need to be reduced, and the efforts of companies are also directed in this direction, and the contribution of AI is directly proportional to the use of equipment of all kinds. On the other hand, the current reality requires the use of equipment in this competition to provide information as quickly as possible. At the same time, the public's level of culture is an element which can influence this decision to use and the level of use of equipment. A public educated in environmental protection can lead to pressure of a different kind when it comes to the providers of these messages, i.e., the media. |
| R.5 | Environmental sustainability has to be seen primarily in terms of the role of the media in society. Namely, it can educate the public to protect nature. On the other hand, AI is energy-intensive and the use of technology requires additional measures in this respect. This is obviously up to the large companies that operate the technology. I believe that if measures are taken to limit the effects, we will also be able to ensure environmental sustainability. |

When asked, "What do you consider to be the most significant risk associated with the integration of artificial intelligence in the media sector?", most respondents indicated the erosion of public trust in journalism and the veracity of information. They emphasized that, with the increasing involvement of AI in the process of creating and distributing content, there is a risk that information will be perceived as less credible, especially in the absence of transparency regarding the use of these technologies. The possibility of disinformation was also mentioned, through the automatic generation of false content.

The results obtained from the qualitative analysis confirm the initial hypothesis that artificial intelligence influences the sustainability of the media in its economic, social, and environmental dimensions. Respondents highlighted both clear advantages and serious challenges related to workforce restructuring, the risks of misinformation, and loss of public trust. These findings support the idea that the integration of AI into the media sector is not just a technological innovation, but a process with profound structural implications that requires a strategic, ethical, and adaptive approach to support long-term sustainability.

## 7 Conclusions

This study has highlighted the complexity of artificial intelligence's impact on the sustainability of the media, emphasizing its potential to bring significant improvements and the major risks associated with improper use. AI stands at the intersection of technological innovation and journalism, offering unprecedented opportunities for efficiency and innovation and serious challenges that require a strategic and responsible approach.

Artificial intelligence has the ability to radically transform the processes of producing and distributing media content. Among the identified benefits are the improvement of newsrooms' operational and economic efficiency, particularly through the automation of repetitive tasks and the optimization of SEO. AI can help media organizations save time and resources, allowing journalists to focus more on creative work and in-depth investigations.

Additionally, AI can contribute to a more environmentally sustainable approach by reducing the consumption of material resources and optimizing energy use in digital production and distribution processes. Advanced technologies enable the digitization and automation of processes, which can reduce dependence on physical materials and help decrease the industry's carbon footprint.

On the other hand, AI also presents significant risks, especially in the social and ethical context. One of the greatest dangers identified is the amplification of disinformation and the spread of fake news. AI algorithms can be exploited to create fabricated content, such as deep fakes, which can deceive the public and provoke negative social reactions, including polarization and social destabilization.

The erosion of public trust in the media is another significant risk. As AI becomes increasingly involved in generating and distributing news, there is a risk that the public may perceive information as less credible, especially if there needs to be more transparency in how these technologies are used. This decline in trust can have severe consequences for the role of the media as a guardian of democracy and can lead to an information deficit in society.

In conclusion, AI can bring significant benefits to the sustainability of the media, but this potential can only be fully realized with careful and responsible management. The risks associated with misinformation, algorithmic biases, and the loss of public trust are real and can have long-term consequences on the democratic health of society. Therefore, media organizations must approach the integration of AI with caution, ensuring that it is used in ways that support and strengthen the journalistic mission while also protecting the rights and interests of the general public.

This study is based on structured interviews with media professionals, which provides valuable insight but also has certain limitations. First, the number and diversity of respondents may influence the generalizability of the conclusions, and the opinions expressed may reflect specific experiences and contexts that are not representative of the entire media industry.

Furthermore, given the qualitative nature of the method, the results are subjective and may be influenced by the interpretation of the participants or the researcher.

Future research could explore concrete models of best practice in the responsible integration of AI into newsrooms, as well as the development of ethical frameworks and regulations that minimize associated risks, particularly with regard to misinformation and algorithmic transparency.

## References

Anderson, C. (2012). Towards a sociology of computational and algorithmic journalism. *New Media & Society, 15*(7), 1005–1021. https://doi.org/10.1177/1461444812465137

Baraniuk, C. (2018). *China unveils AI news presenter*. Accessed Mar 21, 2024, from https://www.bbc.com/news/technology-46136504

Broussard, M., Diakopoulos, N., Guzman, A. L., Abebe, R., Dupagne, M., & Chuan, C. H. (2019). Artificial intelligence and journalism. *Journalism & Mass Communication Quarterly, 96*(3), 673–695. https://doi.org/10.1177/1077699019859901

Choudhary, Kamal, Brian DeCost, Chi Chen, Anubhav Jain, Francesca Tavazza, Ryan Cohn, Cheol Woo Park et al. (2021). *Recent Advances and Applications of Deep Learning Methods in Materials Science. ArXiv*.

Coddington, M. (2015). Clarifying journalism's quantitative turn. *Digital Journalism, 3*(3), 331–348. https://doi.org/10.1080/21670811.2014.976400

Cox, M. (2000). *The development of computer-assisted reporting*. In: Newspaper Division, Association for Edu-cation in Journalism and Mass Communication, pp. 1–22.

Chandler, S. (2020, February). Reuters uses AI to prototype first ever automated video reports. *Forbes*. Accessed Jun 09, 2024, from https://www.forbes.com/sites/simonchandler/2020/02/07/reuters-uses-ai-to-prototype-first-ever-automated-videoreports/?sh=1831d2f07a2a#2636cd0a7a2a?

Chen, C., & Chekam, G. A. (2021). Algorithms and media ethics in the AI age. *Handbook of Global Media Ethics*, 301–328.

Cham, T. H., Cheah, J. H., Memon, M. A., Fam, K. S., & László, J. (2022). Digitalization and its impact on contemporary marketing strategies and practices. *Journal of Marketing Analytics, 10*(2), 103–105. https://doi.org/10.1057/s41270-022-00167-6

Couldry, N., & Turow, J. (2014). Big data, big questionsl advertising, big data and the clearance of the public realm: Marketers' new approaches to the content subsidy. *International Journal of Communication, 8*, 17. Retrieved from https://ijoc.org/index.php/ijoc/article/view/2166/1161

Diakopoulos, N. (2019). *Automating the news: How algorithms are rewiring the media*. Harvard University Press.
Dörr, K. N. (2015). Mapping the field of algorithmic journalism. *Digital Journalism, 4*(6), 700–722. https://doi.org/10.1080/21670811.2015.1096748
Elkington, J. (1997). *Cannibals with forks – Triple bottom line of 21st century business*. StoneyCreek, CT: New Society Publishers
Florea, N. M., Bădîrcea, R. M., Meghisan-Toma, G. M., Puiu, S., Manta, A. G., & Berceanu, D. (2021). Linking public finances' performance to renewable-energy consumption in emerging economies of the European Union. *Sustainability, 13*(11), 6344.
Gondwe, G. (2023). Mapping AI arguments in journalism studies. *SSRN Electronic Journal*. https://doi.org/10.2139/ssrn.4573697
Grewal, D., Hulland, J., Kopalle, P. K., & Karahanna, E. (2020). The future of technology and marketing: a multidisciplinary perspective. *Journal of the Academy of Marketing Science, 48*(1), 1–8. https://doi.org/10.1007/s11747-019-00711-4
Gutiérrez-Caneda, B., Vázquez-Herrero, J., & López-García, X. (2023). AI application in journalism: ChatGPT and the uses and risks of an emergent technology. *El Profesional De La Informacion*. https://doi.org/10.3145/epi.2023.sep.14
Haţieganu, A. (2024). Accessed Jun 10, 2024, from https://www.sport.ro/fotbal-intern/sport-ro-il-prezinta-pe-e-mil-primul-prezentator-ai-din-romania.html
Hansen, A. (2018). Environment, media and communication. In *Routledge eBooks*. https://doi.org/10.4324/9781315625317.
Henestrosa, A. L., Greving, H., & Kimmerle, J. (2022). Automated journalism: The effects of AI authorship and evaluative information on the perception of a science journalism article. *Computers in Human Behavior, 138*, 107445. https://doi.org/10.1016/j.chb.2022.107445.
Hermann, E. (2021). Artificial intelligence and mass personalisation of communication content—An ethical and literacy perspective. *New Media & Society, 24*(5), 1258–1277. https://doi.org/10.1177/14614448211022702
Hermida, A., & Young, M. L. (2019). *Data journalism and the regeneration of news* (1st ed.). Routledge. https://doi.org/10.4324/9781315163895
Huang, Y., & Yu, Z. (2023). Understanding the continuance intention for artificial intelligence news anchor: Based on the expectation confirmation theory. *Systems, 11*(9), 438. https://doi.org/10.3390/systems11090438
Larsson, S., & Heintz, F. (2020). Transparency in artificial intelligence. *Internet Policy Review, 9*(2). https://doi.org/10.14763/2020.2.1469
Léchenet, A. (2014). *Global database investigations: The role of the computer-assisted reporter*. Reuters Insti-tute fellowship paper. https://cutt.ly/FkNG99C
Lewis, S. C., Guzman, A. L., & Schmidt, T. R. (2019a). Automation, Journalism, and Human–Machine Communication: Rethinking Roles and Relationships of Humans and Machines in News. Digital Journalism, 7(4), 409–427. doi:https://doi.org/10.1080/21670811.2019.1577147.
Lewis, S. C., Sanders, A.-K., & Carmody, C. (2019b). Libel by algorithm? Automated journalism and the threat of legal liability. *Journalism & Mass Communication Quarterly, 96*(1), 60–81. https://doi.org/10.1177/1077699018755983
Márquez, J. (2023). *GPT-4: así es la IA más avanzada de OpenAI, cómo funciona y todas las novedades*.
McCarthy, J. (1998). *What is artificial intelligence? CogPrints*; Stanford. Accessed Sep 30, 2024, from http://cogprints.org/412/2/whatisai.ps
McGregor, S. (2013, March 18). CAR hits the mainstream: Computer-assisted reporting is being recognized as an important journalistic discipline. *Columbia Journalism Review*. Retrieved from https://www.cjr.org/data_points/computer_assisted_reporting.php
Meyer, P. (2002). *Precision journalism: A Reporter's Introduction to Social Science Methods*. Bloomsbury Publishing PLC.

Nah, F. F. H., Zheng, R., Cai, J., Siau, K., & Chen, L. (2023). Generative AI and ChatGPT: Applications, challenges, and AI-human collaboration. *Journal of Information Technology Case and Application Research, 25*(3), 277–304. https://doi.org/10.1080/15228053.2023.2233814

O'Neil, C. (2016). *Weapons of Math Destruction.* Penguin UK. http://books.google.ie/books?id=60n0DAAAQBAJ&printsec=frontcover&dq=Weapons+of+Math+Destruction:+How+Big+Data+Increases+Inequality+and+Threatens+Democracy&hl=&cd=1&source=gbs_api

Pantic, M. (2021). Local Media in a Digital Market: Establishing niche and promoting original reporting to ensure sustainability. *Journalism Practice, 16*(8), 1736–1752. https://doi.org/10.1080/17512786.2021.1874483

Parratt-Fernández, S., Mayoral-Sánchez, J., & Mera-Fernández, M. (2021). Aplicación de la inteligencia artificial al periodismo: análisis de la producción académica. *El Profesional De La Informacion.* https://doi.org/10.3145/epi.2021.may.17

Purvis, B., Mao, Y., & Robinson, D. (2019) Three pillars of sustainability: in search of conceptual origins. *Sustainability Science, 14*(3), 681–695. ISSN 1862-4065

Reiter, E. (2010). Natural language generation. In A. Clark, C. Fox, & S. Lappin (Eds.), *The handbook of computational linguistics and natural language processing* (pp. 574–598). Wiley-Blackwell.

Rooth, E. (2024). *OpenAI's video generator Sora is launching today.* Accessed Jan 22, 2025, from https://www.theverge.com/2024/12/9/24317090/openai-sora-text-to-video-ai-generator-confirmed-release

Serrano, M. J. H., Greenhill, A., & Graham, G. (2015). Transforming the news value chain in the social era: A community perspective. *Supply Chain Management an International Journal, 20*(3), 313–326. https://doi.org/10.1108/scm-05-2014-0147

Stănescu, G. C. (2023). The impact of artificial intelligence on journalism. adverse effects vs. benefits. *Social Sciences and Education Research Review, 10*(1), 258–261. https://doi.org/10.5281/zenodo.8151135

Thäsler-Kordonouri, S., & Barling, K. (2023). Automated journalism in UK local newsrooms: Attitudes, integration, impact. Journalism Practice, 1–18. https://doi.org/10.1080/17512786.2023.2184413

Underwood, C. (2019). Automated journalism–AI applications at New York times, Reuters, and other media giants. https://emerj.com/ai-sector-overviews/automated-journalism-applications/.

Voinea, D. V. (2025). Reconceptualizing gatekeeping in the age of artificial intelligence: A theoretical exploration of artificial intelligence-driven news curation and automated journalism. *Journalism and Media, 6*(2), 68. https://doi.org/10.3390/journalmedia6020068

World Business Council for Sustainable Development. (2002). *Investing for sustainable development.* WBCSD. https://docs.wbcsd.org/2002/08/Investing_For_SD.pdf

Wu, H., Chen, Z., Huang, Y., & Tu, H. (2024). Research on the uncanny valley effect in artificial intelligence news anchors. *Multimedia Tools and Applications., 83.* https://doi.org/10.1007/s11042-023-18073-z

Wang, X., & Zhu, F. (2022). The application of artificial intelligence in AI news anchor. In M. Atiquzzaman, N. Yen, & Z. Xu (Eds.), *2021 international conference on big data analytics for cyber-physical system in Smart City. BDCPS 2021* (Lecture notes on data engineering and communications technologies) (Vol. 102). Springer. https://doi.org/10.1007/978-981-16-7466-2_121

Xue, K., Li, Y., & Jin, H. (2022). What do you think of AI? Research on the influence of AI news anchor image on watching intention. *Behavioral Sciences, 12*(11), 465. https://doi.org/10.3390/bs12110465

Yeh, S. C., Wu, A. W., Yu, H. C., Wu, H. C., Kuo, Y. P., & Chen, P. X. (2021). Public Perception of Artificial Intelligence and Its Connections to the Sustainable Development Goals. *Sustainability, 13*(16), 9165. https://doi.org/10.3390/su13169165

**Georgiana Camelia Stănescu, PhD,** teaches in the Audiovisual Communication specialization within the Faculty of Letters at the University of Craiova. With a PhD in media law at the same institution, she published about 20 studies and articles in communication and media, as well as a book presenting the research on media law, the topic of her PhD thesis. Her more recent research approached topics such as fake news, artificial intelligence, virtual reality, and immersive journalism, and the results of these studies were discussed at international conferences or published in indexed scientific journals. Georgiana Camelia Stănescu is a member of CCSCMOP's Scientific Committee (Based on the document published on December 25, 2019) and an editorial board member of the Social Sciences and Education Research Review. https://sserr.ro/editorial-board/. The editorial team of the *Annals of the University of Craiova for Journalism, Communication, and Management* is also available: https://www.aucjc.ro/.

# A Rigid Dichotomy: Artificial Intelligence and Ethical Responsibility: Comparative Assessment of EU, UK, and US AI Policy Framework

**Adebimpe Adesua Lincoln and Christian J. Lincoln**

**Abstract** This research provides a comparative assessment of Artificial Intelligence governance mechanisms adopted in the European Union, United Kingdom, and the United States. The rationale for the study stems from the fact that Artificial Intelligent technologies currently pose significant ethical risks to individuals, businesses, and the wider society. Furthermore, there is a dearth of legal regulatory framework in many countries, and where such frameworks exist, they lag behind advancements in technology due to rapid technological innovation. This research therefore highlights the importance of ensuring risks posed by Artificial Intelligence are identified and that strategies are proffered to mitigate these risks, thereby ensuring global justice in the development and deployment of Artificial Intelligent technologies.

**Keywords** Algorithms · Computational thinking · Machine consciousness · Machine learning · Artificial intelligence · Social responsibility · Governance · Human rights · Ethics · Morality · Fairness · Cybersecurity · Data and privacy protection · Safety · Security · United States · United Kingdom · European Union · Ethical responsibility · Policy · Artificial intelligence hallucination · Digital technology · Media · Traceability · Transparency · General purpose artificial intelligence · Explainability · Equality · Artificial intelligence risk systems · Artificial intelligence technology · Algorithmic discrimination · Stakeholder engagement · Equitability · Competitive artificially intelligent ecosystem

---

A. A. Lincoln (✉)
Robert Kennedy College, Zurich, Switzerland

C. J. Lincoln
Atlantic Wharf, Cardiff Bay, UK



M. Todorovic et al. (eds.), *Sustainability and Social Responsibility of the Media and in the Media*, CSR, Sustainability, Ethics & Governance,
https://doi.org/10.1007/978-3-032-00086-6_6

# 1 Introduction

Artificial Intelligence[1] (AI) focuses on the intelligence of machines and software. In the last century, Alan Turing the architect of the "Turing test," an experiment that measures the ability of a machine to simulate human conversation stated that the question of whether machines can think is baseless. He states, "*The original question, 'Can machines think!' I believe it to be too meaningless to deserve discussion. Nevertheless, I believe that at the end of the century, the use of words and general educated opinion will have altered so much that one will be able to speak of machines thinking without expecting to be contradicted. I believe further that no useful purpose is served by concealing these beliefs*" (Turing, 1950).[2] Since then, AI has become a fast-growing trend. Increasingly, organizational use of AI technologies now influences how people work and how and whether they experience meaningfulness in their work (Bankins & Formosa, 2023). Furthermore, AI is becoming more prevalent in every aspect of our daily lives, for decision-making in private and public affairs. The UK DSIT (2024) recognizes this fast growth in its 2024 policy proposal wherein it states that the UK AI market is witnessing exponential advancement—unlocking novel skills, creating new employment opportunities, and lifesaving treatments for terminal diseases. The industry is projected to grow to over $1 trillion by 2035. Globally, AI is now actively used in medicine and medical research to accurately diagnose and treat patients; in game creation and the development of generalized AI and computer playing systems; in the military application of threat detection and identification, sensors, command and control, communications, and integration, and military logistics. In addition, there is widespread use of generative AI—such as the prominence of ChatGPT, based on GPT-3, and other large language models. AI is utilized by various industries, governmental institutions, and in science and academia[3] (see Fig. 1).

---

[1] Collins et al. (2021) identified a large variety of different definitions. According to their findings, apart from the lack of cohesion, a noteworthy observation is that most definitions tend to focus on what AI systems are capable of, instead of what AI is. According to Koniakou (2023), the term can be traced back to 1956 when AI was described by John McCarthy as "the science and engineering of making intelligent machines" (McCarthy, 2018). The European Commission's Communication on AI for Europe defines AI as "Systems that display intelligent behavior by analyzing their environment and taking actions—with some degree of autonomy—to achieve specific goals. AI-based systems can be purely software-based, acting in the virtual world (e.g. voice assistants, image analysis software, search engines, speech and face recognition systems) or AI can be embedded in hardware devices for example advanced robots, autonomous cars, drones, or Internet of Things application" (AI HLEG, 2019a, 2019b).

[2] Recent advancements in AI support Turing's sentiments about machines—AI is the ability of computers and other artificial entities to do things typically classified as requiring intelligence were a human to do them, such as reason, plan, problem solve, and learn from experience (Wang, 2019).

[3] Industry-specific examples of AI utilization include: Space exploration and autonomous operations; Energy Sector (energy storage); Legal Sector (applications that predict the result of judicial decisions); Politics (predicts foreign policy); Strategy and logistics (used in supply chain manage-

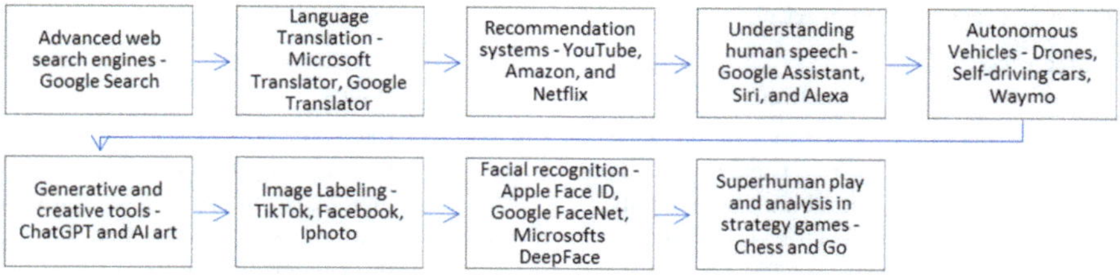


**Fig. 1** Use of AI by various industries, governmental institutions, science, and academia

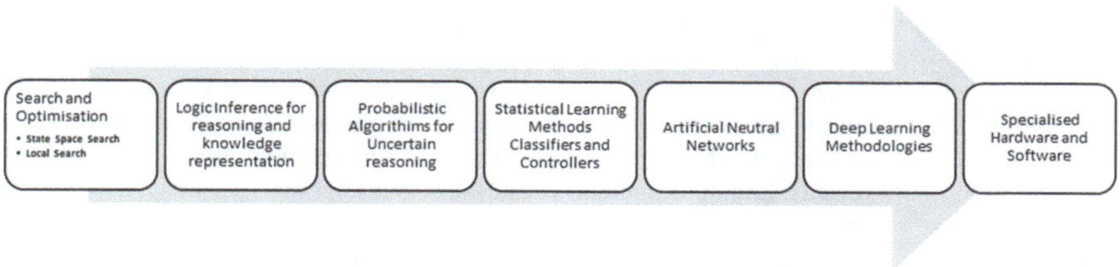


**Fig. 2** AI tools

AI as a construct is centered around specific goals. These goals are essential in simulating or creating intelligence. AI goals comprise specific attributes or skills that researchers expect an intelligent system to exhibit.[4] Russell and Norvig (2021) state that an important long-term vision of AI research is the acquisition of "AI general intelligence." General intelligence is the ability of an AI to complete any human task. AI researchers have designed and amalgamated a range of problem-solving methodologies, for example, statistics, search and mathematical optimization, formal logic, and artificial neural networks (see Fig. 2). They also embrace aspects from divergent disciplines and fields such as psychology, linguistics, philosophy, neuroscience, operations research, and economics in finding workable solutions.

There is no established unifying theoretical perspective to the development of AI research. Symbolism and Connectionism are the two main paradigms in AI research and development with the former taking a systematic and logical approach to discrete symbols whereas the latter relies on neurons inspired from the biology of the human mind to achieve intelligence. AI is also classified into four main distinct categories namely reactive machines; limited memory machines; theory of mind machines and self-aware machines (see Table 1). The categories can also be grouped

---

ment and logistics operations); Agricultural Sector (automate greenhouses, water conservation, identify areas in need of irrigation, fertilization, pesticide treatments, predict the ripening time for crops, monitor soil moisture, operate agricultural robots, detect diseases in animals, detect pests in crops); Astronomy (data analysis and application of data, the discovery of exoplanet and gravitational wave).

[4]These are reasoning and problem-solving skills; knowledge representation; planning and decision-making; learning and natural language processing; perception awareness; social intelligence and general intelligence.

**Table 1** Classification of AI

| Categories/ classification | Function | Issues |
|---|---|---|
| Reactive machines | Used to respond automatically to limited sets of inputs or to a combination of inputs.<br>Reactive to situations as presented to it.<br>It mimic's or copy's the human mind's ability to react to distinct categories of stimuli. Examples include machines that play chess. | Lacks memory-based functionality-cannot rely on memory to improve its operations, learn or use prior experience to inform current actions.<br>Tasks are performed with narrow scope. It is incapable of performing tasks outside of its limited capability. |
| Limited memory machines | Some capacity to react to the world and their environment. Capability to make informed decisions by learning from historical data. Possess some albeit limited understanding of past events. Examples include virtual Assistant; chatbots; self-driving cars | Incapable of forming a robust or comprehensive understanding of the world and their environment.<br>Limited recall of past events.<br>Limited time frame. |
| Theory of mind machines | Awareness and understanding of other actors and entities that exist within the world and in their environment.<br>Ability to discern needs, emotions, and thought. Able to create representations of the world and their environment. | Conceptual and purely theoretical. The reality is yet to be materialised into a functioning machine. |
| Self-aware machines | Possesses self-awareness and desire. Emotionally intelligent and able to evoke emotions.<br>Possess awareness and an understanding of the world, other actors, and entities and of itself.<br>Analogous to the human brain - also referred to as Artificial General Intelligence (AGI). | Hypothetical and purely theoretical. The reality is yet to be materialised into a functioning machine.<br>Capable of having ideas and self-preservation.<br>May be able to exceed human intellect. |
| Artificial Narrow Intelligence (ANI) | Able to perform specific tasks autonomously using human-like capabilities. Encompasses all reactive and limited memory artificial intelligence machines. | Possess a limited and narrow range of competencies based solely on the tasks it has been programmed to accomplish. |
| Artificial General Intelligence (AGI) | Ability to learn and understand like a human being. Capable of perception and replicating human multi-functional competences.<br>Able to form links and generalizations.<br>Reduced training time. | Capable of having ideas and self-preservation.<br>May be able to exceed human intellect. Potential for offsetting jobs and have an impact on economic equilibrium. |
| Artificial Superintelligence (ASI) | Capable of perception and replicating human multi-functional competences.<br>Superior memory and intelligence resulting in its ability to surpass human learning and understanding.<br>High-speed data processing and analysis, and decision-making capabilities. | Capable of threatening human way of life and existence.<br>Has the potential to create systems more intelligent than itself; the 'singularity'. |

into three main classifications namely, Artificial Narrow Intelligence (ANI); Artificial General Intelligence (AGI), and Artificial Superintelligence (ASI). There are a host of benefits and advantages to the use of AI technology, e.g., improved decision-making; cost reduction; and greater speed, accuracy, and efficiency of machines. Davenport et al. (2020) identify various ways in which AI will continue to shape behavior and business processes. They identify examples of AI influence in business models, sales processes, and customer service options, as well as customer behavior.[5] The growing significance of AI technologies is also identified by Gui et al. (2019) who opine that AI has a multifaceted impact in various sectors and industries—AI is said to play a transformative role in the telecommunications sector, optimizing telecommunications networks, boosting predictive maintenance, forecasting potential network problems, and reducing downtime and operational costs.

As the use of AI has become increasingly widespread, several unintended consequences and risks have been identified. Some of the issues identified in various academic discourse include—the use of AI in spreading radical ideologies online, job loss due to increased automation; privacy and cybersecurity concerns; misinformation; potential for bias or discrimination due to the information and type of data the AI is trained on and lack of transparency vis-a-vis decisions reached and system development. For example, the UK AI policy framework recognizes that AI technologies pose "new and amplified risks," the framework cites examples of AI chatbots that promote misinformation by making false information more prominent and highly capable AI systems that may be misused to engage in criminal activities. There are also ethical issues vis-a-vis advancement in AI technologies with AI systems pursuing complex goals with limited human supervision or intervention (DSIT, 2024).

AI ethics is a set of values, principles, and techniques that employ widely accepted standards of "right and wrong" to guide moral conduct in the development and use of AI technologies (Leslie, 2019). Many organizations and governmental institutions have realized that the use of AI technologies raises questions, and that consumers and society at large are increasingly unlikely to engage with unethical companies or overlook instances of unethical practices. Leslie (2019) argues that to manage the impacts of AI responsibly and to direct the development of AI systems toward optimal public benefit, considerations of AI ethics and safety are required as a priority.[6] It is imperative that AI systems are developed and used in a socially

[5] For example, transportation industry—driverless, AI-enabled cars; salespeople assisted by an AI agent that monitors real-time tele-conversations; retailers utilizing AI to identify customers' preferences and shipping items to customers without a formal order, with customers afforded the option to keep or return the orders (Agrawal et al., 2018; Gans et al., 2017).

[6] This will involve integrating considerations of the social and ethical implications of the design and use of AI systems into every stage of the delivery of AI projects. It will also involve a collaborative effort between the data scientists, product managers, data engineers, domain experts, and delivery managers on the team to align the development of AI technologies with ethical values and principles that safeguard and promote the well-being of the communities that these technologies affect (Leslie, 2019).

responsible manner and that ethical guidelines and regulations are devised to prioritize critical quality requirements of AI. AI governance and questions about "AI and algorithmic systems" have attracted significant attention, which continuously grows as the technologies in question become more ubiquitous (Koniakou, 2023). AI is increasingly becoming a key area of deliberation for companies, governments, academics, technologists, and civil society organizations. Furthermore, the ethical significance of the use of AI is widely recognized and discussed in extant academic literature resulting in various organizational, national, and international institutions proposing ethical principles for AI utilization. Article 19 (2019), "Governance with teeth," focuses on how human rights can strengthen fairness, accountability, and transparency (FAT) and ethics initiatives on AI. Some stakeholders such as the European Commission's High-Level Expert Group on AI (AI HLEG, 2019a, 2019b) have focused on building normative ethical principles and guidelines for Responsible AI, while other stakeholders have gravitated towards a technical discussion of how to build fair, accountable, and transparent AI systems. *For instance, facial recognition technologies are being developed by companies and deployed by governments, and there are active attempts at making these systems fair and accurate.* But it is becoming increasingly clear that these approaches are not enough (Article 19, 2019).

Humans are morally and ethically responsible for the AI they create, deploy, and/or use. The question is not if those involved in technological innovation are aware of any potential risks and or the implications and consequences of flawed decision-making, but whether they understand the risk, and that measures are in place for early detection of potential risks (Henz, 2021). Johnson (2006) raises questions vis-à-vis AI training and active human supervision and intervention in AI technologies wherein he states *"Even when it learns, it learns as it was programmed to learn. The fact that the designer and user do not know precisely what the artifact does makes no difference here. It simply means that the designer in creating the program and the user in using the program are engaging in risky behavior. They are facilitating and initiating actions that they may not fully understand, actions with consequences that they cannot foresee. The designer and users of such systems should be careful about the intentionality and efficacy they put into the world"* (Johnson, 2006).

## 2 Ethics and Social Responsibility

Applications of moral concepts such as democracy, equality, fairness, and transparency are integral to real-world situations to better understand and resolve ethical dilemmas presented by AI technologies and their application. The regulatory and policy landscape for AI is an emerging issue in jurisdictions globally. Contemporary advancements in research, development, and application of AI systems have generated a far-reaching discourse on the importance of social responsibility and AI ethics (European Union, 2020a, 2020b, 2020c). These guidelines and frameworks comprise normative principles and recommendations aimed at harnessing the disruptive potentials of novel AI technologies (Floridi et al., 2018; Cath et al., 2018;

Hagendorf, 2020; Jobin et al., 2019). They contain a variety of recommendations on social responsibility issues, providing minimal requirements for building and using ethically sound AI systems. A review of the frameworks shows that ethical values such as bias, explainability, transparency, openness and accountability, fairness, nondiscrimination, justice, data, and privacy protection, safety, and cybersecurity are considered vital in ensuring the effectiveness of AI systems (Mittelstadt et al., 2019; Gebru et al., 2018; Baron & Musolesi, 2017). There is a wide range of ethical aspects that are significantly related to the research, development, and application of AI systems, but are not or very seldomly mentioned in national and international guidelines.[7] Those omissions range from the danger of a malevolent artificial general intelligence, machine consciousness, the reduction of social cohesion by AI ranking and filtering systems on social networking sites, the political abuse of AI systems, a lack of diversity in the AI community, links to robot ethics, weighting between algorithmic or human decision routines, hidden social and ecological costs of AI, and the problem of public-private-partnerships and industry-funded research (Hagendorff, 2020).[8]

Despite the availability of AI guidelines and principles, there is no common or unifying discussion on how to govern ethics in AI implementations, or the ongoing auditability once machine learning is improving without human intervention (Forbes, 2021). Floridi and Cowls (2019) refer to this as "principle proliferation." *There are too many different frameworks available, not only making it difficult to navigate through them but also opening an opportunity for choice when there should not be one* (Forbes, 2021). In practice, AI ethics is often considered as "extraneous," a "surplus" or "add-on" to technical concerns, as an unbinding framework that is imposed on the "technical community" from "outside institutions." As a result, Hagendorff (2020) argues that currently, AI ethics is failing in many cases. This is because generally, ethics lacks a reinforcement mechanism, thus, "deviations from" and "derelictions of" the various codes of ethics have no consequences. Alston (2019) notes that these ethics guidelines contain a reference to human rights, but the substance of human rights law is invariably lacking. Instead, the token reference to

[7] The various organizational, national, and international guidelines and frameworks include the United Nations which launched an advisory body in 2023 to provide recommendations on AI governance; the body comprises technology company executives, government officials, and academics. Other international and country-specific initiatives include the European Commission's High-Level Expert Group on AI; Report on the Future of AI; Beijing AI Principles; AI at Google; Microsoft AI Principles; OpenAI leaders which published a ten-year vision recommendation for the governance of superintelligence in 2023; The OECD Recommendations of the Council on AI. The OECD AI principle addresses various issues such as human-centered values and fairness; transparency and explainability; accountability, robustness, security, and safety. The 2020 Global Partnership on AI also advocates for the development of AI in accordance with human rights and democratic values, to ensure public confidence and trust in the technology.

[8] Significant ethical considerations arise in the adoption of AI and machine learning and continued innovation and advancements in this field. For example, psychologist Carol Gilligan, argues that the way AI ethics is performed and structured constitutes a typical instantiation of a male-dominated justice ethics (Gilligan, 1982).

human rights serves only to enhance claims of legitimacy and universality. The relevant discussions of ethics are based on almost entirely open-ended notions that are not necessarily grounded in legal or even philosophical arguments and can be shaped to suit the needs of the industry.

Article 19 report (2019) found critical gaps in both ethical and technical approaches to AI systems. It states that they are both lacking in enforcement mechanisms and definitions, which institutions they invoke, and how they empower individuals. This view is supported by Weinbaum et al. (2019) who found that the effectiveness of ethical guidelines or ethical codes is almost zero and that they do not change the behavior of professionals from the tech community. They state that "*ethical research also requires internalizing a commitment to it, aided by training and education on codes and appropriate research methods, mentoring and workplace cultures that foster ethics, transparency about how the research was conducted, and forums (in person and writing, local and international) where researchers can share their experiences and the challenges they face.*" In specific cases where ethics is integrated into institutions, it mainly serves as a marketing strategy. Distributed responsibility in conjunction with a lack of knowledge about long-term or broader societal technological consequences causes software developers to lack a feeling of accountability or a view of the moral significance of their work. Especially, economic incentives are easily overriding commitment to ethical principles and values. This implies that the purposes for which AI systems are developed and applied are not in accordance with societal values or fundamental rights such as beneficence, non-maleficence, justice, and explicability (Taddeo & Floridi, 2018; Pekka et al., 2018). AI poses fundamental questions concerning its ethical, social, and economic use and impact. This is especially true when using AI algorithms that are inherently unexplainable in deep learning (Russell & Norvig, 2021; Laskowski, 2023).[9] Examples of ethical issues include:

***Algorithmic Bias and Fairness*** There are examples of AI exhibiting bias and often sensitive predilections due to disparity in sample sizes or sample size bias, inaccurate, or unfair data. For example, AI's mistaking black people as gorillas due to insufficient training. Russell and Norvig (2021) and Christian (2020) opine that machine learning models are designed to make predictions, however, where this training is based on unfair, biased, discriminatory, or inaccurate data it is more probable than not that the application will follow suit. Therefore, if an unfair, biased, discriminatory, and inaccurate algorithm is utilized in making decisions that impact people's lives and well-being, the algorithm may result in prejudice and discrimination. A valid legal argument can be proffered that self-learning and training of AI

[9] Sivarethinamohan et al. (2021) argue that "just because we can—does not mean we should"? They state that some issues for consideration include: *privacy*—consideration of consumers' rights; *errors*—how do we ensure a correct result (e.g., self-driving cars, decision algorithms); *job Loss*—will this replace human staff; *bias*—do the training sets and algorithms promote inherent bias; *security*—can data sets or algorithms be hacked by nefarious sources; *control*—is there a risk of losing control over the algorithm and its results; the *"Creep Factor"*—perhaps it is not illegal or does not break official privacy rules, but does it "feel right."

systems should be based on regulated sources of data and information to foster memory intelligence. In addition, AI and robotics systems should only be considered safe where it can be demonstrated that the system is free of bias. Many AI systems are inherently biased and unfair as a result of systems being developed using training data with a striking underrepresentation of minority groups and women. During development, these factors should be critically taken into consideration to mitigate these problems and create professional, trustworthy, and reliable systems. For example, AI that will assess the trustworthiness of customers at a bank or for a job application may reproduce discrimination by picking up on real-life details and replicating past bias and discriminatory practices in selection criteria that humans have exhibited and in its identification of those deemed trustworthy and mimicking these traits. This can be amended by cleaning the data provided to the model when training. The importance of memory intelligence in AI cannot be overemphasized—memory Intelligence provides AI with a means to adapt to new information that may threaten it and its user's post-deployment and cope with evolving workspaces. For example, self-driving cars and AI involved in statistical predictions may quickly adapt to a scenario unforeseen by its developers, preserving this information for later use and effectively training itself on the spot. AI without memory intelligence can struggle to adapt post-deployment and carries along more risk. For example, chatbots may be easily manipulated in a malicious way where developers may struggle to provide a quick correction without these facilities. For modern Neural network-based systems, once deployed, the efforts required to re-train the system may be significant. The opaqueness of neural networks as it stands may also be a significant problem. It becomes significantly more difficult to review, amend, and predict errors that these systems may create during operation without performing many tests to ensure the AI's safety in real-life scenarios. Consider a self-driving car being presented with an awkward and unexpected scenario on the road and having limited means to mitigate the problem. Here, there will be difficulty in attributing responsibility to a particular part of the neural network and re-training cannot guarantee safety in other unique situations.

***Explainable AI and Algorithmic Transparency*** While popular explainability techniques exist, many AI systems are so complex that their creators are unable to clearly articulate or explain how they reached their decisions. This is particularly true vis-à-vis deep neural networks, involving a large number of nonlinear relationships between inputs and outputs. In addition, there have been instances where a machine learning program learned drastically different information or practice than the developer originally intended for the system, even though the machine learning program passed rigorous initial testing. The scale and advancements of modern neural networks have led to systems that can produce convincing statements, essays, and even entire theses with plausible evidence. However, these systems do not reflect on their own work or consider the task in the same way that a human would. A user could request a large language model to construct a persuasive argument for a completely false premise as though it were correct. These models can "hallucinate" producing convincing results that may trick an unsuspecting user when relied on unwittingly, particularly when prompted with information they have a limited understanding of. A salient example is many models' inability to multiply numbers of a few digits together with accuracy due to the very nature of their intelligence.

***Workplace Impact of AI*** The introduction of technological innovation should enhance work function and result in employee upskilling. In recent times, many of these functions have become obsolete as various human functions have become autonomous using AI technologies—as certain tasks are now carried out by computers. Advances in technology thus resulted in deskilling numerous working-class jobs. McGaughey (2022) identifies a real risk of mass unemployment and redundancies as a result of AI technologies, which is exacerbated by inadequate social policy for full employment.

***Data Protection, Privacy, and Copyright*** Machine learning algorithms require vast amounts of data to function at an optimal level. Techniques adopted by AI developers and technology companies to obtain this data have raised data protection, privacy, surveillance, and copyright concerns. This includes technology companies collecting audio, and video data from their users and monitoring online activity and geolocation data. In addition, Generative AI is usually trained based on unlicensed copyrighted works, images, or computer code. The output generated from the AI is then utilized under the "fair use" umbrella. According to Nicholas et al. (2021), generative AI models are exposed to privacy and data protection violations due to pervasive training on personal data, the memorization of training data, and inversion attacks.

***Weaponized AI*** AI and its use by various institutions and world governments—including use by criminal and terrorist organizations, can result in excessive intrusion into people's lives and their privacy. It can also pose significant threats to human life. Harari (2018) posits that AI technologies provide various tools such as deepfake and generative AI aid to spread misinformation; and promotion of propaganda using extremist memes spyware; and facial and voice recognition which subverts privacy and enables widespread mass surveillance.

***Adversarial Attacks and Fraudulent Use and Abuse*** AI technologies can be susceptible to cyber/adversarial attacks, which destabilize the system, thus rendering it unsafe and subject to manipulation. Cyberattacks are attacks on the operation of information and technology assets that have consequences affecting the availability of information system, their confidentiality, and/or integrity (Cebula et al., 2014). These intrusions corrupt the data or information that the AI has learned resulting in inaccurate outputs and possible data breaches. As the processing power of modern computer systems progresses, as shown by Moore's Law, the number of parameters in Neural Networks will follow a similar trend, leading to more powerful and faster thinking systems. However, this progression in AI technology grants those with malicious intent greater power in deceiving others. The use of generated photorealistic videos and lifelike voices of close family members makes identity fraud and phishing much easier for criminals and can potentially weaken the reliability of existing forms of evidence such as CCTV recordings (a notable example is the May 2024 Schumacher case—in which a newsagent published a fake AI interview of the sports legend. If we do not prepare in advance and quickly consider these changes in technology to have robust legislation and countermeasures "grow" with AI as both a tool and a weapon, it may take us by surprise.

## 3 Research Methodology

The theoretical construct for the study is drawn from the ethics, social responsibility, and information technology literature in order to arrive at a working conceptual framework. An exploratory assessment of three jurisdictions—United Kingdom, the European Union, and the United States is carried out to gain new insights, identify limitations inherent in the approaches adopted by each jurisdiction, and propose ways in which an ethical and socially responsible approach can constructively inform and enhance these efforts. The information presented in this research is based on an in-depth analysis of publicly available government data only. Data was collected through an in-depth review of the UK, EU, and US policy initiatives on AI. Document analysis involving the use of a systematic procedure was adopted in analyzing the documentary evidence obtained from official government websites and publications. Repeated examination and interpretation of the data was undertaken to obtain valuable insight into the objectives of the AI policy frameworks. To ensure the document sources identified yielded the best quality result pertinent to the overriding research objectives, inclusion criteria were adopted to assist in identifying apt sources of data, systematic document selection, and thus ensure authenticity and representativeness of the data source. Two inclusionary criteria were used:

- *Time Limit*: a 2-year limit was placed on the period for the AI policy initiative due to the fast-paced nature of advancements in AI technologies to ensure the approach is current and contextually relevant.
- *Documents*: only official government documents on AI were analyzed and included.[10]

Initial categorization of keywords derived from the IEEE (2018) guidelines was used in aiding the evaluation of key ethical and socially responsible initiatives outlined in the policy framework of each jurisdiction. The IEEE framework contains a comprehensive compilation of pertinent principles that apply to autonomous and intelligent systems, including physical robots such as driverless cars and software

[10]A document analysis of the following government publications was carried out—UK: A pro-innovation approach to AI regulation: government response—2024; EU: European Parliament Laying down harmonized rules on AI—2024; US: Blueprint for an AI Bill of Rights: A Vision for Protecting Our Civil Rights in the Algorithmic Age—2022; Executive Order on the Safe, Secure, and Trustworthy Development and Use of AI—2023.

systems for example algorithmic chatbots. According to the IEEE (2018), the ethical design, development, and implementation of these technologies should be guided by the following FIVE General Principles:

***Human Rights*** Consideration of various international human rights conventions is essential in ensuring that the design, development, and implementation of technologies do not infringe on internationally recognized human rights.[11] The international frameworks encourage governments to ensure adequate protection which requires AI to be designed and operated in a way that both respects and fulfils human rights, freedoms, human dignity, and cultural diversity. That AI is verifiably safe and secure throughout its operational lifetime. Where harm is caused through the design, development, and implementation of AI technologies, it must always be possible to discover the root cause, by assuring traceability for said harm.

***Well-Being*** AI technologies must prioritize metrics of well-being in their design and use. Well-being encompasses human satisfaction with life and the conditions of life as well as an appropriate balance between positive and negative effects (IEEE, 2018).[12] AI technologies can be narrowly conceived from an ethical standpoint; be legal, profitable, and safe in their usage; and yet not positively contribute to human well-being. This means technologies created with the best intentions, but without considering well-being metrics, can still have dramatic negative consequences on people's mental health, emotions, sense of themselves, their autonomy, ability to achieve their goals, and other dimensions of well-being. Key issues of transparency, accountability, and algorithmic bias should be addressed when designing and implementing AI technologies. For AI technologies to advance benefits for humanity, it is essential that society clearly defines and adequately measures the benefits it seeks to increase.[13] Key indicators should ensure measures are in place to maintain "full human autonomy" of the use of AI technology and that personal, environmental, and societal factors are prioritized as highly as fiscal metrics of success such as GDP, to mitigate risks and harms to the planet and population—thereby resulting in positive and holistic impact for humanity (IEEE, 2018).

---

[11] The Universal Declaration of Human Rights; The International Covenant for Civil and Political Rights; The Convention on the Rights of the Child; the Convention on the Elimination of all Forms of Discrimination against Women; the Convention on the Rights of Persons with Disabilities; The Geneva Conventions.

[12] According to the IEEE (2018), well-being is defined based on the Organization for Economic Co-Operation and Development's (OECD) Guidelines on Measuring Subjective Well-being that notes, being able to measure people's quality of life is fundamental when assessing the progress of societies. There is now widespread acknowledgment that measuring subjective well-being is an essential part of measuring quality of life alongside other social and economic dimensions.

[13] Avoiding negative unintended consequences and increasing value for customers and society are often the only indicators utilized in determining the success of AI technologies. Beyond the critical importance of designing and manufacturing these technologies in an ethically driven and responsible manner, determining the key performance indicators (KPIs) of their success once introduced into society is also integral (IEEE, 2018).

***Awareness of Misuse*** The IEEE (2018) advocates for the minimization of the risks inherent in the misuse of AI technology. According to the IEEE (2018), novel technologies give rise to a greater risk of misuse and increase the impact of risks such as hacking, the misuse of personal data, "gaming," or exploitation of vulnerable users by unscrupulous parties. A link is made here with the EU and UK General Data Protection Regulation (GDPR) which provides measures to remedy the misuse of personal data. AI designers must anticipate, reflect, and engage with users on the importance of education in raising awareness of associated AI technology risks. Users, lawyers, and governments have a role to play in developing accountability structures. They also have a role to play in guiding new innovative technologies proactively toward achieving positive benefits for humanity (IEEE, 2018). A thematic analysis which involved the use of pattern recognition, deductive categorization of the data, and coding was then carried out to confirm, expand, or refine the existing understanding of AI applications, ethics, and social responsibility.[14]

***Accountability*** The design, development, and implementation of AI technologies ought to be based on the sociocultural context. In addition, the application, and use of AI, people, and institutions must be clear vis-à-vis the manufacture and deployment of these technologies to establish responsibility and accountability and avoid potential harm (IEEE, 2018). The IEEE (2018) posits that AI designers and operators must act responsibly and should be accountable for the design, development, and implementation of AI technologies. Furthermore, manufacturers of these technologies must be able to provide programmatic-level accountability proving why specific technologies operate in certain ways to address legal issues of culpability and where required apportion culpability among designers, manufacturers, owners, and/or operators, to avoid confusion or fear within the general public.

***Transparency*** Transparent AI technologies are ones in which it is possible to discover how and why an AI technology made a particular decision, or in the case of a robot, acted in a particular way (IEEE, 2018). AI technologies must operate and function transparently to a wide range of stakeholders for different reasons. This is particularly important for safety-critical AI technologies which have significant consequences for human safety or well-being, such as medical diagnosis AI systems, or driverless car autopilots (IEEE, 2018). Lack of transparency increases the risk and magnitude of harm because users do not understand the systems they are using. In addition, lack of transparency also increases the difficulty of ensuring accountability, traceability, explicability, and interpretability. In addition, the complexity of AI technologies compounds the challenges users face when operating these technologies—many users do not fully appreciate the risks or understand the capabilities and limitations of the AI technologies that they use. This complexity, combined with the often-decentralized manner in which AI technologies are devel-

[14] The following SEVEN patterns and categorizations were identified: scope and flexibility; safety and security; equity and civil rights; innovation and competition; international cooperation; government use of AI and penalty for noncompliance.

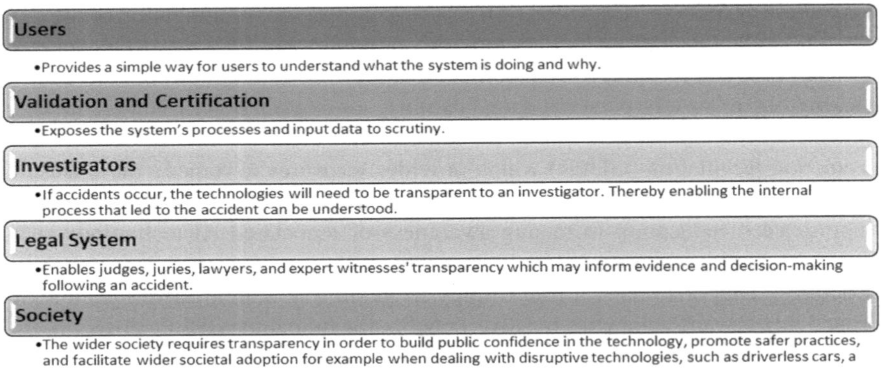


**Fig. 3** Importance of transparency for diverse stakeholder groups

oped, complicates efforts to determine and allocate responsibility when something goes wrong with the technology (IEEE, 2018). Figure 3 depicts the importance of transparency for diverse stakeholder groups.

# 4 Comparative Assessment of AI Policy Frameworks

Recently, the EU, UK, and USA implemented foundational policies of AI risk management to cater to the fast-paced nature of technological innovation and advancements. While the policies have a shared emphasis on safety, equity, and fundamental rights, there are however significant differences. The EU and US approach is a "pro-consumer"—"fundamental rights" approach, while the UK's approach is a "pro-innovation"—"context-based approach." The EU's hard rule-based legislative approach is prescriptive and sets out key requirements for AI design and deployment, especially high-risk AI technologies. The USA on the other hand adopts a more pragmatic approach in focusing on US leadership in AI advancement both on the domestic level and internationally as a global partner. It however proposes mandatory reporting requirements vis-à-vis foundational models that pose significant national or economic security risks. The UK's preference is for a much softer principle-based approach which stresses the importance of not inhibiting potential benefits that AI innovation presents.

## *4.1 European Union*

The EU recognizes the importance of AI in ensuring better healthcare; safer and cleaner transport; more efficient manufacturing; and cheaper and more sustainable energy (European Parliament, 2023). The EU's approach to AI is programmatically based on the safeguard of fundamental rights, whose common core is represented by the concept of human dignity, upon which the anthropocentric European architecture is grounded.[15] In May 2016, the EU released its first document addressing the issue of AI governance: a draft report, by the European Parliament's Committee on Legal Affairs, entitled "Civil Law Rules on Robotics," which called for a coordinated European approach that would employ a mix of hard and soft laws to guard against possible AI risks. This was followed by the Declaration of Cooperation on AI in April 2018, signed by 25 European countries.[16] In June 2018, the EU Commission appointed the High-Level Expert Group on AI (HLEG), a group of 52 experts—amongst other outputs, they produced Ethics Guidelines for AI which focus on the EU Requirements for trustworthy AI (European Commission, 2020). The EU proposed requirements for trustworthy AI—a list of key requirements that must be met continuously, evaluated, and addressed throughout the AI systems life cycle (see Table 2):

Although the existing EU legislation provides some level of protection, the EU recognizes the inability of these provisions to provide a robust level of protection and to address the specific challenges that AI systems and models bring.

**Table 2** Requirements for trustworthy AI (European Commission, 2020)

| Specific issues | Fundamental rights, human agency, and human oversight |
|---|---|
| Technical robustness and safety | Resilience to attack and security, fall-back plan and general safety, accuracy, reliability, and reproducibility |
| Privacy and data governance | Respect for privacy, quality and integrity of data, and access to data |
| Transparency | Traceability, explainability, and communication |
| Diversity, non-discrimination, and fairness | The avoidance of unfair bias, accessibility and universal design, and stakeholder participation |
| Societal and environmental well-being | Sustainability and environmental friendliness, social impact, society, and democracy |
| Accountability | Auditability, minimisation and reporting of negative impact, trade-offs, and redress AI |

[15] The ethical guidelines refer to fundamental rights in the framework of democracy and the rule of law, stating that "these rights are described in the EU Charter by reference to dignity, freedoms, equality and solidarity, citizens' rights and justice. The common foundation that unites these rights can be understood as rooted in respect for human dignity—thereby reflecting what we describe as a 'human-centric approach' in which the human being enjoys a unique and inalienable moral status of primacy in the civil, political, economic and social fields."

[16] Shortly after, the European Commission published a Communication on the European Approach to AI, which outlines a coordinated path forward, centering on the following three priorities:

(1) Boosting the EU's technological and industrial capacity across the economy, in private and public sectors.
(2) Preparing for the changes brought about by AI through anticipating market change, modernizing education and training, and adapting social protection systems.
(3) Ensuring that there is an appropriate legal and ethical framework that is in line with the EU's values.

Consequently, in April 2024, the EU adopted the world's first comprehensive AI law.[17] The EU AI Act which applies to private organizations as well as to public authorities represents the first binding worldwide horizontal regulation on AI. It sets a common framework and establishes a "risk-based approach" vis-à-vis legal obligations and classification for the development and use of AI systems within the EU. The Act promotes a uniform application of rules and seeks to ensure safe, transparent, traceable, nondiscriminatory, and environmentally friendly use of AI systems across EU Member States. To prevent potentially harmful outcomes, the EU advocates for AI systems to be administered and managed by people, rather than being automated.[18] The Preamble to the Act states that: "*The purpose of this Regulation is to improve the functioning of the internal market by laying down a uniform legal framework in particular for the development, the placing on the market, the putting into service and the use of AI systems in the Union, based on Union values, to promote the uptake of human-centric and trustworthy AI while ensuring a high level of protection of health, safety, fundamental rights as enshrined in the Charter of Fundamental Rights of the European Union, including democracy, the rule of law and environmental protection, to protect against the harmful effects of AI systems in the Union, and to support innovation. This Regulation ensures the free movement, cross-border, of AI-based goods and services, thus preventing Member States from imposing restrictions on the development, marketing, and use of AI systems, unless explicitly authorized by this Regulation*" (European Union AI Act, 2024).

The EU AI Act is a direct response to the 2022 citizens' proposals from the Conference on the Future of Europe (COFE). In particular, the proposals relating to:

- Enhancing EU's competitiveness and deepening the Single Market. Proposal 12(10) recognizes the need to "Identify and develop strategic sectors, including space, robotics, and AI."
- Safe and trustworthy digital society—cybersecurity and disinformation. Proposal 33(5) recognizes the need to counter disinformation and ensure humans are ultimately in control of the use of algorithms.
- Digital innovation to strengthen the social and sustainable economy. Proposal 35(3) and 35(8) recognize the need to ensure human oversight and trustworthy and responsible use of AI, setting safeguards that "ensure transparency, interoperability, generate trust, enhance ease of use and avoiding any discriminatory or biased algorithms."
- Citizen's information, participation, and youth. Proposal 37(3) recognizes the need to use "AI systems and translation technologies to circumvent language

[17] Based on the recommendations proposed by the European Commission (EC) in April 2021.

[18] The EU AI Act (2024) defines AI systems as "a machine-based system designed to operate with varying levels of autonomy and that may exhibit adaptiveness after deployment and that, for explicit or implicit objectives, infers, from the input it receives, how to generate outputs such as predictions, content, recommendations, or decisions that can influence physical or virtual environments."

barriers, ensuring the accessibility and usability of all the digital tools for people with disability" (Conference on the Future of Europe, 2022).

The Act proposes a risk-based approach to regulating AI and outlines FOUR categories of risk: unacceptable, high-risk, limited risk, and minimal/no risk (see Table 3). For systems that are not high-risk, voluntary codes of conduct are encouraged.

Violating this Regulation leads to fines of up to 6% of global turnover or 30 million euros, whichever is greater. The Act also provides specific rules for "General Purpose" AI (GPAI) models,[19] setting out stringent requirements for GPAI models

**Table 3** EU Act categories of risk

| | |
|---|---|
| Unacceptable risks | Systems that represent 'unacceptable risks' that violate EU fundamental rights and values are prohibited. These risks include subliminal, manipulative, or deceptive techniques; exploiting mental or physical vulnerabilities in specific groups; biometric categorization systems; social scoring; assessing the risk of an individual committing criminal offenses; compiling facial recognition databases; inferring emotions in workplaces or educational institutions; real-time remote biometric identification in publicly accessible spaces for law enforcement. |
| High risk | These include systems that are safety-critical components and those AI systems that can have a detrimental impact on people's safety, health (e.g. toys, aviation, cars, medical devices, lifts) and their fundamental rights. A wide range of 'high-risk' AI systems are authorized but are subject to conformity assessments and post-market monitoring. The Act places specific obligations on providers, importers, distributors, users, and authorized representatives to ensure compliance. High-risk AI providers must ensure that they establish a risk management system with appropriate levels of accuracy, robustness, and cybersecurity; conduct data governance and ensure training, validation, and testing datasets are relevant and free of error; device technical documents that evidence compliance with the Act; keep adequate and transparent records; provide instructions for the use of AI systems; ensure their AI systems are designed in a manner that allows human oversight; establish a quality management system. |
| Limited or transparency risk | AI systems that interact with humans. These systems pose risks of manipulation, impersonation, and deception. These systems are used for biometric categorization or generate manipulative content for example chatbots, emotion recognition systems, biometric categorisation systems, and AI-generated content - deepfakes (AI systems that generate or manipulate image, audio, or video content). These systems have specific information and transparency requirements and obligations. |
| Low or minimal risk | These are AI systems presenting 'minimal risks' - these systems will not be subject to further legal requirements or obligations. Minimal risks are unregulated, they include the majority of AI applications available in the EU single market. Common AI systems that could pose low or minimal risks include spam filters, recommender systems, and AI-enabled video games. |

[19] GPAI models are models that can serve a variety of purposes, both for direct use as well as for integration into other AI systems (European Union Briefing, 2024). The Act defines GPAI as models that "are trained with a large amount of data using self-supervision at scale," that display "sig-

that pose a systemic risk due to their "high-impact capabilities" on the internal market. All GPAI model providers are subjected to two obligations.

A. GPAI system transparency requirements: devise and maintain technical documentation and make information and documentation available to supply to downstream providers of AI systems; ensure adherence to the copyright directive and ensure they make available a sufficiently detailed summary of the content used for training the GPAI model.
B. Systemic-risk GPAI obligations: perform model evaluations and adversarial testing to identify and mitigate systemic risk; assess and mitigate possible systemic risks; track, document, and report serious incidents and ensure cybersecurity protection.

The Act also proposes measures to support innovation and the Small and Medium Enterprise (SME) sector. It advocates for the provision of a "testing environment" by national authorities within Member States which stimulates "real-world conditions" to be provided and accessible to SMEs and start-ups which allows these enterprises to develop and train innovative AI models before their placement on the market and release to the public. Table 4 outlines the proposed rules:

A significant shortcoming of the EU approach is that the Act applies equally to all entities operating in EU Member States regardless of organizational sector or size. This approach could potentially place a disproportionate obligation on smaller enterprises and smaller service providers. The Act will enter into force 21 days after

**Table 4** Rules for AI models before placement on the market

| |
|---|
| • Addresses risks specifically created by AI applications |
| • Prohibits AI practices that pose 'unacceptable risks' |
| • Determines a list of high-risk AI applications |
| • Sets clear requirements for AI systems for high-risk AI applications. |
| • Defines specific obligations for providers of high-risk AI applications |
| • Requires conformity assessment before AI systems can placed on the market |
| • Puts in place enforcement mechanisms after AI systems have been placed into the market |
| • Establishes a governance structure at EU and Member State level. |

nificant generality" and are "capable of competently performing a wide range of distinct tasks," and "can be integrated into a variety of downstream systems or applications" (European Union AI Act, 2024).

its publication in the Official Journal and will be fully applicable 24 months after it enters into force. Obligations for high-risk systems will have more time to comply—these systems will have 36 months after the entry into force to comply. However, some provisions of the Act will be applicable sooner.[20]

## 4.2 *United Kingdom*

According to the UK National AI Strategy (2021), AI is the fastest-growing deep technology in the world, with huge potential to rewrite the rules of entire industries, drive substantial economic growth, and transform all areas of life. The UK does not put forward a formal definition of AI, rather, it adopts an outcome-based approach in guiding sectoral interpretations. This focuses on two distinguishing characteristics namely "*adaptivity and autonomy*." According to the DSIT (2024), adaptivity is "*the ability to see patterns and make decisions in ways not directly envisioned by human programmers*." Autonomy is the "*capability to operate, take action, or make decisions without the express intent or oversight of a human*." The UK National AI Strategy[21] builds on the UK's strengths as a global power in AI technological advancements—"*recognizing the power of AI to increase resilience, productivity, growth, and innovation across the private and public sectors*." The UK's ten years strategic vision for AI is built on three assumptions namely: global competition for access to people, data, and finance; AI technological advancement will become mainstream and ensure UK sectors and regions are strategically positioned to benefit from the transition; ensuring excellence in technology and innovation, protecting safety and security for all must be benchmarked against the need to keep abreast of innovation and developments vis-à-vis AI technology and ensuring that governance and regulatory landscape align with the fast-paced advancement of AI to effectively maximize growth and ensure competitive advantage. On the 6th of February 2024, the UK Government unveiled its long-awaited response to the March 2023 AI Regulation White Paper (DSIT, 2023). The AI White Paper seeks to address regulatory challenges and gaps posed by the emergence of AI technologies. The policy framework identifies THREE broad categories of AI risk: societal harms; misuse risks; and autonomy risks (see Table 5). The UK government initiative is a

[20] For example: the bans on prohibited practices and AI systems posing unacceptable risks will be applicable 6 months after the Act enters into force; codes of practice will be applicable 9 months after entry into force; rules on General Purpose AI (GPAI) systems that need to comply with transparency requirements including governance will apply 12 months after the entry into force.

[21] The UK's National AI Strategy through broad public trust and support, and by the involvement of the diverse talents and views of society seeks to achieve the following aims: Invest and plan for the long-term needs of the AI ecosystem to continue our leadership as a science and AI superpower; Support the transition to an AI-enabled economy, capturing the benefits of innovation in the UK, and ensuring AI benefit all sectors and regions; Ensure the UK gets the national and international governance of AI technologies right to encourage innovation, investment, and protect the public and our fundamental values.

**Table 5** UK AI White Paper categories of AI risks

| Societal harm | Misuse risks | Autonomy risks |
|---|---|---|
| Protection of workers and employment rights. | Safeguarding democracy from electoral interference. | Accountability, governance, and oversight for developers. |
| Protection of intellectual property rights. Protection against bias and discrimination. Reforming data protection law to support innovation and privacy. | Preventing the misuse of AI technologies—crime prevention, cyberattacks or weapons design. | Testing and benchmarking powerful AI systems to address risks. |
| Ensuring AI generated online content is trusted/safe. | | |
| Ensuring AI driven digital markets are competitive. | | |
| Ensuring AI best practice in the public sector. | | |

pro-innovation sector-based approach[22] that embeds a pro-safety action that will eventually form the basis of future regulatory approaches. The AI policy framework sets out how the UK government proposes to deliver a proportionate, context-based approach to regulate the use of AI, it identifies how regulators and the government are addressing some of the most important issues that arise from using AI and sets out regulatory challenges posed by the rapid advancement and development of "*highly capable general-purpose systems*" and how the UK is currently tackling these through voluntary measures. Particular emphasis is placed on the challenges of "*Frontier AI*"—defined "*as highly capable general-purpose AI models that can perform a wide variety of tasks and match or exceed the capabilities present in today's most advanced models*" (DSIT, 2024).

The policy framework recognizes the substantial risks that highly capable general-purpose AI systems (GPAI) pose to the context-based framework and the recognition that as a result of its general application if there is a flaw in design and implementation, a GPAI could potentially pose significant risks across multiple sectors and applications in the economy. To ensure accountability and that safety measures are in place the UK AI policy framework seeks to devise effective rules and adopt a "*targeted binding requirement*" aimed at organizations developing GPAI (DSIT, 2024). As shown in Fig. 4, the proportionate regulatory approach effectively addresses the risks posed by AI systems by distinguishing "*Foundation Models*" across three broad categories which are often integrated into numerous downstream AI systems. Foundation models are defined in the UK AI Regulation White Paper as "*a type of AI model that is trained on a vast quantity of data and is adaptable for use on a wide range of tasks. Foundation models can be used as a base for building more specific AI models*" (DSIT, 2023).

[22] The UK AI regulatory policy framework is tailored to the needs and risks posed by different sections of the UK economy.

The initiative proposes principles for existing UK regulators to follow and set out the UK government's expectations for responsible AI innovation (DSIT, 2024). The AI White Paper is premised on "*an initial assessment of AI-specific risks and their potential to cause harm, with reference to the values that they threaten if left unaddressed. These values include safety, security, fairness, privacy and agency, human rights, societal well-being, and prosperity*" (DSIT, 2024). The framework is underpinned by FIVE cross-sectoral principles to guide and inform the responsible development and use of AI in all sectors of the economy as depicted in Fig. 5.

The UK Government seeks to implement the FIVE principles by leveraging existing regulatory authorities and frameworks: establishing a central function to facilitate effective risk monitoring and regulatory coordination and supporting innovation by piloting a multi-agency advisory service—the AI and Digital Hub. The Department for Science, Innovation and Technology (DSIT) is the government department responsible for overall responsibility for AI policy in the UK, including regulation. The principles-based, non-statutory, and cross-sectoral initiative will not be codified into law for the time being, because this strategy enables the UK government to keep abreast of the fast-paced nature of the technology sector and rapid change and uncertain advances in AI technologies. Adopting this approach enables the government to devise future legislative intervention if required as the risks and challenges inherent in utilizing AI technologies become more apparent. There are criticisms of the approach adopted by the UK government. Firstly, a shortcoming of the UK approach is the ambiguity it creates and a lack of clarity on how AI is defined, the White Paper merely distinguishes between two characteristics namely—adaptivity and autonomy. The UK leaves the interpretation of adaptivity and autonomy to regulators such as the Office of Communication (Ofcom) and the Financial Conduct Authority (FCA). Secondly, fundamental principles of "*fairness, appropriate transparency and explainability, and contestability and redress*" which

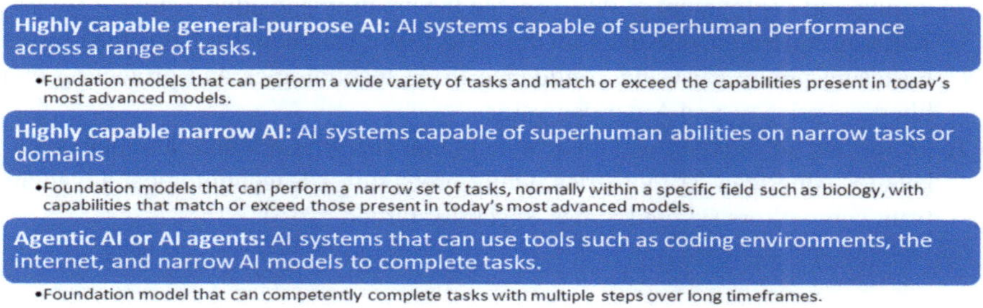


**Fig. 4** UK AI White Paper categories of "Foundation Models"

**Fig. 5** UK AI FIVE cross-sectoral principles

underscore the use of AI systems in recruitment and employment identified in the White Paper are left under the purview of the Equality and Human Rights Commission (EHRC); Information Commissioner Office (ICO); Employment Agency Standards Inspectorate (EASI) and other regulators. There are significant challenges to adopting this approach, it has the potential to create inconsistencies—the interpretations adopted by various regulators can result in a lack of clarity thereby creating uncertainties for businesses operating across sectors. In addition, as UK regulators look to consider the impact of AI technologies in their various domains, the potential for overlapping initiatives may be exacerbated.

## *4.3 United States*

The term AI is defined by the United States National AI Act (2020) under Section 5002 (3) as a "machine-based system"[23] that can, for a given set of human-defined objectives, make predictions, recommendations, or decisions influencing real or virtual environments (United States National AI Act, 2020). In October 2016, the White House Office of Science and Technology Policy (OSTP) released the United States (US) first reports focusing specifically on AI: "*Preparing for the Future of AI*" *and the* "*National AI Research and Development Strategic Plan*." This report defined the government's role in the development of AI as a facilitator of innovation and a minimalist regulator and outlined how federal research and development investments would guide the "*long-term transformational impact of AI.*" FIVE important principles were emphasized:

(1) "Driving technological breakthroughs in AI to promote scientific discovery, economic competitiveness and national security.
(2) Developing appropriate technical standards and reducing barriers to the safe testing and deployment of AI to enable the creation and adoption.
(3) Training American workers with the skills to prepare them for jobs of the future.
(4) Fostering public trust and confidence in AI technologies and protecting civil liberties, privacy, and American values.
(5) Promoting an international environment that supports research and opens markets for American AI industries, while protecting US technological advantage from competitors (United States Executive Order, 2016)."

[23] Section 5002 (11) defines machine learning as an "application of AI that is characterized by providing systems the ability to automatically learn and improve based on data or experience, without being explicitly programmed. AI systems use machine and human-based inputs to—perceive real and virtual environments; abstract such perceptions into models through analysis in an automated manner; and use model inference to formulate options for information or action."

In November 2020, the White House released guidance for government agencies proposing new AI regulations for the private sector.[24] Table 6 presents the US Guidance for Regulation of AI Principles (United States Executive Order, 2020) which contains ten principles for the stewardship of AI applications namely:

Recently, the National AI Initiative Act of 2020,[25] passed in 2021 identifies the need for US leadership in AI technology advancement and governance and the development of trustworthy AI systems. In addition, the Act recognizes the importance of ensuring effective preparation for employment disruptions and coordination of the military sector and the society at large. The Act mandates the establishment of government bodies and parastatals to guide the Federal level. This includes the creation of the "*National AI Office*," tasked with "supporting AI R&D, educational initiatives, interagency planning, and international cooperation." The creation of an

**Table 6** US guidance for regulation of AI principles

| | |
|---|---|
| Public trust in AI | The government must promote reliable, robust and trustworthy AI applications |
| Public participation | The public should have a chance to participate in all stages of the rule-making process |
| Scientific integrity and information quality | Policy decisions should be based on science |
| Risk assessment and management | Agencies should decide which risks are unacceptable |
| Benefits and costs | Agencies should select approaches that maximise net benefits |
| Flexibility | Agencies should pursue a technology-neutral, flexible approach |
| Fairness and non-discrimination | Agencies should make sure AI systems do not discriminate illegally |
| Disclosure and transparency | Context-specific transparency measures are necessary for public trust |
| Safety and security | Agencies should promote AI systems that are safe, secure and operate as intended |
| Interagency coordination | Interagency cooperation and coordination are necessary for consistent policies |

[24] The Regulations center around three themes namely, limiting regulatory overreach; Ensuring public engagement; and promoting trustworthy AI that is fair, transparent, and safe.

[25] A key aspect of the 2020 Act under Section 5103 is the promotion of a coordinated effort by an interagency committee tasked with the following responsibilities: (A) "determine and prioritize areas of AI research, development, and demonstration requiring Federal Government leadership and investment; (B) support long-term funding for interdisciplinary AI research, development, demonstration, and education; (C) support research and other activities on ethical, legal, environmental, safety, security, bias, and other appropriate societal issues related to AI; (D) provide or facilitate the availability of curated, standardized, secure, representative, aggregate, and privacy-protected data sets for AI research and development; (E) provide or facilitate the necessary computing, networking, and data facilities for AI research and development; (F) support and coordinate Federal education and workforce training activities related to AI; and (G) support and coordinate the network of AI research institutes described in Section 5201(b)(7)(B)."

expert National AI Advisory Committee for assessing the degree to which the USA is fulfilling its aims and a subcommittee for AI and law enforcement that advises on issues such as legality, biases, and data security. The latest US AI initiative was published on October 30, 2023—the White House Executive Order on the "*Safe, Secure, and Trustworthy Development and Use of AI*." An important focal point of the US AI policy framework is the need to evaluate the potential for AI technologies and advancement to "pose a threat" or be "misused." Some of the key areas of priority identified by the Executive Order include:

- *"Ensuring safe, secure and reliable AI.*
- *Managing AI in critical infrastructure and cybersecurity.*
- *Reducing risks at the intersection of AI and chemical, biological, radiological, and nuclear (CBRN) threats*—e.g. *ways in which AI can increase biosecurity risks, including risks from generative AI models trained on biological data.*
- *Reducing the risks posed by synthetic content—including labeling, detecting, testing software, auditing, and maintaining synthetic content.*
- *Soliciting input on dual use 'Foundation Models' with widely available model weight.*
- *Safe release and preventing the malicious use of federal data for AI training*" (United States Executive Order, 2023).

The Executive Order recognizes the potential of AI in advancing various aspects of life and work and advocates for the need for AI to be used responsibly in "*making our world more prosperous, productive, innovative, and secure*" (United States Executive Order, 2023). It also identifies some of the consequences of the irresponsible use of AI systems[26]—and how these could potentially be utilized in exacerbating "*societal harms such as fraud, discrimination, bias, and disinformation; displace and disempower workers; stifle competition; and pose risks to national security*" (United States Executive Order, 2023). In addition, in 2022 the USA introduced the AI Bill of Rights. The framework uses a two-part test to determine AI systems that fall within the scope of the proposed Bill, i.e., "*automated systems that have the potential to meaningfully impact the American public's rights, opportunities, or access to critical resources or services. These Rights, opportunities, and access to critical resources of services should be enjoyed equally and be fully protected, regardless of the changing role that automated systems may play in our lives*" (United States Blueprint for an AI Bill of Rights, 2022). The AI Bill of Rights proposes FIVE public and economic safeguards:

- Protection from unsafe or ineffective systems: AI "Systems should undergo pre-deployment testing, risk identification, and mitigation, and ongoing monitoring that demonstrate they are safe and effective based on their intended use, mitigation of unsafe outcomes including those beyond the intended use, and adherence to domain-specific standards."

[26] Such as data systems, software, hardware, applications, tools, or utilities that operate in whole or in part using AI.

- Protection from algorithmic discrimination: AI systems should be used and designed equitably.
- Protection from abusive data practices: Building on current data protection legislation to ensure greater security and agency over how data is used and processed by AI systems.
- Protection from AI systems acting without people understanding how or why—transparency on AI decision-making process.
- Protection from AI having the final say, with opt-out and human remedies to overrule AI systems.

A major criticism of the US approach is that whilst the AI policy initiative recognizes the need for legislative intervention in tackling some issues vis-à-vis data protection and privacy legislation and the need for Congress to pass data privacy legislation—it fails to provide a legislative framework which offers protection or enable strict enforcement measures. Another criticism of the approach adopted in the USA is that the AI policy framework is in the form of an Executive Order. Executive Orders are insufficient in tackling the problems and challenges posed by advancements in AI technologies. This is because Executive orders are inherently restricted in their authority and easily reversed.[27] It remains to be seen if the initiatives contained in the current Executive Order are retained when the transition occurs between President Biden and Trump. The US and UK approach differ significantly from the hard-line regulatory approach adopted by the EU (see Table 7). In evaluating the effectiveness of a country's preference for AI policing (i.e., whether a rule-based regulatory response or a principle-based approach is more palatable), Forbes (2021) argues that regulation might not necessarily be the right response; in fact, concerns that poorly designed regulation could slow down innovation are well sustained. Roger (2019) on the other hand opines that an approach that considers regulatory alternatives (e.g., self and industry regulation, co-regulatory arrangements, and formal laws) should be carefully designed and evaluated given the technical and political complexities.

Considering the wide-reaching impact of AI technologies, some form of mandatory measures is required to address the potential harm that AI technologies pose to the public and to ensure public safety. While not discounting the need for a voluntary principle-based approach, this approach may prove ineffective in ensuring compliance and protecting the public in the long run. We argue that the rule-based approach adopted by the EU should be favored in ensuring the effective protection of citizens and in fully realizing the transformative opportunities that AI technologies offer. An argument can also be proffered vis-à-vis engaging AI companies that can demonstrate an internal commitment to AI ethics, for example, if they have adopted an internal policy overseeing these ethical developments. This enables trust, which in exchange facilitates innovation (Forbes, 2021). Koniakou (2023) however argues against the efficacy of such a strategy—that such self-declared

[27] As evidenced by the December 2020 Executive Order proposed by Donald Trump on Promoting the Use of Trustworthy AI in the Federal Government.

**Table 7** Contextualizing AI policy initiative in UK, EU, and USA

| | 2024 AI White Paper | 2024 AI Act | 2023 Executive Order and Bill of Rights |
|---|---|---|---|
| Context | United Kingdom | European Union | United States |
| Objectives | Adopts a sector-based approach empowering regulators to create bespoke measures tailored to the needs and risks posed by different sections of the economy. | Adopts a prescriptive approach which focuses on the importance of AI technologies in the European Single Market and potential risks these technologies pose to fundamental rights and EU values. | Focuses on the benefits of AI technologies and inherent risks – emphasises safety, equity, civil rights, and US leadership in AI. |
| Scope and Flexibility | A pro-AI innovation approach. It builds on the existing strengths of the UK's thriving AI industry and expert regulatory ecosystem. Combining cross-sectoral principles and a context-specific framework, international leadership and collaboration, and voluntary measures on developers. Distinguishes AI systems as either:<br>– **Highly capable general-purpose AI**<br>– **Highly capable narrow AI**<br>– **Agentic AI or AI agents**<br>It proposes FIVE principles. These are: (1) Safety, security and robustness; (2) Appropriate transparency and explainability; (3) Fairness; (4) Accountability and governance; (5) Contestability and redress. | Clearly defined rules which ensures a level playing field across EU Member States. Recognises the rapidly evolving nature of AI technologies and systems and the need for implementation of adaptable regulatory frameworks. Adopts a tiered approach for risk management, allowing a flexible approach to AI policing. Classifies AI applications into FOUR risk categories: **(1) Unacceptable risk systems:** prohibits real-time remote facial recognition systems used in public spaces; **(2) High-risk systems:** subjecting critical infrastructures to several requirements e.g. conformity assessments; **(3) Limited risk systems:** transparency requirements; **(4) Minimal risk systems:** voluntary guidance. Promotes stakeholder involvement, encourages collaboration with developers/deployers and includes mechanisms which enables periodic review of the AI regulatory landscape and advancements in AI technologies to ensure the law keeps up with the pace of advancement of AI technologies. Advocates for a regulatory Sandboxes for businesses to encourage experimentation in a controlled environment. | Broad coverage of a variety of sectors - healthcare, education and national security. Lays down a robust foundational direction for AI governance and allows for flexible implementation of the directives. Proposes FIVE areas of focus: **(1) Protection from unsafe or ineffective systems; (2) Protection from algorithmic discrimination; (3) Protection from abusive data practices; (4) Protection from AI systems taking action or decisions without people understanding how or why; (5) Protection from AI having the final say, with opt-out and human remedies to overrule AI systems.**<br>Recognises the importance of adopting a cross-sectoral and collaborative approach to AI governance. Identifies the important role that federal agencies play in AI governance, oversight, and advancement. Adopts a flexible approach on implementation entrusting the task of developing best practices guidelines to specific US government departments. It recognises the need for continuous monitoring and calls for development of a 'National Security Memorandum' tasked with monitoring emerging threats from AI technologies. |

| | | | |
|---|---|---|---|
| Safety and Security | Focus on measures which protect people's wellbeing and autonomy against AI-related harms and protecting AI technologies and systems from technical interference and cyber-attacks. | Distinguishes AI applications from a hierarchical perspective based on the level of risk to fundamental rights. Prohibits specific high-risk uses and imposes strict regulations for high-risk applications. | Imposes obligations to protect against AI fraud and cyber-security threats. Developers of robust AI systems required to share safety tests with the government, develop safety and security standards for AI and ensure protective mechanism are in place to prevent the use of AI in creating biological threats. |
| Equity and Civil Rights | Adopts a fairness principle and highlights the need to protect UK citizens from AI-related bias and discrimination. Promotes fundamental rights and related values e.g. freedom, pluralism, privacy, equality, inclusion, and accessibility. Identifies importance of Equality and Human Rights Commission (EHRC) and Information Commissioners Office (ICO) in devising strategies to address inherent bias and discriminatory practices. Introduction of Data Protection and Digital Information Bill (DPDI) to reform data protection laws. | Adopts a strict approach which seeks to protect fundamental rights and EU core tenants – data privacy and protection of personal data. Prohibits applications that could be biased and unfair, harmful or discriminatory. Promotes a system of transparency, accountability and mechanism for redress across EU Member States. Alignment with GDPR in reinforcing the EU's commitments on data protection, data consent, access, and erasure. | Highlights the importance of fundamental rights and calls for AI policy focus to address – privacy issues, biases, discrimination, inclusive stakeholder engagement, equitable access to benefits and impact on employment for underrepresented and marginalised communities. Calls for the enactment of data privacy legislation. |
| Innovation and Competition | Pro-AI innovative approach—fosters competitive edge; creative industries; AI enhancement of creativity and innovation. | Seeks to ensure balance between regulating use of AI systems and innovation | Supports research in AI and promotes innovation and competitive AI ecosystem. |

(continued)

**Table 7** (continued)

| | 2024 AI White Paper | 2024 AI Act | 2023 Executive Order and Bill of Rights |
|---|---|---|---|
| Context | United Kingdom | European Union | United States |
| International Cooperation | Through agreements with multi-stakeholders, it seeks to collaborate with international partners; industry, academia, civil society to promote AI technologies, governance and safety. | Focuses regionally on EU Member States and the EU Single Market. The EU has a global reach and is recognised on an international level as an important blueprint for AI technologies and safety. | Advocates for global alliance and international partnership on AI ethics and governance and in fostering US leadership on AI technologies and safety. |
| Government use of AI | Central Digital & Data Office (CDDO) developed Algorithmic Transparency Recording Standard (ATRS) to help public sector provide information on algorithmic tools used to support decisions. | Does not explicitly provide guidance on government use of AI -this can be inferred. The regulations are applicable to all AI applications, including AI technologies implemented by EU governments. | Provides guidance on use of AI technologies by the government agencies and the Federal Government. Identifies the importance of AI training for government officials. |
| Penalty for Non-Compliance | Penalty is context-based. UK Regulators are empowered to allocate liability within their sector. | Non-compliance attracts penalties of up to 6% of a company's total worldwide annual revenue or €30 million. | Empowers government agencies with the responsibility of allocating liability and taking appropriate action. |

commitments to ethical AI may be utilized merely for "*ethics washing*," promoted to avoid scrutiny/criticism or used for "branding grandiosity." Furthermore, Wagner (2018) states that the "*turn to ethics*" sidelines the role of the States in ensuring that the legal regulatory landscape is robust and effective. Thus, "ethics as self-regulation is used as a form of escape from regulation or a way to delay the debate and work on law for AI." However, anecdotal evidence exists to support the argument vis-à-vis the efficacy of ethically motivated efforts where technical solutions exist for specific ethical dilemmas such as safety, security, accountability, discrimination, and privacy protection.

# 5 Conclusion

A rigid dichotomy exists whereby, even though AI is becoming an increasingly prevalent feature in social settings, workplaces, and governmental affairs, there remains a lack of understanding of the extent to which AI can revolutionize society and the ethical implications of such changes. This research therefore highlights the importance of ensuring the risks posed by AI are identified and that strategies are proffered to mitigate them, thereby ensuring global justice in the development and deployment of AI technologies. The risk that AI poses is borderless, consequently, international cooperation is required as well as a concerted effort from governments, researchers, academics, the private sector as well as members of society. There is a dearth of efficient AI legal frameworks in many countries and where AI regulatory frameworks exist, they lag behind advancement in technology due to rapid changes in technological innovation. Consequently, until the legal regulatory landscape measures up, ethics must drive behavior and determine what is "right" (Henz, 2021). As the adoption of AI becomes more commonplace, so do questions vis-à-vis the ethical implications of AI technologies and their application (European Union, 2020a, 2020b, 2020c). While the policy approach adopted by countries defines various constellations of responsibility, emphasizes the importance of cooperation, and mentions specific areas of concern and values to be upheld, it is becoming increasingly clear that these approaches are not enough. To harness AI for societal good and be able to realize the myriad potential that AI offers, there is a need to mitigate the substantial risks that AI technologies present. Figure 6 depicts a comprehensive ethics framework derived from this study.

In seeking a common ground, there is a need for the AI community to be educated on ethical principles encouraging the development of ethical AI. For AI deployment, development, and application to be carried out in an ethical and socially responsible manner, "*it must fulfil its functionally sourced obligations, where the latter can be either design sourced, or etiologically determined via the benefit that the relevant AI brings, and which contributes to the explanation of its continuous existence*" (Simion & Kelp, 2023). Those looking to harness the benefits of AI must first ensure that a holistic understanding of its effects on states, businesses, and

**Fig. 6** AI ethical and social responsibility framework

society is in place. It is also essential to have a clear system of accountability in place. An accountability system would encourage a practice of openness, promote data security, and ensure that AI systems are designed to be transparent and free from bias. To mitigate AI risks, a sound foundation is required that fundamentally adopts a responsible innovative, and collaborative approach to AI guidelines and rules. In particular, areas in the AI policy framework requiring improvement include how AI creates inequalities of opportunities, resources, and power in business and society and how this perpetuates historic injustice. In addition, specific attention is required vis-à-vis risks of algorithmic bias and discrimination, and unequal opportunities created by machine learning. A gap exists in some of the AI policy initiatives reviewed vis-à-vis explicit discourse of opportunities and challenges presented by AI in terms of value alignment in sustainable development goals; use of AI technologies by businesses in exploiting market failures, and the difficulties of addressing these failures on a global level.

## References

Agrawal, A., Gans, J. S., & Goldfarb, A. (2018). *Prediction machines: The simple economics of artificial intelligence*. Harvard Business School Press.

Alston, P. (2019). *Report of the special rapporteur on extreme poverty and human rights*. www.ohchr.org/Documents/Issues/Pov erty/A_74_48037_AdvanceUneditedVersion.docx

Article 19. (2019). *Governance with teeth: How human rights can strengthen FAT and ethics initiatives on AI*. Free Word Centre. https://www.article19.org/wp-content/uploads/2019/04/Governance-with-teeth_A19_April_2019.pdf.

Bankins, S., & Formosa, P. (2023). The ethical implications of AI for meaningful work. *Journal of Business Ethics, 185*, 725–740.

Baron, B., & Musolesi, M. (2017). Interpretable machine learning for privacy-preserving pervasive systems. *arXiv* 1–10.

Cath, C., Wachter, S., Mittelstadt, B., Taddeo, M., & Floridi, L. (2018). Artificial intelligence and the 'good society': The US, EU, and UK approach. *Science Engineering Ethics, 24*, 505–528.

Cebula, J. J., Popeck, M. E., & Young, L. R. (2014). *A taxonomy of operational cyber security risks version 2*.

Christian, B. (2020). *The alignment problem: Machine learning and human values*. W. W. Norton & Company. ISBN 978-0-393-86833-3.

Collins, C., Dennehy, D., Conboy, K., & Mikalef, P. (2021). Artificial intelligence in information systems research: A systematic literature review and research agenda. *International Journal of Information Management, 60*, 102383.

Conference on the Future of Europe. (2022). *Report on the final outcome*. European Union. https://conference-followup.europarl.europa.eu/cmsdata/267078/Report_EN.pdf#page=80

Davenport, T., Guha, A., Grewal, D., & Bressgott, T. (2020). How AI will change the future of marketing. *Journal of the Academy of Marketing Science, 48*, 24–42.

DSIT. (2023). *A pro-innovation approach to AI regulation*. Department for Science, Innovation and Technology (DSIT). https://assets.publishing.service.gov.uk/media/64cb71a547915a00142a91c4/a-pro-innovation-approach-to-ai-regulation-amended-web-ready.pdf.

DSIT. (2024). *AI regulation: A pro-innovation approach—Policy proposals*. Department for Science, Innovation & Technology (DSIT). https://www.gov.uk/government/consultations/ai-regulation-a-pro-innovation-approach-policy-proposals/outcome/a-pro-innovation-approach-to-ai-regulation-government-response

European Commission. (2020). *White paper on AI - A European approach to excellence and trust. Brussels, COM (2020) 65*. https://commission.europa.eu/system/files/2020-02/commission-white-paper-artificial-intelligence-feb2020_en.pdf

European Parliament. (2023). *EU AI Act: First regulation on AI. European Parliament Directorate general for communication*. Available EU AI Act: first regulation on AI | News | European Parliament (Europa.eu)

European Union. (2020a). *European framework on ethical aspects of AI, robotics, and related technologies*. https://www.europarl.europa.eu/RegData/etudes/STUD/2020/654179/EPRS_STU(2020)654179_EN.pdf

European Union. (2020b). *Ethics of AI: Issues and initiatives*. https://www.europarl.europa.eu/RegData/etudes/STUD/2020/634452/EPRS_STU(2020)634452_EN.pdf

European Union. (2020c). *Artificial Intelligence: From ethics to policy*. https://www.europarl.europa.eu/RegData/etudes/STUD/2020/641507/EPRSSTU(2020)641507EN.pdf

European Union Artificial Intelligence Act. (2024). *European Parliament*. https://www.europarl.europa.eu/doceo/document/TA-9-2024-0138-FNL-COR01_EN.pdf

European Union Briefing. (2024). *Artificial intelligence act. European Union legislation in progress*. European Parliament Publication. https://www.europarl.europa.eu/RegData/etudes/BRIE/2021/698792/EPRS_BRI(2021)698792_EN.pdf

Floridi, L., & Cowls, J. (2019). A unified framework of five principles for AI in society. *HDSR*. https://doi.org/10.1162/99608f92.8cd550d1

Floridi, L., Cowls, J., Beltrametti, M., Chatila, R., Chazerand, P., Dignum, V., et al. (2018). AI4People—An ethical framework for a good AI society: Opportunities, risks, principles, and recommendations. *Minds and Machines, 28*(4), 689–707.

Forbes, K. (2021). Opening the path to ethics in AI. *AI and Ethics, 1*, 297–300. https://doi.org/10.1007/s43681-020-00031-2

Gans, J., Agrawal, A., & Goldfarb, A. (2017). How AI will change strategy: A thought experiment. *Harvard Business Review*. Retrieved February 11, 2019 from https://hbr.org/product/how-aiwill-change-strategy-a-thought-experiment/H03XDI-PDF-ENG

Gebru, T., Morgenstern, J., Vecchione, B., Vaughan, J. W., Wallach, H., Daumeé, III, H., & Crawford, K. (2018). Datasheets for datasets. *arXiv* 1–17.

Gilligan, C. (1982). *In a different voice: Psychological theory and women's development*. Harvard University Press.

Gui, T., Ma, R., Zhang, Q., Zhao, L., Jiang, Y. G., & Huang, X. (2019, August). CNN-based Chinese NER with lexicon rethinking. In *IJCAI international joint conference on artificial intelligence*, pp. 4982–4988. International Joint Conferences on AI. https://doi.org/10.24963/ijcai.2019/692

Hagendorff, T. (2020). The ethics of AI ethics: An evaluation of guidelines. *Minds and Machines, 30*, 99–120.

Harari, N. (2018). *Why technology favors Tyranny*. The Atlantic. https://www.theatlantic.com/magazine/archive/2018/10/yuval-noah-harari-technology-tyranny/568330/

Henz, P. (2021). Ethical and legal responsibility for AI. *Discover Artificial Intelligence*. https://doi.org/10.1007/s44163-021-00002-4

High-Level Independent Group on Artificial Intelligence (AI HLEG). (2019a). *A definition of AI: Main capabilities and disciplines*. European Commission. https://ec.europa.eu/digital-single-%0A10.1145/3301275.3302317

High-Level Independent Group on Artificial Intelligence (AI HLEG). (2019b). *Ethics guidelines for trustworthy AI* (Issue December). https://digital-strategy.ec.europa.eu/en/library/ethicsguidelines-trustworthy-ai.

IEEE. (2018). *Global initiative on ethics of autonomous and intelligent systems. Ethically aligned design: A vision for prioritizing human well-being with autonomous and intelligent systems, version 2*. IEEE. http://standards.ieee.org/develop/indconn/ec/autonomous_systems.html

Jobin, A., Ienca, M., & Vayena, E. (2019). The global landscape of AI ethics guidelines. *Nature Machine Intelligence, 1*(9), 389–399.

Johnson, D. (2006). Computer systems: Moral entities but not moral agents. *Ethics Information Technology, 8*, 195–204.

Koniakou, V. (2023). From the "rush to ethics" to the "race for governance" in AI. *Information Systems Frontiers, 25*, 71–102.

Laskowski, N. (2023). *What is AI and how does AI work? TechTarget*. Enterprise AI.

Leslie, D. (2019). *Understanding AI ethics and safety: A guide for the responsible design and implementation of AI systems in the public sector*. The Alan Turing Institute.

McCarthy, J. (2018). *What is artificial intelligence?* http://www-formal.stanford.edu/jmc/whatisai.pdf

McGaughey, E. (2022). Will robots automate your job away? Full employment, basic income, and economic democracy. *Industrial Law Journal, 51*(3), 511–559.

Mittelstadt, B., Russell, C., & Wachter, S. (2019). Explaining explanations in AI. In *Proceedings of the conference on fairness, accountability, and transparency—FAT'19* (pp. 1–10).

Nicholas, C., Tramer, F., Wallace, E., Jagielski, M., Herbert-Voss, A., Lee, K., & Roberts, A. (2021). *Extracting training data from large language models*. https://doi.org/10.48550/arXiv.2012.07805

Pekka, A.-P., Bauer, W., Bergmann, U., Bieliková, M., Bonefeld-Dahl, C., Bonnet, Y., Bouarfa, L., et al. (2018). *The European Commission's high-level expert group on AI: Ethics guidelines for trustworthy AI* (pp. 1–37). Working Document for Stakeholders' Consultation.

Roger, C. (2019). *Regulatory alternatives for AI*. ScienceDirect. http://www.rogerclarke.com/EC/AIR-Final.pdf

Russell, S., & Norvig, P. (2021). *Artificial intelligence: A modern approach* (4th ed.). Pearson. ISBN 978-0134610993.

Simion, M., & Kelp, C. (2023). Trustworthy AI. *Asian Journal of Philosophy, 2*, 8. https://doi.org/10.1007/s44204-023-00063-5

Taddeo, M., & Floridi, L. (2018). How AI can be a force for good. *Science, 361*(6404), 751–752.

Turing, A. (1950). Computing machinery and intelligence. *Mind, LIX*(236), 433–460. https://doi.org/10.1093/mind/LIX.236.433. ISSN 0026-4423.

United Kingdom National AI Strategy. (2021). *Presented to parliament by the secretary of state for digital, culture, media and sport by the command of her Majesty*. https://assets.publishing.service.gov.uk/media/614db4d1e90e077a2cbdf3c4/National_AI_Strategy_-_PDF_version.pdf

United States Executive Order. (2016). *'Preparing for the future of AI' and the 'National Artificial Intelligence Research and development strategic plan'*. Executive Office of the President National Science and Technology Council Committee on Technology. https://obamawhitehouse.archives.gov/sites/default/files/whitehouse_files/microsites/ostp/NSTC/preparing_for_the_future_of_ai.pdf

United States National AI Act. (2020). *National AI Initiative Act. Division E of P.L. 116–283*. United States.

United States Executive Order. (2020). *Executive Order on Promoting the use of trustworthy AI in the Federal Government*. White House Publication, United States. https://trumpwhitehouse.archives.gov/presidential-actions/executive-order-promoting-use-trustworthy-artificial-intelligence-federal-government

United States Executive Order. (2023). *Executive order on the safe, secure, and trustworthy development and use of AI*. White House Publication. https://www.whitehouse.gov/briefing-room/presidential-actions/2023/10/30/executive-order-on-the-safe-secure-and-trustworthy-development-and-use-of-artificial-intelligence/

United States Blueprint for an AI Bill of Rights. (2022). *Making automated systems work for the American people*. White House Publication. https://www.whitehouse.gov/ostp/ai-bill-of-rights/

Wagner, B. (2018). Ethics as an escape from regulation: From ethics-washing to ethics-shopping. In M. Hildebrandt (Ed.), *Being profiled: Cogitas ergo sum* (pp. 86–90). Amsterdam University Press.

Wang, P. (2019). On defining AI. *Journal of Artificial General Intelligence, 10*(2), 1–37.

Weinbaum, C., Landree, E., Blumenthal, M. S., Piquado, T., & Gutierrez, C. I. (2019). *Ethics in scientific research*. RAND Corporation.

**Adebimpe Adesua Lincoln** holds an LLB and an LLM in Commercial Law from Cardiff University and an MBA in International Business and PhD in Entrepreneurship and Business Development from the University of South Wales. She holds a Postgraduate Diploma in Legal Practice and also holds various teaching qualifications in Higher Education. She is Fellow of the Higher Education academy and Associate Member of the Chartered Institute of Personnel Development. She has over 20 years' experience in Higher Education. She has held positions in Cardiff University and the University of South Wales and was Senior Lecturer in Law at Cardiff Metropolitan University before taking up a position as Assistant Professor in Saudi Arabia. Adebimpe Lincoln is a legal practitioner. She also works with University of Liverpool online Master's in Law Programmes and Kaplan Online Learning and is Adjunct Professor with Robert Kennedy College Switzerland. Adebimpe Lincoln has vast experience supervising research students. She has supervised a wide range of Master's level research including MSc, MBA, and LLM dissertations. She has also supervised students at Doctoral Level in the area of Entrepreneurship and Corporate Governance. Her research interests lie in Female Entrepreneurship, Small and Medium-Sized Enterprise (SME) sector, Leadership Practices of Nigerian Entrepreneurs, Corporate Governance and Board Diversity, Circularity, Sustainability, and Corporate Social Responsibility among Nigerian SMEs. She has published and presented several articles on Leadership, Sustainability, CSR, Governance, and Entrepreneurship.

**Christian J. Lincoln's** research focuses on Artificial Intelligence, Computational Thinking, Game Development, and Game theory. Currently, he is working on the development of neuro-symbolic system aiming to fill in the gap from the connectionist and symbolic frameworks for applications on more human and transparent intelligence. He is passionate about advancements in computational thinking and fostering a socially responsible and ethical approach in the development of applications and utilization of artificial intelligence. His current research seeks to contribute to global discuss in this area.

# Journalistic Content in the Digital Age: Loss of Supremacy, Redefining Gatekeepers, and the Rise of Artificial Intelligence

**Predrag Đ. Bajić and Zoran B. Jevtović**

**Abstract** While changes are a common part of the development of human civilization, including the media ecosystem, the entry into the digital age and the revolutionary turn that has experienced global expansion in the twenty-first century have redefined the concept of media and journalistic professions. Through a theoretical perspective, questions arise about trends in setting the media agenda, changes regarding "gatekeepers" and, above all, the ways in which journalistic content is changing..

This paper problematizes the issue of creators of journalistic content and what is today implied by that term, especially considering the introduction of artificial intelligence into journalism and media editing, leading to questions related to media ethics and media literacy. Additionally, questions are raised about the role of journalists in the digital age in the face of new technologies and enabled communication and information without such intermediaries, primarily through social networks. This leads to questions about the content itself that is available to the public through various media platforms, as well as topics addressing the sustainability of journalistic content in ways that are traditionally known.

Moreover, if journalistic content is considered as something that represents the first draft of history, it is relevant to question what is now recognized as such a form. Furthermore, considerations are made about the historical legacy of journalistic content and its archiving, which is particularly problematic due to its nature and the volume on digital platforms.

---

P. Đ. Bajić (✉)
Faculty of Sport, University "Union – Nikola Tesla", Belgrade, Serbia
e-mail: predrag.bajic@fzs.edu.rs

Z. B. Jevtović
Department of Communicology and Journalism, Faculty of Philosophy, University of Niš, Niš, Serbia
e-mail: zoran.jevtovic@filfak.ni.ac.rs



M. Todorovic et al. (eds.), *Sustainability and Social Responsibility of the Media and in the Media*, CSR, Sustainability, Ethics & Governance,
https://doi.org/10.1007/978-3-032-00086-6_7

**Keywords** Agenda-setting · Algorithm · Archiving · Artificial intelligence (AI) · Communication · Computer · Content · Context · Democracy · Digital age · Digital revolution · Ethical · Filter bubble · Gatekeepers · Information · Internet · Journalism · Journalist · Life cycle patterns · Mass media · Media literacy · Media organizations · Network · New media · New technologies · News organizations · News websites · Newspaper · Person of the year · Personalization · Platform · Print media · Professional communicators · Radio · Self-communication · Social media · Supremacy · Television · Traditional media · User · Web portals

## 1 Introduction

Is journalism endangered? Or is it not quite so? Can the traditional role of journalists as guardians of democracy and mirrors of society be defended today? And who are the journalists today anyway? What exactly constitutes mass media today?[1]

The postmodern era has opened a series of questions. We live in a digital age, with more and more generations that do not even know the world as it existed before. Increasingly, generations are growing up in a digital environment as their only known environment, making comparative approaches and their critical thinking about the positive and negative aspects more challenging, as they lack personal experience from both periods.

Today's abundance of information, ideas, products, content, and other things simultaneously offers far greater possibilities than ever before—and confusion. The value of or the very existence of former leading elements, role models, and key points that were or still are the foundation of a given society and that help in the orientation of individuals and society is diminished or collapsed.

Such an understanding is exciting, intriguing, as it represents the possibility of creating something new, a stronger improvement of the existing or a complete change if it was not good. At the same time, all mentioned can lead to increased insecurity, greater disorder in any system we focus on in the analysis, where enduring values are reduced, and almost everything is given an expiration date.

Therefore, all mentioned relates to the topics that are the focus of this work. Although changes are a common part of the development of human civilization, including the media ecosystem, the entry into the digital age and the revolutionary turn that experienced global expansion in the twenty-first century has redefined the concept of media and the journalistic profession.

[1] The term "mass media" is traditionally used "to describe organizations that transmit information to many people, such as those that create web pages, news portals, or blogs on the Internet, newspapers, television and radio companies, as well as magazines" (Shoemaker & Vos, 2009).

## 2 "The Computer Moves In"

The 1990s have been recorded as a period of tectonic changes in various segments of society, including mass media. Throughout the twentieth century, the print media gradually faced competition from audio and audiovisual media, primarily radio and television. However, when it came to journalism, it was still discussed in a traditional manner, albeit with the specificities of these platforms.

A significant shaking of the mass media landscape occurred in the aforementioned last decade of the century, with technological innovations from the 1980s foreshadowing this shift, reflected in numerous popular culture themes. For instance, the American news magazine "*Time*" named the Computer as the "Machine of the Year" instead of a Person of the Year for 1982. The cover featured the Machine of the Year, "with a plaster person by sculptor George Segal contemplating a concept computer which TIME commissioned from a design firm" (McCracken, 2013).

This was the initial phase of something new for the general population, as indicated by the opening segment of the cover story titled "The Computer Moves In"—with the addition "By the millions, it is beeping its way into offices, schools and homes" (Friedrich, 1983). The excitement for what was coming was evident, as was the uncertainty, so innovations were tested through various contexts. Therefore, the call for caution was not omitted, with messages that these innovations could have both good and bad sides, like previous ones, including television.

In the end, the author of the text, Otto Friedrich, concludes:

> Perhaps the revolution will fulfill itself only when people no longer see anything unusual in the brave New World, when they see their computer not as a fearsome challenger to their intelligence but as a useful linkup of some everyday gadgets: the calculator, the TV and the typewriter. Or as Osborne's Adam Osborne puts it: "The future lies in designing and selling computers that people don't realize are computers at all." (Friedrich, 1983)

Sounds familiar, doesn't it? It can be said that for a significant part of the world's population this is precisely the present age, what is happening today, the world in which we live and work. Numerous generations have already grown up that can be classified as Digital Natives, in the way Marc Prensky defined them at the beginning of this century (2001).[2] All others, namely Digital Immigrants, accept new technologies because they want to or have to, in order to keep pace with others.

All the changes we are talking about have happened in less than half a century, probably faster than expected. Certainly, this represents an extremely short period in a historical context. According to the Life Cycle Patterns[3] discussed by W. James

---

[2] They are "'native speakers' of the digital language of computers, video games and the Internet" (Prensky, 2001).

[3] "Some of the mass media industries are relatively old and have been around for more than two centuries (e.g., books, newspapers, and magazines) while others are relatively young (e.g., cable television and especially the Internet). Although each medium has been shaped by different historical influences, technologies, regulations, and audience needs, all of these media have followed a similar pattern in the way they have evolved over time. This evolution is composed of stages in a

Potter (2022), these are the years of digital media, which are at their peak, while others are adapting to it to survive. However, the question remains, what is the peak of digital media? Maybe it is precisely this time, or perhaps in the future we will be able to classify the entire new media spectrum in the digital age into more detailed subgroups, thereby leaving some of these segments at their peak and hinting at the arrival of others.

## 3 The Loss of Supremacy

Why is all of this important for mass media and the journalism profession? Print media entered the twentieth century as undisputed rulers, practically without serious competition. Then, in the 1940s, radio was at its peak. Subsequently, broadcast TV took the lead until the 1980s, when it entered the Decline phase, followed by the Adaptation phase (Potter, 2022). Thus, this was a time when platforms were changing as the leading media. Naturally, the journalism profession adapted to the new circumstances. However, the foundations of traditional media and journalism remained unshaken.

Then came the digital revolution, in which "text became the first cultural medium to be subjected to digitization in a massive way" (Manovich, 2001: 73).[4] This wave particularly hit the newspaper industry in the last decade of the past century with the popularization of the World Wide Web and the emergence of graphical web browsers. "This global network united the capabilities of traditional media, so previously often divided readers, listeners, and viewers gained a revolutionary platform without divisions, enriched with numerous possibilities for rapid feedback" (Bajić, 2017: 58).[5] Nevertheless, in the early years, the online content of many media organizations was still created from the content of print editions transferred to the new environment without utilizing the potential for content enrichment (Craig, 2005). Later, with higher Internet speeds, new possibilities opened up, including more space for audiovisual content. More and more people are using online content, and "being on the Web has become an obsession, first partially, depending on the

life cycle pattern of innovation (or birth), penetration (or growth), peak (maturity), decline, and adaptation" (Potter, 2022: 74).

[4] "Text is unique among media types. It plays a privileged role in computer culture. On the one hand, it is one media type among others. But, on the other hand, it is a metalanguage of computer media, a code in which all other media are represented: coordinates of 3-D objects, pixel values of digital images, the formatting of a page in HTML. It is also the primary means of communication between a computer and a user: One types single line commands or runs computer programs written in a subset of English; the other responds by displaying error codes or text messages" (Manovich, 2001: 73).

[5] "The role of the media to inform, entertain, and educate is transforming into the role of the media to inform, entertain, and connect," while "the media audience is transforming into a large, interactive, mobile, and individualized actor on the media scene" (Milivojević, 2011: 78).

number of available hours, and then permanently as technology advanced" (Bajić, 2020: 20).

In this initial period, traditional media were still dominant, but significant changes in the journalism profession were already visible. This was evident through the redefinition of the concept of timeliness when it came to publishing news, as well as the notions of proximity and prominence (Craig, 2005). Additionally, there was increasing talk about "the blurring of media boundaries and the power of conglomerates" (Turow, 2009: 580). The development of computers and the online world led to "huge transformation in the way mass communication is taking place" (Turow, 2009: 579) and new dimensions of the concept of "society of the screen."[6] It has long been warned that "The Digital Revolution isn't just some news-magazine cover headline. It's an actual revolution, yielding revolutionary changes, thousands or millions of victims, and an entirely new way of life" (Garfield, 2009). In such an environment, journalism and traditional mass media needed to adapt, including audio and audiovisual media such as radio and television, considering that "the technological revolution is directing broadcasting towards a 'museum of antiquities'" (Radojković, 2012: 26).

However, the key change occurred in the first decade of the twenty-first century, when "the explosion of user-created media content on the web (2005-) has unleashed a new media universe" (Manovich, 2008). And, "this new universe was not simply a scaled up version of twentieth century media culture. Instead, we moved from media to social media" (Manovich, 2008).

Lev Manovich (2008) highlighted two significant changes in the 2000s. First, there was a gradual shift from the majority of Internet users accessing content created by a relatively small number of professional producers to increasingly consuming content generated by other nonprofessional users. Second, the web transitioned from being primarily a publishing medium to becoming predominantly a communication medium.

In such communication, variations exist, and Manuel Castells (2009: 54–55) describes a new form—"mass self-communication." This type of communication enables the transmission of messages from many to many in real time or at a chosen moment, thereby dismantling the previous order dictated by traditional mass communication, which was predominantly one-directional.

Furthermore, explaining mass self-communication, Castells highlights the following:

> It is mass communication because it can potentially reach a global audience, as in the posting of a video on YouTube, a blog with RSS links to a number of web sources, or a message to a massive e-mail list. At the same time, it is self-communication because the production of the message is self-generated, the definition of the potential receiver(s) is self-directed,

[6] Manovich (2001: 94) highlighted that "we may debate whether our society is a society of spectacle or of simulation, but, undoubtedly, it is a society of the screen." The "screen" he refers to "has been used to present visual information for centuries—from Renaissance painting to twentieth-century cinema" (Manovich, 2001: 94).

and the retrieval of specific messages or content from the World Wide Web and electronic communication networks is self-selected. (Castells, 2009: 55)

How does this affect journalism and the journalism profession? Initially, professional media editorial offices known from the world of print media, radio, and television published certain content on online platforms (content already distributed on traditional platforms, and later new content as well). Also, new media editorial offices specialized for the online platform were formed, which also functioned traditionally in terms of journalism and media. The key content was still published from certain centers where professional communicators worked. The novelty was that two-way communication was more open than ever before, through controlled interactivity (e.g., through comments).

However, the key novelty arose when numerous online participants became content creators, that is, when they gained the ability to publish in ways previously exclusive to media professionals. "The monopoly of professional communicators and groups (journalistic editorial offices, PRs, information services, marketing experts, etc.) has disappeared forever, as a large part of the functions of institutionalized communication centers can now be performed by individual citizens and civil society organizations" (Jevtović & Vulić, 2019: 332).

For this "new media universe," the previously mentioned technological development is crucial: "On a practical level, this universe was made possible by free web platforms and inexpensive software tools which enable people to share their media and easily access media produced by others" (Manovich, 2008). In this way, the nature of information is changing globally, whether it is citizen journalism[7] or other forms of content creation and distribution. On one hand, such competition can contribute to quality, while on the other hand it raises the question of the relevance of numerous sources and how one can navigate the abundance of information and content.

In any case, traditional media collectives (and the professional journalists working in them), as we have known them, no longer have supremacy, meaning they are no longer the absolute rulers of the media space. Of course, this does not mean they no longer have a significant role, but they now face strong competition that prevents them from having a practical monopoly in the media space. Above all, there are the possibilities of previously mentioned mass self-communication and the power of social media and networks (today it is not uncommon for professional journalists to operate outside the collectives in which they work, as individuals, for example on social media).

Moreover, this significance was recognized by the American news magazine "*Time*," which declared the Person of the Year 2006 as—You (representing individual content creators on the World Wide Web). In the explanation (Grossman,

---

[7] Citizen journalism is "journalism that is conducted by people who are not professional journalists but who disseminate information using Web sites, blogs, and social media. Citizen journalism has expanded its worldwide influence despite continuing concerns over whether citizen journalists are as reliable as trained professionals" (Albarado).

2006), it was precisely mentioned the impact on the global media scene: "And for seizing the reins of the global media, for founding and framing the new digital democracy, for working for nothing and beating the pros at their own game, TIME's Person of the Year for 2006 is you."

Also, despite the romantic approach outlined in the previous sentence, the same text highlights that there are both positive and negative sides, with the hope that such an experiment will be successful.

And not only in that year, but the influence of social media has also been shown in the years that followed. Thus, in 2010, the Person of the Year was the founder of the social-networking website Facebook (Grossman, 2010): "For connecting more than half a billion people and mapping the social relations among them; for creating a new system of exchanging information; and for changing how we all live our lives, Mark Elliot Zuckerberg is TIME's 2010 Person of the Year."

Then, in 2011, it was "The Protester" (Andersen, 2011). In the protests in different parts of the world during that period, online communication through social media was very significant. Castells (2015) discussed this in his book "*Networks of Outrage and Hope*," where he presented research on new forms of social movements and protests associated with the creation of autonomous communication networks based on the Internet and mobile networks. He emphasized how all this "began on the Internet social networks, as these are spaces of autonomy, largely beyond the control of governments and corporations that had monopolized the channels of communication as the foundation of their power, throughout history" (Castells, 2015: 2). He also added that "the movements spread by contagion in a world networked by the wireless Internet and marked by fast, viral diffusion of images and ideas," and then "from the safety of cyberspace, people from all ages and conditions moved toward occupying urban space" (Castells, 2015: 2).

And, on social media, both media companies and individuals have accounts and use them in various ways. When it comes to media professionals, these platforms have become dominant in promoting content since more and more users do not directly visit news websites but do so through links on social media. This is also discussed in the Reuters Institute for the Study of Journalism's research titled "Digital news report 2023" (Newman et al., 2023: 11):

> Every year, we see direct access to apps and websites becoming less important and social media becoming more important due to their ubiquity and convenience. At an aggregate level, we reached a tipping point in the last few years, with social media preference (30%) now stretching its lead over direct access (22%).

Also, the main ways of getting news online are search (25%), mobile alerts (9%), aggregators (8%), and email (5%).[8]

[8] The conclusions also highlight the following: "Although search and social media play different roles, news access has for some time been dominated by two giant companies: Google and Facebook (now Meta), who at their height accounted for just under half of online traffic to news sites. Although the so-called 'duopoly' remains hugely consequential, our report this year shows how this platform position is becoming a little less concentrated in many markets, with more providers competing" (Newman et al., 2023: 11).

In addition to promoting content from news websites, newspaper companies adapt the forms of their content to social media or create new content specifically for social media to be as competitive as possible on these platforms, given their increasing significance. They also create content that does not belong to traditional journalism and engage actors who are not journalists but contribute to greater visibility among social media users. In this competition, according to the same research, it is noted that "while mainstream journalists often lead conversations around news in Twitter and Facebook, they struggle to get attention in newer networks like Instagram, Snapchat, and TikTok, where personalities, influencers, and ordinary people are often more prominent, even when it comes to conversations around news" (Newman et al., 2023: 13).

## 4 Changes in Media Theory and Journalism

Today's professional journalists work in an environment where traditional media organizations no longer hold complete dominance, as many other organizations and individuals in online media have become competitors. Consequently, the content they distribute can also be competitive and available on the global market through online platforms. While professional journalists affiliated with certain media organizations have a specific moral responsibility prescribed by their profession and organization,[9] those who work outside such a system do not have norms defined in that way. However, this does not mean that one group will definitely act in one way while the other will not; it only means that in an environment where clear norms exist, there is a greater possibility of identifying, resolving, and sanctioning anomalies.[10]

Another question arises: how do readers, listeners, viewers, or users navigate the overwhelming amount of information presented daily, given that more information has been generated since our birth than the total sum of all recorded information before that time—and it continues to grow rapidly (Potter, 2022: 4).

Zygmunt Bauman (2005) spoke about today's society as a "fluid society," where sublimity and duration are irrelevant, as everything is transient. In such a society, instant gratification is the individual's primary goal. In the context of daily life, the overabundance of information can lead to confusion and increase the possibility of

---

[9] When discussing to whom a journalist has moral responsibility, Clifford Christians et al. (2020: 29–30) identify five groups: duty to ourselves, duty to clients/subscribers/supporters, duty to our organization or firm, duty to professional colleagues, and duty to society.

[10] Speaking about journalism today, some authors point out certain challenges for this profession: "The old journalistic paradigm of one-way information dissemination is being relegated to the margins of history with democratic changes, globalization and digitalization offer citizens the opportunity not only to consume information but also to create it, while the crisis of print media, along with the increasingly pronounced race for ratings in electronic media, leads to the lowering of professional standards and the loss of ethical standards in the profession" (Jevtović et al., 2014: 7).

misinformation. Questions arise about relevance and how to stay well-informed. And, who determines which content reaches users and how is it ranked?

When it comes to the media's agenda, it is "the pattern of news coverage for the major issues of the day," and "the agenda-setting role of the news media is the transmission of issue salience from the media's agenda to the public's agenda" (McCombs & Guo, 2014: 251–252). Thus, this theoretical foundation speaks about the power of the media to direct public attention to specific topics—"The basic agenda-setting role of the news media is to focus public attention on a small number of key issues and topics" (McCombs & Guo, 2014: 252).[11] In this regard, media professionals are crucial on informative web portals, in print media, radio, and television. On the other hand, through search engines, content aggregators, and social media, the power of selecting which content will be highlighted and which will be less visible shifts to algorithms.[12] These algorithms are controlled not by professional communicators from traditional media institutions but by those who control specific platforms and software.

When discussing the revolution that the digital age has brought to media and journalism through a theoretical lens, significant changes occur regarding "gatekeepers." Specifically, this theory suggests that messages are created from information about events that have passed through a series of gates and been altered in the process (Shoemaker & Vos, 2009). Additionally, some information does not pass through this process and, as a result, remains invisible. Furthermore, some information may be relevant to certain news media but not to others—or it may be more or less emphasized in various news media outlets.

Thus, the quality selection of content is the ultimate requirement in the vast amount of content in which the average user can get lost. It is impossible for a person to accumulate all the information, and the change has occurred in who selects the content that will reach them. Previously, professional communicators in traditional media organizations held the keys, directly selecting and determining importance. Today, for many users, this is done by algorithms. This shift is not simple and affects the essence of "gatekeepers." "Basic conceptual categories, the gate and the keeper, are also modified. There is a noticeable trend towards the transformation of

[11] Additionally, "beyond these basic, first-level agenda-setting effects on the public's focus of attention and its perception of what the most important issues of the day are, there is a second level of agenda-setting effects: attribute agenda setting. If we think of the items on the agenda at the first level as a set of objects (analogous to attitude objects), then it is clear that these objects have certain attributes that are emphasized to varying degrees on the media's and the public's agendas"—"at the second level of agenda setting, the theoretical proposition is that the media's attribute agenda influences the salience of these attributes on the public's agenda" (McCombs & Guo, 2014: 254).

[12] Speaking about algorithms, David R. Brake highlighted the following: "Algorithms in their broadest sense are ubiquitous—as Gillespie puts it, they are 'encoded procedures for transforming input data into a desired output, based on specified calculations' (2014: 1), and as such do not even necessarily involve computers—a cooking recipe is also an algorithm." Focusing on a specific segment, he emphasized "public relevance algorithms," which "search engines such as Google and social media companies such as Facebook use 'to select what is most relevant from a corpus of data composed of traces of our activities, preferences, and expressions' (2014: 2)" (Brake, 2017: 26).

gatekeeping towards gatewatching, in which social media users do not create their own gates but observe and use already existing gates" (Pałka-Suchojad, 2021: 91).

Observing these changes, it is clear that in the era of traditional media dominance, editors selected certain information and controlled its flow, thus being able to create and direct certain narratives. Today, a significant portion of that power belongs to algorithms, which select what will be shown to the user and what they will be able to access.

> Algorithms distribute and filter news based on preprogrammed criteria, aimed at selecting content that best fits readers' interests and prior behaviors (Thurman et al., 2019; Wallace, 2018). In contrast to professional gatekeeping where editorial processes are involved in selecting and creating content, algorithms choose from an already existing corpus of news and distribute those to the users (Wallace, 2018) (Scheffauer et al., 2024: 756).

Additionally, users on online platforms already narrow the information space by following topics and views that suit them, thus personalizing content,[13] including news.[14]

> The sophistication of these gatekeeping rules is shown by the ability of readers to tailor the content to their own interests, for example by asking for more world news and less sports information. These 'more or less' rules and placement of categories on the web page turn into algorithms created by individual readers and are remembered by the news portal's servers, so that when the reader enters the site again, the cookies (audience-defined sets of information and rules) left behind by the website on the person's computer allow the server to recognize the person and his preferences. (Shoemaker & Vos, 2009)

In these ways, the abundance of content and plurality of opinions in the online world suddenly narrow significantly. The key issue is the narrowing of diversity, focusing on selected topics from selected sources instead of broader informativeness. Previously, before the global development of the Internet, when print media, radio, and television dominated, media professionals were responsible for selecting and editing what would be presented to the public. However, these "gatekeepers" no longer have supremacy as different possibilities for connecting and disseminating information have emerged through independent action or synergy with traditional methods.

These changes have brought new questions and dilemmas. For example, unlike people who can react differently in the same situation when it comes to ethical principles, the question is how algorithms apply them, as they do not have the same level of sensitivity for each individual situation. Selection algorithms primarily use user

---

[13] "Personalization of the newsfeed is, however, usually more rudimentary on the websites of news organizations than it is for social media outlets or news aggregators" (Scheffauer et al., 2024: 757).

[14] Nicholas Negroponte spoke about this as early as the mid-1990s (1995: 153): "There is another way to look at a newspaper, and that is as an interface to news. Instead of reading what other people think is news and what other people justify as worthy of the space it takes, being digital will change the economic model of news selections, make your interests play a bigger role, and, in fact, use pieces from the cutting-room floor that did not make the cut on popular demand. Imagine a future in which your interface agent can read every newswire and newspaper and catch every TV and radio broadcast on the planet, and then construct a personalized summary."

profiles and activities. Pinja Hokkanen (2019) wrote about this, discussing filter bubbles and relying on Eli Pariser's (2011)[15] thoughts:

> The more personalized the information feed is for the unique user, the more relevant the platform seems to be for the individual. So again, the information shown to us is chosen, edited and tailored, but this time on an individual level. And the filtering is done by the algorithms owned by these massive companies instead of real people. The algorithms have become information gatekeepers, which create a filtered information bubble around every one of us. A filter bubble is one's own, unique universe of information in which they have access to online, isolating the person from other kinds of information. It is caused by digital services trying to personalize their experience by providing you information that they think is relevant and especially what they think you would like to see. This information shown to you is determined by an algorithm that is specific for that exact online service. Thus, you are in a filter bubble when an algorithm chooses the information you can see and explore online, which is based on what the algorithm thinks you need and want to see.

Additionally, the same author (Hokkanen, 2019) spoke about another phenomenon—online echo chambers:

> The other problem with filter bubbles is that together with the fact that our networks are mostly composed of similar-minded people, it creates an online echo chamber. An echo chamber is a closed system, like our mind, social circle or facebook feed, where the same ideas, beliefs, and values are communicated, repeated and through confirmation bias, amplified. They polarize us to see only one viewpoint, which distorts our whole world view. We don't even know that we don't know something, which makes us even more sure that our opinion is the only correct one. No differing viewpoints exist in our online platforms, so why would they exist in real life?

## 5 Questions About Content

If the reader/listener/viewer, or user, accepts the previously stated information without further contemplation, they can easily encounter problems regarding being informed and even become a victim of manipulation. In this vast number of information and content, constant attention is necessary—and recognizing the importance of media literacy.[16] It is a continuous process that lasts throughout one's entire life and applies to both those who consume content and those who create it. Therefore, it also applies to professional communicators, to maintain ethical standards of the profession and adhere to other journalistic rules (e.g., verifying sources, avoiding sensationalist headlines and content manipulation for more clicks, etc.).

This adherence to the rules of the journalistic profession has historically distinguished true journalists from propagandists. This is crucial considering that journalism is often referred to as "the first draft of history,"[17] in the sense that it is written

[15] More on https://www.ted.com/talks/eli_pariser_beware_online_filter_bubbles

[16] "Media literacy is a set of perspectives that we actively use to expose ourselves to the media and interpret the meaning of the messages we encounter. It is multidimensional and a continuum" (Potter, 2022: 13).

[17] Kari Andén-Papadopoulos (2014: 149) suggests that "these shifts necessitate a critical re-evaluation of the role of journalism in shaping 'the first draft of history' which, in effect, tends also to

immediately after events occur. Furthermore, a higher level of media literacy helps in better understanding the context and time in which certain content was created.

Considering the aforementioned, journalistic content represents valuable material and legacy for future generations. Hence, the question of archiving content is significant, especially for those found on the Internet, given their easy accessibility. As easily accessible as it is, it can also easily disappear, as shown by a Pew Research Center study: "A quarter of all webpages that existed at one point between 2013 and 2023 are no longer accessible, as of October 2023. In most cases, this is because an individual page was deleted or removed on an otherwise functional website" (Chapekis et al., 2024). The same study states that the percentage is even higher for pages older than a decade: "For older content, this trend is even starker. Some 38% of webpages that existed in 2013 are not available today, compared with 8% of pages that existed in 2023."

Regarding links on news websites, the study showed the following:

> When we tracked these links to their destination, we found that 5% of all links on news site pages are no longer accessible. And 23% of all the pages we sampled contained at least one broken link. Broken links are about as prevalent on the most-trafficked news websites as they are on the least-trafficked sites. Some 25% of pages on news websites in the top 20% by site traffic have at least one broken link. That is nearly identical to the 26% of sites in the bottom 20% by site traffic. (Chapekis et al., 2024)

Therefore, this raises the question of the adequate preservation of one's journalistic work, into which a certain amount of time has been invested, considering that such content can easily be lost to future generations. However, this doesn't always have to be the case, as some content is stored on Internet archive services. Nevertheless, in such cases, they are no longer available at their original addresses and, consequently, are less recognizable to the general public, remaining on the margins. This issue becomes significant considering that an increasing amount of content is being created exclusively for online platforms, and much of the content from traditional media is being published through online platforms.

Social media poses a particular issue, given the previously mentioned fact that professional news organizations do not have the final word there, but rather those who manage those social media platforms. Although account management largely depends on individuals delegated by these professional news organizations, this does not represent the ultimate order, unlike news websites (both in terms of media content placement and interaction with the audience, or set boundaries in that communication on news websites and social media).

---

become a final one (Berkowitz, 2011: 203)." Additionally, Andén-Papadopoulos adds the following: "As Carolyn Kitch writes, the initial draft that journalism provides of current events is 'also the first draft of memory, a statement about what should be considered, in the future, as having mattered today' (2008: 312)."

## 6 The Revolution Is Yet to Come

Previous examples have shown the revolutionary changes that the digital age has brought to the media and journalism, and thus to their products. All these changes have led to the redefinition of these concepts, but journalistic content created by media professionals still has a significant impact. The reason for this can also be found in the fact that people find it difficult to navigate the overabundance of information, so they trust an editorially shaped agenda (as known from the pre-digital age) in their desire for information.[18] In this way, they prioritize professional communicators, people whose main job and life calling is journalism, and their content over the countless other actors.

However, the freedom that publishing content outside of journalistic norms provides also brings advantages in communication and information dissemination, but only with a clear understanding of both the good and the bad sides. Without this, one becomes an easy prey for manipulation in terms of what and how to think. This applies to every form of information, whether it concerns professional communicators or individuals, on informative web portals or social media, etc. Nevertheless, on traditionally media-regulated informative web portals, it is clear who is responsible for published content, whereas the situation is different on social media.

The challenge for professional journalists and media organizations in the future lies precisely in the fact that more and more traffic to their content on informative web portals comes from social media. On these social media platforms, they are not the complete rulers of communication, despite what might seem through account administration. After all, this also appears to be the case for other users. In problematic situations, as mentioned earlier, those who manage these social media platforms ultimately decide.

In any case, caution is implied both when trusting well-known journalistic names and professional media editorials and when trusting other known and unknown actors. The pluralism of information, opinions, and attitudes brought about by the digital age certainly has its advantages as it allows for a multitude of content that has not gone through traditional "gatekeepers." However, as previously mentioned, for one to be able to properly process all this information, it is essential to understand the content, which is facilitated by media literacy. In this way, today's reader/listener/viewer, or user, can distinguish true journalism from a propagandist model

[18] In a report on how audiences in six countries (Argentina, Denmark, France, Japan, the UK, and the USA) view the rise of generative AI and its possible impact on society and on the news landscape, it is concluded that people prefer news produced by human journalists. Additionally, most people are comfortable with using generative AI for spelling and grammar, translations, and creative charts, but they are very uncomfortable with using it for recreating images and creating AI authors and AI presenters. Moreover, many believe that AI will make news less trustworthy, cheaper to make, and less worth paying for. More on https://reutersinstitute.politics.ox.ac.uk/sites/default/files/2024-05/Fletcher_and_Nielsen_Generative_AI_and_News_Audiences.pdf (Fletcher & Nielsen, 2024).

and recognize those who stand on the barricade of the traditional role of journalists as democracy's watchdogs from those who degrade this profession.

It seems that the media world has become very challenging for the average user, who is caught between trust and distrust in humans and algorithms that select content based on predefined guidelines. But what happens when the competition to human-created journalistic content also becomes content created by artificial intelligence? Hence, the real challenges are yet to come. In fact, it's too late to exclaim "Hannibal ante portas" because artificial intelligence is already among us. Of course, although such an exclamation points to great danger, in this case it is more a call for necessary caution. Like any invention, artificial intelligence can be both—an opportunity and a danger, depending on how people use and manage it. And, the importance of this topic has been recognized by the institutions of the European Union through The AI Act.[19]

The key moment for global recognition of the potential of generative AI[20] was the introduction of ChatGPT by OpenAI in late 2022, and soon other technology companies followed that way. All this quickly reflected on the everyday life of media and journalism. This is also reflected in the conclusions of the study "Generative AI in Journalism: The Evolution of Newswork and Ethics in a Generative Information Ecosystem" (Diakopoulos et al., 2024: 7): "Based on participants' responses and our analysis, we find that generative AI is already changing work structure and organization, even as it triggers ethical concerns around use."[21]

The era of significant changes is already upon us, as evidenced by the news that "German magazine 'Bild' announced that they will reduce their editorial staff and increase dependencies on AI." Additionally, "news organizations such as Buzzfeed, Daily Express and others are exploring ways that the new technology can be used by, or instead of, journalists." The same news also mentions that "AI-generated disinformation is a growing concern, especially in an age of general mistrust in the news."[22]

One of the most intriguing examples comes from Poland, in the fall of 2024. Specifically, OFF Radio Krakow fired real journalists and replaced them with AI-generated "presenters," but soon abandoned this experiment following a strong public backlash. People expressed fears that humans were being replaced by AI.

---

[19] It is a European regulation on artificial intelligence (AI)—the first comprehensive regulation on AI by a major regulator anywhere. More on https://artificialintelligenceact.eu/ai-act-explorer (The AI Act Explorer).

[20] "Generative AI is a type of artificial intelligence technology that can create new content, such as text, images, audio, video, or other media, based on the data it has been trained on and according to written prompts provided by users" (Diakopoulos et al., 2024: 4).

[21] Those changes relate to various segments: Applications in News Production; Changing Work Structure and Organization; Work Redesign; Ethical Concerns and Responsibility; Strategies for Responsible Use; Ambivalence in Content Rights (Diakopoulos et al., 2024: 7–8).

[22] More on https://europeanjournalists.org/blog/2023/08/18/feature-media-literacy-tools-in-the-age-of-ai-generated-disinformation (Feature: Media literacy tools in the age of AI-generated disinformation).

The magnitude of this change and the challenges it brings can be seen in the example of this radio station, which broadcast an "interview" conducted by an AI-generated presenter with a voice pretending to be Wisława Szymborska, a Polish poet, essayist, translator, and Nobel Prize winner in Literature who died in 2012. Additionally, as reported by The Associated Press, before canceling the experiment, the radio station had planned an "interview" with Polish statesman Józef Piłsudski, who died in 1935.[23]

Such occurrences are more of an exception than part of daily life. As such, it is something that is (still) far from the average listener. However, one must remain constantly vigilant when it comes to the use of new technologies, given the controversies involved. In this way, the boundaries between the real person and their work, as well as the actual views of certain figures and the way they are constructed using artificial intelligence, are blurred. The ability to "revive" a voice in this manner carries numerous possibilities for misuse.

On the other hand, it is a fact that some media organizations already publish texts where AI is used for automated writing. After all, the possibilities offered by generative AI have become so vast that *The New York Times* publicly stated the following: "We don't use A.I. to write articles, and journalists are ultimately responsible for everything that we publish."[24]

All these rapidly occurring changes have alarmed other organizations as well, leading to the initiative by Reporters Without Borders (RSF) and chaired by journalist and Nobel Peace Prize laureate Maria Ressa, unveiling the Paris Charter on AI and Journalism—"it defines ethics and principles that journalists, newsrooms and media outlets around the world will be able to appropriate and apply in their work with artificial intelligence" (Boering, 2023).[25]

Thus, the potential danger posed by the new actor in the system has been recognized if its role is not regulated by clear normative acts and controlled. Although manipulation are not new, the actor capable of doing that is new. At the same time, generative AI has a destructive potential for media, journalism and society, but also a different potential that could aid development. Artificial intelligence appears as a news aggregator, news creator, assistant to media professionals, and more. One thing is certain—it is both, the present and the "next big thing," considering that the

---

[23] More on https://apnews.com/article/poland-media-radio-ai-bba6beb01d523c6727d650c69da14960 (Gera, 2024).

[24] More on https://www.nytimes.com/2024/10/07/reader-center/how-new-york-times-uses-ai-journalism.html (The New York Times, 2024).

[25] The result is a set of ten fundamental ethical principles (Paris Charter on AI and Journalism): Journalism ethics guide the way media outlets and journalists use technology; Media outlets prioritize human agency; AI systems used in journalism undergo prior, independent evaluation; Media outlets are always accountable for the content they publish; Media outlets maintain transparency in their use of AI systems; Media outlets ensure content origin and traceability; Journalism draws a clear line between authentic and synthetic content; AI-driven content personalization and recommendation upholds diversity and the integrity of information; Journalists, media outlets and journalism support groups engage in the governance of AI; Journalism upholds its ethical and economic foundation in engagements with AI organizations.

full extent of it is yet to be seen. Questions remain as to where the boundaries lie and whether there are any when it comes to developing artificial intelligence, given their rapid learning and advancement through mutual interaction. Therefore, it is essential for humans to retain the role of regulator and direct new technologies towards their own advancement, rather than becoming slaves to technology.

After all, when considering journalistic content, generative AI can be very helpful and shorten the process of preparation for journalists, data processing, etc. Of course, always with clear caution as there is a possibility of error, although generative AI models will improve. Economically, of course, it is always important to reduce costs as much as possible while maintaining quality, and technologies can help here. There are tendencies for increasing use of programs and applications aimed at generating text, images, videos, and other content, as well as editing content, i.e., automating certain processes.

However, it should be noted that such content is generated from available content, which requires special attention as it raises the question of what influences this process—which content is selected in creating new content and how it is even possible to separate information from misinformation. Also, the question of context remains, how AI creates narratives, and more. The inviolability of journalists (and people in general) lies in the fact that they bring news, i.e., new information, which they place in a certain context, and which will only then become possible new raw material for other content.

Moreover, an important topic of discussion is precisely the labeling of content created with the help of generative artificial intelligence, so that readers/listeners/viewers, or users, are aware of it. Thus, they can better understand these contents.

Changes are an inevitable part of the development of any society, and how that society relates to changes, how it understands them, shows the degree of its development. Responsibility lies with regulators and technological companies. And with journalists, of course, with professional communicators in general. What is certain is that for sustainability in the media and of the media adaptation to new technologies is necessary. After all, referring back to James Potter's Life Cycle Patterns, no medium has disappeared with the advent of a new one, but many representatives of those media who did not adapt to the changes have. In these changes, the journalist who respects ethical standards and creates quality (new) content still plays an irreplaceable role. And this is their strength, as well as the potential of humans in general to properly manage changes and utilize them for the benefit of society, through understanding changes and respecting ethical norms.

## References

Albarado, S. *Citizen journalism*. Britannica. https://www.britannica.com/topic/citizen-journalism

Andén-Papadopoulos, K. (2014). Journalism, memory and the "crowd-sourced video revolution". In B. Zelizer & K. Tenenboim-Weinblatt (Eds.), *Journalism and memory* (pp. 148–163). Palgrave Macmillan.

Andersen, K. (2011, December 14). *Cover story: The protester*. Time. https://content.time.com/time/specials/packages/article/0,28804,2101745_2102132_2102373,00.html
Bajić, P. (2017). Online press: The relations between the content in print and online editions of daily newspapers. *CM: Communication and Media, XII*(39), 57–82. https://doi.org/10.5937/comman12-9129. [In Serbian].
Bajić, P. (2020). Theoretical and practical aspects of the communication theory of power in the global network society. *Sport i biznis, 6*, 19–25. [In Serbian].
Bauman, Z. (2005). *Liquid life*. Polity Press.
Berkowitz, D. (2011). Telling the unknown through the familiar: Collective memory as journalistic device in a changing media environment. In M. Neiger, O. Meyers, & E. Zandberg (Eds.), *On media memory: Collective memory in a new media age* (pp. 201–212). Palgrave Macmillan.
Boering, L. (2023, November 10). *Why the Paris charter on AI and journalism is important*. European Journalism Centre. https://ejc.net/news/why-the-paris-charter-on-ai-and-journalism-is-important
Brake, D. R. (2017). The invisible hand of the unaccountable algorithm: How Google, Facebook and other tech companies are changing journalism. In J. Tong & S. H. Lo (Eds.), *Digital technology and journalism: An international comparative perspective* (pp. 25–46). Palgrave Macmillan. https://doi.org/10.1007/978-3-319-55026-8_2
Castells, M. (2009). *Communication power*. Oxford University Press.
Castells, M. (2015). *Networks of outrage and hope* (2nd ed.). Polity Press.
Chapekis, A., Bestvater, S., Remy, E., & Rivero, G. (2024, May 17). *When online content disappears*. Pew Research Center. https://www.pewresearch.org/data-labs/2024/05/17/when-online-content-disappears.
Christians, C. G., Fackler, M., Richardson, K. B., & Kreshel, P. J. (2020). *Media ethics: Cases and moral reasoning* (11th ed.). Routledge.
Craig, R. (2005). *Online journalism: Reporting, writing, and editing for new media*. Thomson Wadsworth.
Diakopoulos, N., Cools, H., Li, C., Helberger, N., Kung, E., & Rinehart, A. (2024, April). *Generative AI in journalism: The evolution of newswork and ethics in a generative information ecosystem*. https://www.researchgate.net/publication/379668724_Generative_AI_in_Journalism_The_Evolution_of_Newswork_and_Ethics_in_a_Generative_Information_Ecosystem. https://doi.org/10.13140/RG.2.2.31540.05765
Feature: Media literacy tools in the age of AI-generated disinformation. (2023, August 18). *European Federation of Journalists*. https://europeanjournalists.org/blog/2023/08/18/feature-media-literacy-tools-in-the-age-of-ai-generated-disinformation
Fletcher, R., & Nielsen, R. K. (2024, May). *What does the public in six countries think of generative AI in news?* Reuters Institute for the Study of Journalism. https://reutersinstitute.politics.ox.ac.uk/sites/default/files/2024-05/Fletcher_and_Nielsen_Generative_AI_and_News_Audiences.pdf. https://doi.org/10.60625/risj-4zb8-cg87
Friedrich, O. (1983, January 3). *The computer moves in*. Time. https://content.time.com/time/subscriber/article/0,33009,953632,00.html
Garfield, B. (2009). *The chaos scenario*. Stielstra Publishing.
Gera, V. (2024, October 28). *Polish radio station abandons use of AI 'presenters' following outcry*. The Associated Press. https://apnews.com/article/poland-media-radio-ai-bba6beb01d523c6727d650c69da14960
Gillespie, T. (2014). The relevance of algorithms. In T. Gillespie, P. J. Boczkowski, & K. A. Foot (Eds.), *Media technologies: Essays on communication, materiality, and society*. MIT Press.
Grossman, L. (2006, December 17). *You — Yes, you — Are TIME's person of the year*. Time. https://web.archive.org/web/20070112060107. http://www.time.com/time/magazine/article/0,9171,1570810,00.html
Grossman, L. (2010, December 15). *Person of the year 2010: Mark Zuckerberg*. Time. https://web.archive.org/web/20101215133743/http://www.time.com/time/specials/packages/article/0,28804,2036683_2037183,00.html.

Hokkanen, P. (2019, September 26). *Filter bubbles: Algorithms as information gatekeepers*. Medium. https://medium.com/social-media-writings/filter-bubbles-algorithms-as-information-gatekeepers-58fd53680c22
Jevtović, Z., & Vulić, T. (2019). Change in journalistic discourse and media theory. *DHS – Društvene i humanističke studije: časopis Filozofskog fakulteta u Tuzli, 1*(7), 331–346. [In Serbian].
Jevtović, Z., Petrović, R., & Aracki, Z. (2014). *Genres in contemporary journalism*. Jasen. [In Serbian].
Kitch, C. (2008). Placing journalism inside memory—And memory studies. *Memory Studies, 1*(3), 311–320. https://doi.org/10.1177/1750698008093796
Manovich, L. (2001). *The language of new media*. The MIT Press.
Manovich, L. (2008). *The practice of everyday (media) life*. http://manovich.net/content/04-projects/059-the-practice-of-everyday-media-life/56_article_2008.pdf
McCombs, M. E., & Guo, L. (2014). Agenda-setting influence of the media in the public sphere. In R. S. Fortner & P. M. Fackler (Eds.), *The handbook of media and mass communication theory* (pp. 251–268). Wiley Blackwell.
McCracken, H. (2013, January 04). *TIME's machine of the year, 30 years later*. Time. https://techland.time.com/2013/01/04/times-machine-of-the-year-30-years-later
Milivojević, S. (2011). *Profession at the crossroads—Journalism at the treshold of information society*. Center for Media and Media Research of the Faculty of Political Sciences of the University of Belgrade. https://www.fpn.bg.ac.rs/wp-content/uploads/2017/01/Profesija-na-Raskr%C3%85%C2%A1%C3%84%E2%80%A1u.pdf. [In Serbian]
Negroponte, N. (1995). *Being digital*. Hodder & Stoughton.
Newman, N., Fletcher, R., Eddy, K., Robertson, C. T., & Nielsen, R. K. (2023). *Digital news report 2023*. Reuters Institute for the Study of Journalism.
Pałka-Suchojad, K. (2021). Who keeps the gate? Digital gatekeeping in new media. *Zeszyty Prasoznawcze, 64*(2), 91–99. https://doi.org/10.4467/22996362PZ.21.012.13477
Paris Charter on AI and Journalism. (2023, November 10). https://www.thomsonfoundation.org/media/252844/paris-charter-on-ai-and-journalism.pdf.
Pariser, E. (2011, March). *Beware online "filter bubbles"*. TED2011. https://www.ted.com/talks/eli_pariser_beware_online_filter_bubbles
Potter, W. J. (2022). *Media literacy* (10th ed.). Sage.
Prensky, M. (2001). Digital natives, digital immigrants. *On the Horizon, 9*(5). https://www.marcprensky.com/writing/Prensky%20-%20Digital%20Natives,%20Digital%20Immigrants%20-%20Part1.pdf
Radojković, M. (2012). The implementation of "chaos scenario" in broadcasting in Serbia. In R. Veljanovski (Ed.), *Broadcasting in Serbia: Present and future* (pp. 25–34). University of Belgrade – Faculty of Political Sciences & Čigoja štampa. [In Serbian].
Scheffauer, R., Goyanes, M., & Gil de Zúñiga, H. (2024). Social media algorithmic versus professional journalists' news selection: Effects of gate keeping on traditional and social media news trust. *Journalism, 25*(4), 755–778. https://doi.org/10.1177/14648849231179804
Shoemaker, P. J., & Vos, T. P. (2009). *Gatekeeping theory*. Routledge.
The AI Act Explorer. *EU Artificial Intelligence Act*. https://artificialintelligenceact.eu/ai-act-explorer.
The New York Times. (2024, October 7). *How the New York Times uses A.I. for Journalism*. https://www.nytimes.com/2024/10/07/reader-center/how-new-york-times-uses-ai-journalism.html.
Thurman, N., Moeller, J., Helberger, N., & Trilling, D. (2019). My friends, editors, algorithms, and I. *Digital Journalism, 7*(4), 447–469. https://doi.org/10.1080/21670811.2018.1493936
Turow, J. (2009). *Media today: An introduction to mass communication* (3rd ed.). Routledge.
Wallace, J. (2018). Modelling contemporary gatekeeping. *Digital Journalism, 6*(3), 274–293. https://doi.org/10.1080/21670811.2017.1343648

**Predrag Đ. Bajić** is an assistant professor at the Faculty of Sport, Union—Nikola Tesla University in Belgrade (Serbia). Previously, he spent more than a decade and a half working in journalism. He earned his PhD at the Faculty of Political Sciences, University of Belgrade, in the field of culture and media. His primary interests are related to topics in the areas of communication studies, media, and journalism. He has published around 30 articles in scientific journals, thematic collections, and conference proceedings.

He is the author of the monograph *Being Fourth: Champions Without an Olympic Medal* (2021), which explores the phenomenon of fourth-place finishes in sports through interviews and life stories of athletes who placed fourth in the Olympic and Paralympic Games. The monograph was produced in collaboration with journalism students from the Faculty of Sport and received an award for Publishing Venture of the Year in Print Journalism by the Association of Sports Journalists of Serbia. He is also the editor of the monograph "Sergije Lukač—The Prometheus of Journalism" (authors Radivoje S. Petrović and Slobodan M. Penezić with journalism students from the Faculty of Sport, 2023), which discusses the life and work of one of the pioneers of academic journalism studies in former Yugoslavia.

**Zoran B. Jevtović** is a full professor at the Faculty of Philosophy, University of Niš, and is one of the leading theorists in media and communication studies in Serbia. He earned his Master's and PhD degrees at the Faculty of Political Sciences, University of Belgrade, specializing in radio journalism.

In his extensive academic career, he has been engaged with the Journalists' Association of Serbia and the Center for Contemporary Journalism in Belgrade. He has worked as a tutor at the Belgrade Open School and served as a professor at the Faculty of Performing and Applied Arts—Academy of Fine Arts and the Faculty of Culture and Media. As a visiting professor, he has been involved for several years with the Master's program in communication studies at the Faculty of Philosophy in Novi Sad, and the doctoral program in sociology at the Faculty of Philosophy in Kosovska Mitrovica. He is also a member of the editorial boards of several leading national journals in political and communication sciences.

He has authored more than two hundred scientific articles, as well as several scientific monographs, including: *Totalitarianism and Mass Media* (2000), *Public Opinion and Politics* (2003), *History of Journalism* (with Mihailo Bjelica, 2006), *Politics, Media, Security* (with Srđan Milašinović and Ljubiša Despotović, 2013), and *Geopolitics of Media* (with Ljubiša Despotović, 2019).

# Responsible Media Dialectics: Communities, Sustainability, Education, and Ethical Voices

# In the Media We Trust: The Crave for Media Adjudication on Social and Quasi-Legal Issues in Ghana

**Sam Sarpong**

**Abstract** 'Trial by media' is a popular phrase that has been used against the media for commenting on pending cases. This is often deemed to be the case when the media thoroughly analyse cases just about to be heard or pending before the court. Currently, a novelty 'media court' show, made up of lay people, aptly fits this description in Ghana. The shows generally adjudicate on socio-cultural and quasi-legal issues and have continued to attract people from all walks of life. The objective of these shows is partly to circumvent the lengthy court processes and to provide justice for the vulnerable, especially people who feel they cannot have justice at the courts or have socio-cultural issues to deal with as well as the settlement of family or paternity suits. The chapter examines the allure of these shows and also provides a detailed account of why and how the shows operate.

**Keywords** Media · Adjudication, 'trial by media' · 'Media court shows' · Jurisdiction · Mistrials · Social media · Law courts · Moral issues · Conversational analysis · Justice system · Watchdog role · Fair comments · Legal entitlements · Judicial corruption · Law enforcement · Observations · Theory of symbolic self-completion · Motivation · Curses · Cultural values · Law and morality · Societal issues · Lawyers · Radio stations · Public sphere · Marginalised social groups · Litigation · Problem-solving mechanisms · Freedom of the press

S. Sarpong (✉)
University of Central Lancashire, Lancashire, UK
e-mail: SSarpong@uclan.ac.uk



M. Todorovic et al. (eds.), *Sustainability and Social Responsibility of the Media and in the Media*, CSR, Sustainability, Ethics & Governance,
https://doi.org/10.1007/978-3-032-00086-6_8

## 1 Introduction

Traditionally, the media are there to inform, educate and entertain, as many media scholars such as McQuail (2010) and Herman & Chomsky (2002) have argued. The news media, as an institution, are integral to much of politics and public life. This clear mandate has, at times, enabled the media to operate in unfettered territories to the chagrin of certain authorities (Hasell et al., 2024). Although expected to operate within certain parameters, the media sometimes stray into some terrains, which may be deemed to be quite inappropriate (see Koetsenruijter & Vanderveen, 2011). Oftentimes, this draws a wedge between them and the powers that be (Chadwick, 2013). But what has drawn so much concern of late globally is the incisive nature of media analysis and disclosures that come about prior to and during court trials (Sarpong, 2022). Many legal experts have denounced this, as they deem it to be a 'trial by media' (Surette, 1989). This situation arises when there is a massive coverage of a case or on a person's reputation by creating a widespread perception of guilt or innocence before a verdict is given in a court of law.

Such intrusion has been aided by the technological prowess that the media enjoy today, which incidentally has turned them into a powerful social force. In the past decades, the internet has transformed the rules of socialisation, defying notions of space and time and setting itself as a new medium. Today, the media's influence seeps through nearly every aspect of modern life, from personal relationships to political landscapes and from business strategies to cultural trends. Besides, their social and technical accomplishments, their perceived capacity to create, reproduce and deliver content in mind-boggling abundance and near instantaneity and their putative impacts on many aspects of social and cultural life have drawn both praise and criticisms (Birkbeck, 2014).

It is worth noting that lately, news about crime and justice and consequently the entertainment centring on them has become wide-spread (Robinson, 2018). Such news and the dramatisation of crime stories are quite popular and constant in the total amount of news broadcast by the media (Robinson, 2018; Banks, 2005). People's fascination with crime might stem from the desire to explore crime vicariously because perhaps only a few peoples have, and will have, that kind of personal experience. The portrayal of true crime has had widespread implications, impacting opinions on crimes and legal proceedings, desensitising violence and being ethically immoral towards victim's loved ones. Despite this, producers are motivated to produce and commercialise true crime due to continued consumer demand, low production costs, newsworthiness and success with advertisers. As media and entertainment have changed throughout the years, the true crime genre has always remained, seeing immense growth across platforms and captivating audiences across the globe.

In today's interconnected world, media platforms also play a significant role in shaping public opinion, including perceptions related to court cases. From Facebook to Twitter, the content shared can directly influence criminal defence strategies and the collection of evidence (Cooke, 2024). There have been instances too where some jury members have sought information on accused persons from the media as a basis for their decisions in courts (Sarpong, 2022). Some jurors too have also used the social media to announce their selection for jury duty. Meanwhile, posts and comments on social media sites about on-going cases also tend to create widespread publicity, sometimes leading to a form of trial outside the courtroom.

These developments have brought cases to a halt and also led to mistrials in certain situations. As Melnitsky (2023) notes, some high-profile cases are not just decided in the courts but also find expression in the court of public opinion. Indeed, some legal experts even contend that the media sometimes do their own investigations, build public opinion against accused persons even before the court takes cognisance of the case. By this way, they prejudice the outcome of such cases. According to Sastry (2019), some lawyers even believe there have been instances where judges altered their judgements due to the intensive case analyses by the media before the completion of these cases. This phenomenon raises concerns about the ability of the criminal justice system, especially to protect the right to a fair trial amidst intense media coverage and public scrutiny (Cooke, 2024).

Despite the foregoing and the seemingly grave danger that the media are seen to be engaged in, a novel approach to this 'trial by media' concept, is currently making waves in Ghana, albeit in a different direction. Some radio and TV stations in Ghana have opened their doors to vulnerable, disgruntled and underrepresented people who feel let down by the Ghanaian legal system. People who feel they cannot find justice in the law courts, lack the needed resources to engage the services of lawyers, or needless to say, need expeditious trial of their cases, or are drawn into moral issues that require societal attention and or resolution, are now finding solace in 'media court' shows operating as pseudo-courts in Ghana where quick resolution of matters are heard.

These shows have allowed people, especially those in poor communities to seek redress from the ills of society. Not only that, the shows have also been a means to restore community values, ethics, morals and cultures that the younger generation is deemed to be unaware of or have simply discarded. The popularity of these shows has transcended beyond the country's borders through the social media. So, Ghanaians in the diaspora now call in to the shows or even put in an appearance when they are back in the country to seek help over wrongs done to them by their kinsmen or to report on the infidelity of women or men they are attached to or connected with in the country.

The popularity of these 'media court' shows emanates from the fact that Ghanaians remain in an era of a barely functioning court system, greatly reduced access to the legal system and a potential weakening of a rule of law, as cases of bribery and corruption remain rampart within the judicial processes (Osse & Asiamah, 2020). But Ghana's justice system is not just slow, it is also expensive and sometimes even harsh (Atuguba, 2018). Initiation of an action sometimes take too long. This explains why some frustrated people are seeking justice outside the law courts. The 'media court' shows tend to adapt to problem-solving lessons and many of these problem-solving strategies have resulted in more positive outcomes for programme participants.

But there are questions regarding the modus operandi of these programmes. Some legal experts have even questioned the legitimacy of such programmes. Others have also complained about the usurpation of cases that otherwise should have been handled by the law courts. Debates have also raged over how subjects are treated and even whether decisions these shows arrive at, are even lawful or have any legal basis at all. Aside these, it is important to know how people make sense of the current media landscape with the propagation of these 'media courts', the extent to which these pseudo-courts can go and what the authorities make of these shows.

The aforementioned issues constitute the basis for the chapter. The chapter, thus, delves into the concept behind the 'media court' shows. It is underpinned by the uses and gratification theory and the socio-legal theory. It reviews some of the cases that the Ghanaian media houses have dealt with and also examines and evaluates both the moral and legal contents underpinning some of these cases. It also revisits the oft-debated relationship between law, in particular, and morality, as well as the place of both within the realm of the Ghanaian setting. This is because such pseudo-courts tend to oscillate between both legal and moral issues. Furthermore, the chapter interrogates why such programmes are attracting a strong showing within local communities and even Ghanaians in the diaspora. The write-up mainly dwells on both the live streams and recorded versions of the programmes.

The outline for the chapter is as follows. It begins with an introduction that sets the tone for the definition of a 'media trial' and subsequently the so-called 'media courts'. It then provides an insight into the media landscape and its relationship with the law and society. There is also an assessment of the justice system in Ghana before the theoretical underpinning of the study is provided. The methodology adopted for this chapter is then delved into. Following from that, the chapter examines the workings of the 'media courts' and the nature of the cases they handle. There is a section for a discussion of the issues under consideration. This is followed by the conclusion.

## 2 Understanding the Media

The media have a unique identity in the society (O'Shaughnessy & Stadler, 2012). They provide information, education, knowledge, motivation, surveillance and entertainment and also create awareness in people. Generally, they play a critical role in facilitating social change and shaping public opinion and attitudes (Happer & Philo, 2013). The media, through their reporting too, can put a spotlight on critical developments that impact on people's lives, as well as bringing to the fore issues that are often ignored and voices that are marginalised. Their agenda-setting function often influences debate, thinking and priorities within society.

Today, both the traditional and social media offer a wider audience pool (Johnson, 2021). They are an important part of the society (McQuail, 2010). The media wield a lot of power in view of their position in society and, therefore, serve as a conduit through which voices, perspectives and lives are brought into the public sphere

(Herman & Chomsky, 2002). The persuasive power of the media is well known as governments, corporations, nonprofit organisations and political campaigns rely on the media to create messages and to send them to the general public (O'Shaughnessy & Stadler, 2012). They often put a spotlight on critical developments that impact negatively and positively on people's lives, as well as bringing to the fore issues that are often ignored and voices that are marginalised.

The media's agenda-setting function often influences debate, thinking and priorities within society (O'Shaughnessy & Stadler, 2012). But, perhaps, one of the most important roles played by the media within modern democracies is the watchdog role through which they monitor the performance and conduct of governments to ensure that they adhere to the promises and expectations of those who put them into power. Consequently, this role often serves as the source of conflict between media, the state and its institutions (Jurkowitz & Mitchell, 2020).

In general, media performances are based on the guiding principles of fairness and accuracy. However, the ability and efficacy with which the media perform these functions largely depends on the political and legal environments in the countries within which they operate (O'Shaughnessy & Stadler, 2012). Where the legal framework is conducive, there is a proliferation of media organisations and products as well as a plurality of voices and access to information. On the contrary, in situations where the legal framework is stifling, there is a lack of diversity of media as well. So much can be said of the media's usefulness to society as well as its transgressions and for some time now, the searchlight seems to have been thrown on the media's analyses of pending cases in the law courts. This has been seen to be unprofessional, dangerous and unfair within legal circles (Sastry, 2019).

## 3 The Media, Society and the Law

For years, the media have been reconstituting and dramatically extending their power to shape crime consciousness and influence the justice system. Seemingly, this has not gone down with some legal experts, though. Simply put, their fear is that a persistent stream of distorted and sensationalised media portrayals may crowd out sober legal facts and may ultimately prove corrosive of law's authority and autonomy (see Haltom & McCann, 2004; Sherwin, 2000). Judging from this, it seems such concerns are the product of some widely-held ideas about media culture, which some imply distort, influence and act as a poor proxy for a proper understanding of what law is about.

Understanding the ever-complex relationship between the media, society and the law, therefore, requires exploring several fluid discussions that connect these domains. Basically, what people know of the justice system and what they think ought to be done about it are based on content that has been explained in the media. This sometimes reinforces the point that the seeming conflict between media culture and law might be about some rivalry and the concern of legal actors to safeguard the law's own spectacle that they consider instrumental in communicating the legal truth in all its integrity. The potential elimination of the court process, under the influence of a media culture that is intent on promoting direct justice and is impatient with the elaborate proceedings required for the proper administration of justice, is therefore harrowing for those in the legal profession.

Incidentally, the media's facilitation of legal self-help through 'media courts' feeds into a need to feel emotionally supportive of people who need help and can speak from personal experiences. This perhaps explains the appeal of media houses and the use of multimedia acting as virtual meeting points where people can commiserate as well as advice each other on legal matters. In that sense, at least one could argue that self-help media thrive on drama and emotion, often eclipsing a dry legal fact. Besides, an important factor in explaining the media's powerful effects is people's presumed lack of any direct involvement with the law, particularly as regards the actual administration of justice. Indeed, for many people, the only contact they have with the law is through the media. This suggests that law is virtually absent from ordinary lay experience, creating ample opportunity for the media to act as a proxy for personal contact with the legal system.

## 4 Analysing the 'Trial by Media' Phrase

'Trial by media' has become a phrase popularly coined to describe the impact of the media on a person's reputation by creating a widespread perception of guilt or innocence before, or after, a verdict in a court of law. It is a fairly well- known concept that goes straight to the issue of protecting the right to a fair trial (Wright & Ross, 1997). It is when the media either frontally or suggestively pronounce on the merits of a matter that is sub judice or build public opinion on guilt or innocence in criminal trials. Ultimately, this enters into the domain of law courts, and therefore is frowned upon within legal circles.

Incidentally, in today's highly-connected social world, the details of a scandal are often posted on social media accounts almost instantly. The media coverage of an issue can even change the way that judges and juries act in the legal system, especially if and when the media give extra information into the case or the people behind it. It is quite significant to note that the tension between wanting a verdict entirely on the evidence in the individual case and recognising that other conditions affect the outcome, which remains part of the tension that is always going to be present in courtrooms because of the way information flows these days. Some of the media's alleged 'transgressions' include building a sensation around such cases, which can affect the decisions of jurors. Since courtrooms are not hermetically sealed from what is going on in the world, it presupposes that a lot too could filter through to cases under consideration (Wright & Ross, 1997).

Ultimately, the promise of the judicial system is that parties would receive a fair and impartial trial, based on the evidence presented in open court and not by any outside influence. That is the reason why any external influence is always deemed to be inappropriate especially when discussions in the media have or could have a significant effect on a matter before the court. A shortcoming of this is that there is a gap between an ideal situation in which fair comments are made and the reality in which they are generally and often perversely entwined. The disparity between what is said and how that will influence the outcomes of cases, therefore, remains a huge concern.

Like other institutions helping to raise the standard of living in society, the media often emphasise on the unquestionable need for improvements in the social conditions of people. To this extent, it has been their avowed determination to ensure the average person gets the appropriate attention. Despite the debates, arguments and reservations in certain quarters regarding the media's seemingly usurpation of cases deemed to be within the preserve of the law courts, it seems seeking justice from the media is catching on well from the perspective of many litigants in Ghana and for some good reasons too. Obviously, a precise set of institutional factors may explain the great social relevance of the 'trial by media' in the Ghanaian context.

Currently, the media in Ghana have almost revitalised themselves as the 'public' court in the country. They are now emphasising their role as the enablers of people who probably can never be heard, the light to the people who can never see the reality and the guide to the judge making the decisions in the law courts. There is a justification for this too, since many Ghanaians are discontent with the court system and the judiciary in general (Atuguba, 2018). Disputes in public courts generally take too long, costs too much, many blame the system for being antiquated and arcane and many more (Osse & Asiamah, 2020). But whatever explanation is proffered, the unvarnished reality is that many people in Ghana cannot enforce their legal entitlements in public courts for one reason or the other.

## 5 Assessing the Justice System in Ghana

Ghanaians generally contend that the three pillars of democracy—legislature, executive and judiciary have woefully failed to work in their best interest. The country's leadership and other politicians have not been seen in the best possible terms of late (Sarpong, 2017). The media have been rife with reports of politicians amassing wealth to the detriment of the people. Whilst politicians in the country are often accused of not taking the plight of the people into consideration, the judiciary, on the other hand, is also seen as corrupt among ordinary Ghanaians (Amissah, 2019). The perception and reality of judicial corruption have greatly undermined the credibility of courts, eroded public trust and also harmed core judicial functions such as the determination of disputes and law enforcement (Benson, 2023). There have been multiple examples of lower-level issues with corruption throughout the judicial system culminating in a former Chief Justice, Georgina Woode, removing 20 judges during her tenure (Amissah, 2019).

Ghana has different levels in terms of its jurisdictional administration. The Supreme Court is the apex court. There are also the Appeals courts, High Courts, Circuit Courts and the District Courts. The district court, by virtue of it being the lowest in the hierarchy of all courts in the country, has the largest number of courts. A more recent initiative to increase access to justice has been the promotion of Alternative Dispute Resolution (ADR) to divert cases from the courts. The Courts Act, 1993 (Act 459) in sections 72 and 73 encourage the use of ADR to resolve disputes pending before the courts. Whilst parties are not required to have an attorney to participate in arbitration, arbitration is a final and legally-binding process. There is now a strong advocacy for litigants and their lawyers to turn their attention to ADR mechanisms instead of the traditional mode of courtroom litigation. This is in view of the financial costs associated with litigation in the country (The Open Society Initiative for West Africa, 2007).

Although Article 294(1) of the Ghanaian Constitution provides for legal aid for the poor, the Ghana Legal Aid Board (GLAB), which oversees this, lacks the presence in many areas of the country, especially in rural locations. This situation is compounded by the lack of lawyers to provide representation in the rural areas and the absence of a formal requirement for lawyers to do pro bono work. Ghana also has a Commission on Human Rights and Administrative Justice (CHRAG), which is responsible for human rights issues in the country. Despite the challenges of under-resourcing, CHRAG has been able to perform quite satisfactorily by seeing to various petitions lodged with it. The country's constitution also recognises regional and national houses of chiefs in adjudicating chieftaincy disputes.

## 6 Theoretical Underpinnings

The study employs the Uses and Gratification Theory (UGT) and Socio-legal Theory. These theories have a profound explanation for the issue under review.

## 7 Uses and Gratification Theory (UGT)

The UGT remains one of the most prevalent theories used in media effects research (Kasirye, 2022). Scholars leverage this approach to analyse a broad array of media, from traditional sources such as television and newspapers to modern formats, which include social media platforms and streaming services. The UGT is a dynamic approach that has adapted and expanded its scope with the development of technology in the 1980s and 1990s and the internet and has increased programme options and control. It was developed by a group of researchers who were interested in understanding the role of media in people's lives (Kasirye, 2022).

The UGT offers insights into the role of the media in fulfilling the needs of individuals. It proposes that people actively consume specific media channels to satisfy their needs.

It portrays how audience members react to the mass media and seek to use it for specific purposes based on their own individual characteristics, social categories and relationships (Pedersen, 2017). People choose to consume certain kinds of media because they expect to obtain specific gratifications as a result of those selections (Lai, 2023). For instance, many people use the media to seek out entertainment, to find information, to communicate with like-minded individuals or to pursue self-expression.

Media consumption can provide relaxation, social interaction, entertainment, escape and a host of interpersonal and social needs. Each of these uses gratifies a particular need, and the needs determine the way in which media are used. Media use is therefore goal-oriented and the audience is always active in seeking out needs gratifications. By examining the motives behind the consumption of a particular form of media, researchers can better understand both the reasons for that medium's popularity and the roles that the medium fills in society. Through UGT, researchers have explored how individuals use media for entertainment, social interaction, information, emotional release, surveillance, escapism, education, persuasion and a sense of belonging and this constitutes the basis for which Ghanaians are using.

## 8 Socio-Legal Theory

The relationship of the law, in its many aspects to a social situation, should be considered a necessary part of the understanding of that situation. A socio-legal approach is an interdisciplinary approach to analyse the law, legal phenomenon and relationships between these and wider society. According to the socio-legal approach, analysis of law is directly linked to the analysis of the social situation to which the law applies, and should be put into the perspective of that situation by seeing the part the law plays in the creation, maintenance and/or change of the situation.

The interdisciplinary field of socio-legal studies delves into the intricate relationship between society and the law, recognising that legal systems are not static entities existing in isolation but are deeply intertwined with the social fabric. This dynamic interaction between societal norms, values and legal frameworks shapes the contours of justice, governance and individual rights within a given community or nation. By scrutinising this interconnection, socio-legal scholars aim to understand how legal rules are created, interpreted and enforced in response to social realities.

Understanding law and legal system as a social formation is the first part of a law and society approach; the second is the inter-relationship between law and other aspects of society. One issue is how law as a system of social rules interacts with other systems of social rules, such as those of civil associations, religious bodies, private institutions, family networks and so on. What happens when legal rules conflict with or try to change other networks of rules? Here the notion of social spheres is developed and put to use in explaining the inter-relationship.

One of the key pursuits of socio-legal inquiry is to uncover how socio-economic factors influence legal processes and outcomes. Poverty, inequality, discrimination and access to resources all play pivotal roles in shaping individuals' interactions with the legal system. For example, marginalised communities often face systemic barriers to justice, including inadequate legal representation, discriminatory practices and unequal treatment before the law. By examining these disparities, socio-legal research sheds light on the ways in which law can either perpetuate or alleviate social injustice (Hart, 2018). It critiques how laws and legal systems are complicit in creating and perpetuating inequalities, hierarchies and exclusion, often reflecting and reinforcing power imbalance.

At its core, socio-legal dynamics highlight the reciprocal relationship between society and the law, wherein societal norms, values and cultural shifts continually influence the development, interpretation and enforcement of laws. Within this framework, efforts to address social justice concerns and enact legal reforms play a pivotal role in fostering fairness, equity and access to justice for all members of society. Initiatives aimed at rectifying historical injustices, combating discrimination and protecting human rights are crucial steps towards creating a more just and equitable legal landscape, navigating the intersection of morality and legality adds layers of complexity to this landscape, as societal perceptions of right and wrong often collide with legal regulations, sparking debates and shaping legal outcomes.

Understanding and reconciling these moral and legal dimensions is essential for fostering consensus and upholding the principles of justice within Ghana's legal system. Analysis through socio-legal approaches can expose how formal laws practical application can generate injustice, for example through biased enforcement, unequal access to justice and the erosion of social justice.

## 9 Methodological Considerations

The chapter dwells on observations (listening and watching) of these shows over the last 3 years. The shows are aired on both the radio and TV stations of the media houses and are also streamed live on Facebook and YouTube. The exceptional, interesting and controversial ones are also shared by other content creators (bloggers), leading to a large patronage of these shows on the internet. The court shows in this study were watched either during live streaming or on the internet after they had been shown. Our interest in this shows stems from the incredible patronage these shows receive from the general public and the debates some of the issues generate.

The study also used conversation analysis (CA) as a basis for the study. CA is a method for studying social life as it happens, in all its rich, varied and intricate complexity. It is an approach to the study of social interaction that examines the mechanisms by which humans achieve mutual understanding (Flick, 2009). CA is a qualitative approach, often applied to analyse social interactions between people in various settings. The fundamental purpose of conversation analysis is to understand how people use conversation to achieve social goals, called social actions (Have, 2007).

Principally, CA can be used to discover how participants understand and respond to one another in their turns at talk, with a central focus on how sequences of action are generated. It reveals the basic ways in which speakers coordinate their talk; how they manage problems of production and understanding; how they perform offers and requests, complaints and compliments; how they tell stories and indicate what they know and do not know; how emotion is displayed and responded to; how gaze, prosody and gesture help to orchestrate interaction and many more themes. It has over the years been adapted to embrace more task and institution-centred interactions, such as those occurring in the media. It is, therefore possible now to trace a number of conversational techniques, varying subject matters and different types of guests in conversation analysis. Through CA, one can learn to decipher meaning, power dynamics and social structures embedded within spoken language, which are all of relevance to this study.

One of the goals of CA is to describe the procedures that people use to produce and understand conversation. Conversations are used for public discourse and together with real life experiences of ordinary people, the 'media court' shows have occasioned informal performances and a language better suited to their audience. Vernacular has been the preferred choice and which is easily embraced by a number of people in the country.

## 10 Collecting and Analysing the Data

The analyses began with observations on the shows and YouTube/Facebook. Anything that participants treated as relevant for their interactions were considered a candidate phenomenon for investigation. The observations concerned the structure of episodes of interactions.

First, data were collected using live streaming or video recordings. Then the recordings were transcribed to capture a number of occurrences such as intonations and other non-verbal cues. We then went on to identify recurring patterns in the data, notably the use of specific words and phrases to denote the participants' issues. Following from this, a clear and concise analysis was developed, focusing on their stories. Finally, we considered the context of the interaction, that is, by looking at the social and cultural norms that might be influencing their interaction.

Aside that, the study used secondary sources that provide second-hand commentary and information from other researchers. Such sources include newspaper and TV comments on the 'media trial' shows.

## 11 Examining the Modus Operandi of the 'Media Courts'

Some radio-cum-TV stations have lately set up shows that they believe can help with some social and legal matters in view of the challenges inherent in the country's judicial system. The most popular ones, though, are those that feature on *Oyerepa FM/TV*, *Akoma FM*, *Sompa FM* and *Nhyira FM*. The language used is vernacular, which is easily accessed by a large number of people. The 'media court' show can be labelled as a postmodern show because of its extreme intertextuality, its combination of both information and entertainment, personal and public experiences, lay people and experts.

Typically, a person with an issue would approach the show, the producers and their teams would then look into the issue. The complainant would then be given a date to appear on the show if the nature of the case is deemed to be appropriate for broadcast. The show also ensures that both complainants and defendants are invited; however, those who are unable to come on their own, either because they do not have the means to do so, are picked up from their locations by the media stations. Sometimes parties attend a mini-trial, which is off camera to determine the resolution of the matter. It is often after this that they are given an opportunity to state their cases live on air.

The show always has a host and a panel, which is made up of community members who are well-versed in moral principles or beliefs that society values. The duty of the host is to facilitate dialogue between parties to enable them to arrive at a mutually acceptable settlement. The host works with the parties to find the situation that best fits them all. When they are able to resolve their dispute, the parties put the settlement into writing. These written terms are then read over and interpreted to the parties who will then sign the terms of the settlement and same will be entered as consent judgement. It is only when the settlement breaks down that the matter is, at times, referred for trial at the law courts. The agreements are also binding, so when a party reneges, these media organisations can take the matter to court for enforcement, if need be. The shows, despite holding deliberations on their own, also have lawyers who are on retainers, and who occasionally help them with the drafting and representation of some cases at the courts.

In practice, the processes undertaken by these shows have followed the same understanding at the courts. What seemingly counts as quite useful too are the moral suasions applied. Some litigants sometimes reconsider their positions on their cases and then soften their stance when such persuasions are applied. Appealing to the conscience of the litigants can also be seen to be a means to ensure they see the essence of their responsibility to end their litigation through amicable ways. Through their individual conscience, the litigants are made aware of certain moral principles and, therefore, become motivated to act upon them, assess their characters, their behaviours and ultimately themselves against those principles that they hold (Sarpong, 2022).

## 12 The Nature of the Cases

There has been an exponential growth of these 'problem-solving media court shows' in Ghana of late. The concept of these problem-solving media courts has also expanded to address many societal issues. The shows combine a therapeutic model with traditional jurisprudence. Modern means to solve issues are also undertaken. For instance, provisions are made for paternity tests in situations where they are required. Besides, cases that dwell on curses and other customary issues that require traditional interventions are also dealt with, with the involvement of traditional authorities. Basically, behaviours that are not in tandem with societal norms are those that the shows tend to focus on. The basis of these problem-solving media courts too is to address the underlying issues that have brought an individual or individuals into an unwarranted situation.

Some of the cases usually handled by the media include land litigations, paternity issues, rape, defilement and family negligence. There are those who also bring in spiritual problems that they are confronted with. Others include issues that the police have not actually investigated, reporting the conduct of a trial judge, looking for missing relatives and sometimes criminal acts that need the support of the public. Sometimes the producers invite the police in when cases that require the intervention of the law enforcement officers are being deliberated upon. At other times, community or religious elders do call in to arrange to settle some matters that come on the shows. The shows too can sometimes turn into a spectacle as they also have an entertainment value and, therefore, give people strong incentives to watch them.

Apart from issues pertaining to family land disputes and misunderstandings, swindling, incestuous relationships, settling marital issues and so on, cases involving imprecation or curses have also been quite rampant and often take a centre stage. People suffering from the effect of curses are brought on to the show where they publicly ask for forgiveness from those who have invoked curses on them. Curses, without doubt, are one of the known belief practices that are feared by many people in the Ghanaian society (Gyasi, 2017). Curses are believed to have devastating spiritual effects.

In recent times, the use of curses has become common in many parts of Ghana. Some people who are aggrieved by certain acts or feel cheated in certain situations resort to the use of curses when they believe they have been wronged. Such curses come in the form of invocation of supernatural powers or river deities to kill, cause harm or place a misfortune on a person who has wronged another either through stealing, being untruthful or saying something unpalatable to someone. These 'media courts' provide help to those affected by the curses by seeking the intervention of traditional healers or fetish priests to reverse the curses put on the affected persons.

One reason for the increased use of imprecation in the pursuit of justice, especially in domestic cases, is the slow speed of case adjudication in the law courts. The consequences of invoking curses can often be fatal, yet people turn to this means because of the pain of being wrongly accused or desiring retaliation for harm done by some persons, known or unknown. Victims of curses who appear on the show portray thin postures, protruding stomachs, swollen legs and other debilitating illnesses, an indication that the curses they are suffering from have reached a serious point. They come in to beg their cursers to reverse such curses. Some even die before the resolution of their cases, others, meanwhile, get the curses reversed in good time if they come to the show early enough and are able to be sent to the appropriate quarters for the necessary rites to be performed.

The political class, in general, is certainly not exempt as a principal accused on such shows. This is because the state lacks any support system for the poor and vulnerable. Hence, there are those who appear on the shows to solicit for funds usually for health reasons. These are often helped by the generosity of audiences. The audiences spontaneously send in money and other items to people who appear on the shows and are deemed to have unmet needs. A recent case is a hole-in-heart patient who needed financial support for an operation at a state teaching hospital. The boy's relatives came on one of the shows and asked for support since such operations are not free in the country. People, subsequently, contributed massively towards the child's medical expenses. There have been occasions too where women suffering from domestic abuse or others who had acid poured on them by jealous rivals have been helped with financial support by those who watch or listen to the shows. Indeed, these media court shows have, in many instances, settled a number of cases that on the face of it could have persisted for years at the law courts (see Ghanaweb, 22 July 2024).

The shows oscillate between morality and law as the basis for their decision-making. They set the tone for the two to be applied in the best possible terms. Ghanaians place much premium on their cultural values, which they deem virtuous. This accounts for the justification for the condemnation of a conduct that does not meet the standards of their conception of cultural morality. Usually, if a person behaves contrary to certain moral standards, he or she might be seen as isolating himself or herself from the community. The shows clearly implore common sense, daily experiences and scientific philosophy. There is a recognition that society cannot be fully appreciated without, at the same time, appreciating those values, norms, morality and worldview that have enhanced human life and created the necessary relations for survival and development. As Waliggo (2005) rightly put it, oftentimes, the law seems to be divorced from public morality, but it can be said though that this is what the media courts are trying to bridge.

## 13 Infractions of the Show

One of the first challenges for those working from the traditional court process perspective has been whether these media shows are usurping the jurisdiction of the courts. Incidentally, the uniqueness of these media shows reflects deeply on solving societal issues rather than to adjudicate on cases deemed to be suitable for the law courts, but that aspect is not lost on the shows as well. The whole arrangement regarding the media court shows have sparked much debate, and many different opinions have been expressed on that. To some extent, these media courts attempt to take on societal issues that are not truly within the jurisdiction of the court such as appealing for funds, helping those unable to take care of themselves, providing funds for hole-in-heart patients, and so on, which obviously do not fall within the domain of the law courts. In spite of that, they also stray into easily identifiable legal situations that require the courts to handle.

Some legal experts have expressed misgivings about the way the media houses are trying cases and have called for a halt to such shows. In particular, the Ghana Bar Association, the umbrella group of lawyers in the country, has decried the tendency, stating that the media have constituted themselves into law courts and judges and have conducted court cases as if they had the right to administer justice (GNA, 2009). A lawyer and a vehement critic of the 'media court shows', Maurice Ampaw, has even been challenging the media stations for some time now, stressing that the media organisations should be careful for treading on issues and cases they are not destined to make pronouncements on (Ampaw, 2023).

Indeed, some journalists have even expressed dismay at the turn of events chastising some radio/TV stations for the excesses being witnessed on the shows. More recently, some journalists even reprimanded one of the stations that took up the case of a rape victim. The radio station went to the extent of mentioning the name of the victim, her profession and residence, all of which go against the ethics of the journalism profession. There have been occasions too where some of these media organisations have used their medium to intimidate and harass people who they believe have wronged others. Such people are summoned to the radio stations under the guise that their failure to be present, might result in their names and pictures being published.

Meanwhile, apart from the disdain expressed by a section of the populace, nothing by way of stopping the broadcast of such shows has ever come from the National Media Commission (NMC), which is the body established under the constitution to take all appropriate measures to ensure the establishment and maintenance of the highest journalistic standards in the media. The NMC, however in 2023, only rebuked a media house for compelling people to appear before its show. This followed a complaint by an invitee who received 'unguided and derogatory comments' about his person when he declined the media house's invitation to appear before it (Senyah, 2023). In the said reaction from the NMC, the commission stated that parties are not legally bound to submit to the programme's host and the station should avoid resorting to intimidation and threats when parties decline participation. Whilst acknowledging the programmes entertainment value and its role in issue resolution, the NMC underscored the importance of giving fair hearing to all parties involved in cases that come before the stations (Senyah, 2023).

A popular Ghanaian media personality, Okatakie Afrifa, has also indicated that not only are these radio/TV stations courting issues bordering on defamation and contempt of court, they are also undermining and destroying the integrity of the journalism profession and abusing the airwaves (Amoako, 2024). Some Ghanaians, in the light of these happenings, have begun to question the power or authority radio/TV presenters have to summon evidence or go public with any information about people they want to appear on their shows. Indeed some see such acts as blackmail that have nothing to do with press freedom. There is even a belief that the media have arrogated to themselves certain powers to the extent that they can summon people before their shows and even fine them.

These 'media court shows' have also been likened to surroundings which offer an array of a morbid and depressing displays of dysfunctional people whose lives are in already in turmoil. Some invitees on the shows tend to display strong emotions, anger and distress as they mount a strong defence of their cases. Sometimes these lead to chaos as huge disagreements emerge on some of the shows. The shows are at times viewed, in some quarters, as harmful to society (Amoako, 2024). The producers have, in some instances, been accused of allowing and even provoking the use of indecent language, some profanities in their attempt to attract large audiences through staging entertaining and controversial subject matters and guests. Some critics are more cynical about the 'democratic' potential of the shows, arguing that the shows by commodifying human behaviour are simply exploitative and inherently repressive.

It is, however, quite significant to note that in the realities of modern society, the dissemination of the news and programmes is part of an industrial process, which is commonly driven by business concerns. TV and radio programmes often require more adverts and as a result, they try to attract huge numbers of audiences. The likeliness of this can mainly be felt if and when such programmes are of huge interest and very entertaining. Given all the outlets available for entertainment and information in a digitally delivered media world, audience members are more empowered today than at any other time in media history in terms of their consumption pattern. Audience members no longer just consume content—they also make content in a multitude of ways, whether through blogging, podcasting, uploading videos or social networking to just name a few options. Social aspects are yet another force driving the transformation process and these reflects on issues taken up in the media courts.

One line of thought, though, is that these shows have now been positioned at the forefront of a new form of 'participatory citizenship' or a display of 'emotional democracy'. In effect, they have now been conceptualised as a new public sphere in which the old dichotomies of public/private, expert/lay person and rational/emotional are challenged and reconfigured. Promoting a cause, sharing experiences with others in similar situations, or trying to look for relatives, seeking help from state authorities or well-wishers friends, are some of the motivations for participating in these programmes. Nevertheless, people's motivations to appear on such shows reflect a deep desire to find solutions to the myriads of problems confronting them. Many feel very positive about appearing on such shows and acknowledge that the shows have been invaluable and worthwhile.

Arguably, the motivation for someone to agree to be a guest on such a show varies considerably. Some also come in order to gain public attention and media exposure, resolve personal issues or disputes, receive free professional counselling or legal assistance or potentially earn money from appearing on the show, as the case is with those seeking financial support. The decision to appear on such shows can also come with risks and negative consequences for one's personal and professional reputation, though, as the case was when a lady feigned homelessness and had wanted some financial support from the public. She was quickly noticed by a member of the audience as a fraudster.

An important strand of these shows is that they offer an alternative public sphere that provides a platform for the vulnerable and troubled to be heard. They can be considered as 'alternative' because they encompass social groups that are usually excluded or under-represented in the public sphere. In particular, the shows offer opportunities for the participation of marginalised social groups or individuals who probably are not normally heard in the traditional news genres to come on air to discuss issues they are faced with. The shows deal with issues and perspectives that are usually excluded from the public sphere, but are nonetheless relevant in many people's lives, especially the lives of less powerful groups in society.

## 14 Conclusion

This chapter has highlighted the allure of these media shows and their infractions. It finds that the main pull of the shows is about ensuring fair, speedy and cost-effective trial procedures while de-emphasising the adversarial nature of the common law civil process. Evidently, existing litigation processes seem ill-suited for resolving some of the complaints that people have, hence the resort to the media courts. Many participants in these 'media court shows' have found help in some situations and have acknowledged their benefits, yet the shows too have drawn a number of adversaries who believe the media are encouraging some illegalities.

Whatever it is, it seems these shows have come to stay as the sheer numbers they attract either from audiences and litigants remain quite overwhelming. Indeed, few people could have anticipated the direction, velocity and momentum of this development. People now accept the media court shows as the voice of the people because there is a complete loss of faith in the processes of justice within the law courts and from the police. Perhaps, this accounts for the dilemma facing the legal authorities or even government officials who invariably can step in if the shows were too extraneous.

## References

Amissah, P. K. (2019). Amos and 'Ghana in the eyes of God': A public theological response to bribery and corruption. *International Journal of Public Theology, 13*(3), 282–300.

Amoako, G. M. (2024). *This nonsense must stop: She is breaking homes in this country – Okatakyie Afrifa blasts Oyerepa FM's Auntie Naa*. Modern Ghana. Retrieved December 9, 2024, from https://www.modernghana.com/news/1322698/this-nonsense-must-stop-shes-breaking-homes-in.html

Ampaw, M. (2023). *Lawyer Ampaw sends waning to Auntie Naa and Mama Efe*. Retrieved March 30, 2024, from https://www.youtube.com/watch?v=0UTh-Wo3tF8

Atuguba, R. (2018, January 4). *Ghana's justice system needs a major overhaul: Here's what should be done*. The Conversation. Retrieved July 2, 2024, from https://theconversation.com/ghanas-justice-system-needs-a-major-overhaul-heres-what-should-be-done-88724

Banks, M. (2005). Spaces of (in)security: Media and fear of crime in a local context. *Crime, Media, Culture, 1*(2), 169–187.

Benson, G. H. (2023). Justice for gains, a potential threat to national security - A tortuous trip to Ghana. *Journal of Forensic Legal and Investigative Sciences, 9*, 071. https://doi.org/10.24966/FLIS-733X/100071

Birkbeck, C. (2014). Media representations of crime and criminal justice. In *Oxford handbooks online criminology and criminal justice*. Oxford University Press. https://doi.org/10.1093/oxfordhb/9780199935383.013.15. Retrieved December 2, 2024.

Chadwick, A. (2013). *The hybrid media system: Politics and power*. Oxford University Press.

Cooke, B. (2024) *How does social media affect court cases? The impact of social media on criminal defence and evidence gathering*. Retrieved June 24, 2024, from https://stlouiscriminaldefense.com/how-does-social-media-affect-court-cases/

Flick, U. (2009). *An introduction to qualitative research* (4th ed.). Sage.

Ghanaweb. (2024, July 22). *Oyerepa afutuo provides more societal help than most politicains*. Retrieved December 9, 2024, from https://www.ghanaweb.com/GhanaHomePage/entertainment/Oyerepa-Afutuo-provides-more-societal-help-than-most-politicians-Auntie-Naa-s-spiritual-mother-1941659

GNA. (2009, October 23). *GBA cautions radio stations against media trial*. Retrieved November 2, 2024, from https://www.modernghana.com/news/245354/gba-cautions-radio-stations-against-media-trial.html

Gyasi, Z. K. (2017, May 17). *Duabo: A practice travelling through ages*. Daily Graphic. Retrieved June 12, 2024, from https://www.graphic.com.gh/features/features/duab-a-practice-travelling-through-ages.html

Haltom, W., & McCann, M. (2004). *Distorting the law: Politics, media, and the litigation crisis*. University of Chicago Press.

Happer, C., & Philo, G. (2013). The role of the media in the construction of public belief and social change. *Journal of Social and Political Psychology, 1*(1), 321–336.

Hart, C. S. (2018). Education, inequality and social justice: A critical analysis applying the Sen-Bourdieu analytical framework. *Policy Futures in Education, 17*(5), 582–598.

Hasell, A., Halversen, A., & Weeks, B. E. (2024). When social media attack: How exposure to political attacks on social media promotes anger and political cynicism. *International Journal of Press/Politics*. https://doi.org/10.1177/19401612231221806. Retrieved October 2, 2024.

Have, P. T. (2007). *Doing conversation analysis* (2nd ed.). Sage.

Herman, E. S., & Chomsky, N. (2002). *Manufacturing consent: The political economy of the mass media*. New York, Pantheon.

Johnson, K. (2021, May 18). *Why social media vs traditional media is not the debate you should care about*. Sprout Social. Retrieved July 3, 2024, from https://sproutsocial.com/insights/social-media-vs-traditional-media/

Jurkowitz, M., & Mitchell, A. (2020). *Most say journalists should be watchdogs, but views of how well they fill this role vary by party, media diet*. Pew Research Centre. Retrieved July 02, 2024, from https://www.pewresearch.org/journalism/2020/02/26/most-say-journalists-should-be-watchdogs-but-views-of-how-well-they-fill-this-role-vary-by-party-media-diet/

Kasirye, F. (2022). The importance of needs in uses and gratification theory. *Advance*. Retrieved December 7, 2024, from https://doi.org/10.31124/advance.14681667.v2

Koetsenruijter, W., & Vanderveen, G. (2011). The popular virgin and the wolf in sheep's clothing: A case study of the imaging of victims and offenders. In T. van Haaften, H. Jansen, J. de Jong, & W. Koetsenruijter (Eds.), *Bending opinion: Essays on persuasion in the public domain* (pp. 241–270). Leiden University Press.

Lai, P. C. (2023). *Strategies and opportunities for technology in the metaverse world*. IGI Global.
McQuail, D. (2010). *Mass communication theory* (6th ed.). Sage.
Melnitsky, R. (2023, October 24). *Social media can change outcomes in the courtroom*. New York State Bar Association. Retrieved June 27, 2024, from https://nysba.org/social-media-can-change-outcomes-in-the-courtroom/
O'Shaughnessy, M., & Stadler, J. (2012). *Media and society* (5th ed.). Oxford University Press.
Osse, L., & Asiamah, G. B. (2020, February 29). *Ghanaians cite high cost, bias, and long delays as barriers to using formal justice system*. Afrobarometer Dispatch No. 347. Retrieved December 7, 2024, from https://www.afrobarometer.org/wp-content/uploads/2022/02/ab_r8_dispatchno347_high_cost_bias_delays_prevent_ghanaians_from_using_justice_system.pdf
Pedersen, P. M. (2017). *Strategic sport communication*. Human Kinetics.
Robinson, M. B. (2018). *Media coverage of crime and criminal justice* (3rd ed.). Carolina Academic Press.
Sarpong, S. (2017). Pushing the boundary: Are reality TV shows ruffling feathers in Africa? *Society, 55*(1), 53–61.
Sarpong, S. (2022). The jury system in the digital age. In T. P. Podshivalov, E. V. Titova, & E. A. Gromova (Eds.), *Digital environment law* (pp. 787–792). South Ural State University (SUSU) Department of Business, Competition, Ecological Law Institute, Prospekt.
Sastry, V. V. L. N. (2019) *Influence of trial by media on the criminal justice system in India*. Walden Dissertations and Doctoral Studies. 6805, Unpublished thesis, Walden University. Retrieved November 12, 2024, from https://scholarworks.waldenu.edu/dissertations/6805
Senyah, Y. K. (2023). *You have no authority to compel anyone to appear on your programmes-NMC to Oyerepa FM*. Graphiconline. Retrieved December 3, 2024, from https://www.graphic.com.gh/news/general-news/you-have-no-authority-to-compel-anyone-to-appear-on-your-programmes-nmc-to-oyerepa-fm.html
Sherwin, R. K. (2000). *When law goes pop: The vanishing lines between law and popular culture*. University of Chicago Press.
Surette, R. (1989). Media trials. *Journal of Criminal Justice, 17*, 293–308.
The Open Society Initiative for West Africa. (2007). *Ghana justice sector and the rule of law – A discussion paper*. The Open Society Network Publication.
Waliggo, J. M. (2005, September 4–8). *Law and public morality in Africa: Legal, philosophical and cultural issues. A paper discussed at the ALRAESA Annual Conference, Imperial Resort Beach Hotel, Entebbe*. Retrieved May 2, 2024, from https://www.justice.gov.za/alraesa/conferences/2005uganda/ent_s3_waliggo.pdf
Wright, J. W., & Ross, S. D. (1997). Trial by media?: Media reliance, knowledge of crime and perception of criminal defendants. *Communication Law and Policy, 2*(4), 397–416.

**Sam Sarpong's** research interests lie in the relationship between society, economy, institutions, and markets. He tends to explore the nature and ethical implications of social-economic problems. His publications span a variety of fields including sustainability issues, management practices, CSR/Business Ethics, entrepreneurship, and global marketing strategy. He is currently affiliated to the School of Business, University of Central Lancashire (UCLan), UK. Prior to joining UCLan, Sam held academic positions at Xiamen University, Malaysia; Narxoz University, Kazakhstan; University of Mines and Technology, Ghana; Swansea Met University, University of London (Birkbeck College); and Cardiff University, respectively. He obtained his PhD from Cardiff University and his MBA from the University of South Wales. He is a reviewer for many leading journals and also serves as Editor (NEP/AFR – RePEc) and as Associate Editor for the *International Journal of Corporate Social Responsibility* (Springer).

# The Media Impact on the Economic News Consumer

**Suzana Demyen, Adrian-Costinel Tănase, and Florența-Diana Tănase**

**Abstract** In today's information-driven society, accurate and timely information has become as valuable as any other organizational resource. The historical need for information evolved significantly with the advent of printing, and today, individuals and organizations face an overwhelming flood of information from various media sources. This chapter examines the critical role of media in shaping consumers' understanding and responses to economic news in a context where verifying accuracy is increasingly challenging. It explores how the accessibility, framing, and perceived credibility of economic information influence consumer behavior and, by extension, economic decisions at both individual and organizational levels. Additionally, the chapter considers how educational background and media literacy affect a consumer's ability to assess the reliability of economic information, ultimately impacting economic choices and quality of life. Through this analysis, we aim to underscore the importance of media literacy and credible information sources in enabling informed economic decision-making in an age of rapid information dissemination.

**Keywords** Media · Impact · News · Economic news · Consumer · Information · Resource · Social · Fake news · Education · Manipulation · Challenge · Organization · Opinion · Articles · Belief · Freedom of expression · Networking · Social networks · Mass media · Image · Brand · Consequence · Influence · Well-being · Behavior · Accuracy · Media consumer · Newspapers · Internet

S. Demyen (✉) · A.-C. Tănase
Department of Business Administration—Reşita, Faculty of Economics and Business Administration, Babeş—Bolyai University, Resita, Romania
e-mail: suzana.demyen@ubbcluj.ro; adrian.tanase@ubbcluj.ro

F.-D. Tănase
Faculty of Political, Administrative and Communication Sciences, Babeş—Bolyai University, Resita, Romania
e-mail: diana.tanase@ubbcluj.ro



M. Todorovic et al. (eds.), *Sustainability and Social Responsibility of the Media and in the Media*, CSR, Sustainability, Ethics & Governance,
https://doi.org/10.1007/978-3-032-00086-6_9

## 1 Introduction

Man is a social being. Communication is part of his life, being an essential component of any relationship, be it professional, personal, or simply social. Since ancient times, man has communicated with his peers through various means, adapted to the development level of the referred period. The forms of communication and the means used have also evolved, and therefore in today's society, we are witnessing the spread of more and more advanced methods and means of communication.

At every step, we find that information is found in abundance around us, enveloping us from various sources, but it depends on us what we select to carry out our activity, to train, or to pass on. Just like the products or services, the information also targets certain consumer segments, being transmitted through those means and appealing to those psychological springs that trigger the "consumer decision."

According to Kelman (1961), a source of information must be strong, attractive, and credible; in other words, we can say that it must meet some of the most important characteristics of a successful person.

The media plays an important role in society, influencing public perceptions and thus shaping, to a considerable extent, people's behavior in various fields, including the economic one.

Mass media is often considered the "watchdog" of democracy, having the capacity to inform and mobilize citizens (McQuail, 2010), but also to defend the rule of law. It serves not only as a channel for transmitting information but also as a tool of influence, having the power to shape opinions (Arnold-Forster, 2023) and build new paradigms.

The media has kept pace with the dynamics of society. It has always had an important role, but in this interconnected world of the last decades, its role has grown significantly. Looking back, history shows that mass media has played and continues to play an important role in shaping public opinion and influencing various events. Thus, perceptions, rebellions, and economic crises have been influenced by it. At other times, hope has been strengthened or restored with the help of mass media. Radio was used to broadcast various news during the Great Depression (Sterling, 2002). The global news press spread information during the 2008 financial crisis; it created panic when the news was about bankruptcies or the collapse of renowned financial institutions, or it led to paradigm shifts regarding these markets by highlighting the causes and consequences of the crisis (Lewis, 2010). Additionally, media coverage of future economic changes influenced public perceptions and was itself influenced by them during periods of economic recession and beyond (Boomgaarden et al., 2011).

Not only have economic and financial crises been influenced by mass media, but rebellions as well. The Civil Rights Movement in the United States and the Arab Spring were influenced by television networks and social media platforms (Facebook, Twitter, YouTube), respectively (Anderson, 2003).

The way news is reported significantly affects public support for Turkey's accession to the European Union, with negative framing having stronger effects (Vreese et al., 2011). The intense promotion on social media platforms (TikTok) influenced the results of the first round of elections in Romania in 2024.

Economic information may slow down or speed up the economy, with accurate and up-to-date information having an impact on investment decisions and market volatility. This happens because financial markets and economic actors rely heavily on information to make predictions and adjust their behavior accordingly. For instance, if media reports suggest an impending economic downturn, investors might pull back on spending, companies might delay expansion plans, and consumers may reduce consumption, thereby amplifying the predicted slowdown. Conversely, positive economic news—such as strong job growth, rising consumer confidence, or favorable monetary policies—can boost investment, increase spending, and stimulate growth.

Additionally, the speed and accuracy of economic reporting play a very important role. Delayed or misleading economic news can create uncertainty, leading to increased market volatility. Rapid dissemination of credible economic indicators allows businesses and policymakers to make informed decisions, minimizing risks and fostering economic stability. A striking example of this dynamic was the 2008 financial crisis, where media coverage of bank failures and recession forecasts led to panic-driven market reactions, accelerating the downturn.

The influence on the financial markets is unquestionable, a single piece of news can shake empires, such as in the case of the Volkswagen scandal that began in September 2015 (ZF, 2015) and led to a significant drop in the value of the automotive group's shares in a very short time. Other examples, in this regard, would be the cyberattack on the Associated Press Twitter account in 2013 (BBC, https://www.bbc.com/news/world-us-canada-21508660), fake news about Bitcoin payment possibilities. Thus, one can grasp the undeniable relationship of economic information in the media with the capital market and, ultimately, with economic policies. In this respect, Stiglitz (2002) underlines that mass media can even influence government decisions, thus changing the economic big picture.

Especially in a society that works day by day, more and more, with data and information, in an increasingly digitized world, information means power. However, as stated by Konstankevych et al. (n.d.) "mankind has begun to use the advantages offered by technological progress for negative purposes," which may subsequently reflect negatively on many aspects of human life. At the same time, in the midst of an information overload, we notice that it is much more difficult to retain a large amount of details, which underscores the need for accurate filtering and selection of important information (Boukes et al., 2021).

Information is a major resource category of an organization, a precious resource that, if used properly, can change the dynamics of that organization. It is easy to find sources of information in today's society, but it is more difficult to be able to select which one is the right one to use in order to reach the set goals.

Unfortunately, very often, fake news, especially in the economic field, is dressed up in such a way that few manage to see beyond what is conveyed to them. As detailed by Damstra and Boukes (2021), in terms of the economic content of the news, an "asymmetrical" attitude be observed, with negative news being exaggerated and positive news being neglected.

In this context, we remark that the media sometimes provides multiple perspectives on a subject. It becomes necessary for a person to be able to discern between what is real and what is false. Tandoc et al. (2017) mention that the term "fake news" has already entered the current language, therefore it is essential to comprehend the correct meaning of this term. Wardle (2017) associates this concept with the intention to spread disinformation, while the range of tools associated with the concept of fake news includes, in the view of Tandoc et al. (2017), "satire, parody, news fabrication, propaganda, as well as advertising," all this contributing to the manipulation of the general public. Jaster and Lanius (2018), on the other hand, define the concept of "fake news" from the perspective of the action of making "mischief with the truth," indicating that news of this nature is characterized by the absence of truth or by a character that distorts it on purpose. This perspective is also shared by Allcott and Gentzkow (2017), who point out the intention of a message sender to falsify an item of information, with the aim of misleading the recipient. Fârte and Obadă (2018) complete these ideas, proposing a conceptual model indicating the following as types of fake news: "clickbait, satire, impostor content, disinformation, erroneous content, false connections, false context, manipulative content, disinformation and fabricated content."

The relationship between fake news and economic news is complex, as misinformation can significantly distort economic perceptions, influence market behavior, and alter policy decisions. Fake economic news can lead to panic, speculative bubbles, or even economic downturns. For example, misleading reports about a company's financial health can trigger stock sell-offs, affecting not just investors but the broader market sentiment.

At the same time, economic news can also be a target of manipulation for financial gains. Certain actors may intentionally spread misinformation to manipulate stock prices, currency values, or public opinion about economic policies.

When fake news and economic news align, they can create powerful, yet misleading, narratives that shape public perception and investor behavior.

Real, in-depth economic news includes analysis, data, indicators that are obtained from painstaking studies needing time, effort, and other mobilized resources. At the same time, news assumes a certain level of preparation to process it. Inserting numbers and using widely circulated economic concepts makes it easier to believe because few individuals, except those in the field or directly interested, have the necessary training, read specialized literature, or listen/watch specialized channels. In other words, it is quite easy and much more credible to associate numbers or other less real data with news and transmit them to targeted consumer categories than to express oneself at a general level, and the impact of what is transmitted will undoubtedly be stronger. At the same time, the information thus seems studied and, therefore, much more credible.

The effects of the propagation of information that is not real can be manifested not only on individuals but also on companies on the market, even in the long term, we may even say there is an increased risk from this perspective that the consumer's perception is wrongly channeled in terms of a brand or product. False advertising campaigns are just one example thereof, while a few words dropped online, falsely presenting a negative message, can influence the big picture in the long run. Baltezarevic (2024) even uses the term “infodemic” in his research as an element to characterize news media in recent years.

On the other hand, the propagation of false information in the media may even endanger the health and safety of the consumers of a product, at the same time affecting their confidence (Fârte & Obadă, 2021).

## 2 Theoretical Background

Fake news has become a problem over time and represents a real challenge that requires a stance on the part of individuals, companies, the state or media organizations. We want a more informed and sustainable society, but there is a need for prioritizing digital literacy, investment in technology-based solutions, all in conjunction with appropriate legislation and adherence to journalistic standards. Thus, the impact of fake news would be lessened, which would lead to increased trust in viable information, in order to protect the integrity of our socio-economic and political prospects, especially nowadays when everything is marked by technological progress, but also by the rapid rise of fake news (Mwangi, 2023).

False information generates consequences on all areas, whether we are talking about social, cultural, political, economic, or religious aspects. Whether purely commercial, political, or otherwise, fake news has a common element: the deliberate deception of news consumers (Hirst, 2017). The dissemination of fake news is problematic and widespread, but it is necessary to make a correct assessment of the strength of its impact and effects, seen through the prism of companies or politicians (Christov, 2019).

The dividing line between fake news and expressing opinions in a particular way is sometimes extremely fine, with some authors considering that jokes are also fake news because “detecting the humor within a truth news story is challenging,” while others, on the contrary, remove satire, pranks, irony from the category of fake news (Özgöbek & Gulla 2017).

When defining the concept of “fake news,” which has already become a popular term (Corbu et al., 2021), certain authors emphasize the intentional nature of disinformation, others emphasize the crisis of values or the desire for enrichment, even by putting other people, entities, organizations, or communities at risk. Thus, if for Özgöbek and Gulla (2017) “fake news articles are intentionally fabricated to be deceptive and can be proven that they are false,” for Arencibia et al. (2023) “fake news is an expression of the crisis of norms, beliefs and values of mercantile societies, endangering the credibility of individuals, organizations, States and communities.”

What is, in fact, false news and what does fake news stand for? According to Dentith (2017), fake news is “designed to deceive its intended audience.” Fake news is dangerous because it misleads by its resemblance with real news and is also has an attractive design (Dentith, 2018). Jaster and Lanius (2018) believe that it omits parts of the truth or may be characterized by a partial absence thereof. Lazer et al. (2018) consider fake news to be fabricated information similar to news in the press in terms of form, but not intent. According to Tandoc (2019), fake news is "a form of falsehood intended to primarily deceive people by mimicking the look and feel of real news."

False news is more comprehensive and may refer to inaccurate information, subjective or erroneous interpretations, which are not based on bad intent, but which, in the end, influence the recipient. Therefore, we can say that fake news is correlated with disinformation, and false news with misinformation. "Disinformation is fabricated or purported" (Dentith, 2017).

Regardless of whether we refer to economic or political benefits, fake news is not a new element, characteristic only nowadays, as it is, as a rule, the result of professionals of persuasion (Bakir & McStay, 2018). The costs of creating false or at least confusing information are, as a rule, very low, being encouraged by "freedom of expression," correlated with numerous legislative loopholes in this area and, in some countries, even with the absence of legal regulations, which gives newsmakers the opportunity to express themselves in almost any way and in any field, in many cases, without violating any of the laws in their countries (Kshetri & Voas, 2017).

These recent years we have been witnessing a "boom" in terms of social networking, which has led to changes in human interactions (Almeida & Mortari, 2021) and, at the same time, to an obvious increase in the consumption of information from social media, in that each individual can be both a consumer and a publisher of content within these environments (Özgöbek & Gulla 2017). Present in these contexts, many people do not have the ability to evaluate sources and they accept information very easily, especially if it corresponds to preconceived ideas, people tending to believe what they would like to be true and thus they "can easily be duped." (Kshetri & Voas, 2017). Among the general population, there are few who have knowledge of economic issues (Eggers & Fouirnaies, 2014), most exhibiting a modest understanding of economic phenomena and processes.

With all the advantages of social networks, the quality of news is lower compared to traditional organizations in the field, which have been professionally active for years (Shu et al., 2017). The same is stated by Allcott and Gentzkow (2017) or Di Domenico et al. (2021) according to which individuals spread the fake news, usually through social media, where there are no too severe barriers to stop the easy propagation of some confusing, misleading, or even false ideas.

In the context of Internet use in almost all fields, blogs are becoming increasingly accessed news sources, with considerable impact on news consumers. However, this truly novel source of information does not enjoy, in many cases, the trust of the public, credibility being one of the fundamental principles of news, regardless of whether we refer to traditional or online journalism (Gunter et al., 2009).

On the other hand, certain blogs are created by research departments, universities, or various professional organizations, and in this case they are important sources for disseminating research and increasing the visibility of those organizations (Fraumann & Colavizza, 2022).

According to Prueksuralai (2014), news blogs provided by individuals were categorized as less credible compared to reports and information provided by news professionals.

The great advantage of blogs is that they are a real "round table" around which users can gather to express their opinions, they can publish their comments right below each post. Thus, the blog represents a new source of news using both the journalist and its target public, the latter being an active participant in the creation and dissemination of news (Dimitrakopoulou, 2010).

In today's society, driven by digitalization, social media has become a primary source of information as well as a fertile ground for the spread of disinformation—an aspect that leads to polarizing effects and leaves its mark on social and political perceptions and behaviors. Thus, disinformation through various platforms can lead to offline violence, serving as a "strategic weapon" for any state, making government control over the virtual space indispensable (Arayankalam & Krishnan, 2021).

On the other hand, the impact of social media also extends to how digital platforms influence behaviors, attitudes, and organizational performance: it improves customer relationships, makes certain information more accessible, and reduces costs while also contributing to the entrepreneurial orientation of various organizations in the competitive market struggle (Parveen et al., 2016).

Disinformation is influenced by a series of factors that depend either on the source or the recipient. When the characteristics of the message (source authority, consensus indicators) align with the characteristics of the user (digital literacy, personality traits), the intended outcome is achieved (Buchanan, 2020). Thus, the persuasiveness of a message and the perceived support are closely linked to changes in the recipients' attitudes (Chang et al., 2018).

Unlike the mass media of other times, nowadays, especially in social media, there is no recourse to rigorous editors who evaluate the quality and veracity of information before its publication, so that users of various platforms can post, appreciate, and share news, images, videos, and other content without being verified by a specialist. In this context, some may have as many followers as famous media organizations such as CNN or the *New York Times* (Kshetri & Voas, 2017). Although in these past years the phenomenon of fact-checkers was born, the very existence and their objectivity and good faith may sometimes be questioned.

Özgöbek and Gulla (2017) note that according to some studies, about 75% of respondents consider fake news to be, in fact, correct news. The question thus arises, what can be done so that the percentage of those who make such confusions becomes lower? (RQ2).

Due to its main characteristics—speed, aggregation, monetizing the clickstream (Hirst, 2017)—the digital environment is a fertile ground for the creation and development of fake news, an environment where creators of fake content, consumers, and various arbiters form a vicious circle by supporting each other. The great challenge is, in this context, to identify the mechanisms for breaking this circle, so that we can more easily distinguish "between facts and non-facts" (Kshetri & Voas, 2017).

In the context where content creators may be both professionals in a certain field and simple citizens, the big challenge is how we manage to distinguish an accurate item of information from a false one?

Besides the previously mentioned issues, in the fight against false information, artificial intelligence can play both the role of an ally and that of an enemy, in many situations the presence of a vigilant and well-intentioned human eye is absolutely necessary to detect the fake.

Currently, research in the AI field leads to a controversy regarding the increase or, on the contrary, the decrease of the trust of users in the information provided by means of AI. Detecting fake news through AI tools becomes essential "in the context of fostering critical media literacy and trust among social media users." (Bezzaoui et al., 2022).

Montoro-Montarroso et al. (2023) emphasize the importance of using AI-based systems in the process of identifying and effectively combating disinformation but, at the same time, the authors highlight the fact that AI is used, on the other hand, to an increasing extent large, for automatic generation of fake content.

According to Demartini et al. (2020), human-in-the-loop AI (HAI) systems represent an effective means of capitalizing on the advantages offered by AI, but based on human intelligence, somehow irreplaceable, in solving extremely complex tasks, these systems being, thus, "more powerful than pure machine-based AI methods."

In this context, finding a balance between what AI means and human intelligence is essential in order to have a properly informed society, where AI systems—capable of handling large amounts of data and time-consuming tasks—should represent only responsibly used tools.

In the economic field, the spread of fake news may lead to changes in the behavior of consumers, entrepreneurs, or investors who are faced with decisions based on incomplete, inaccurate, or completely false data (Mwangi, 2023). Depending on the information circulated in the various media, the activity of the companies may change drastically.

Among the negative effects of fake news in business, Christov (2019) lists:

- decrease of confidence in certain markets;
- increasing the number of restrictions and regulations;
- decrease in a company's sales;
- causing damage to the image of the respective company;
- financial losses and the need to supplement resources.

Mwangi (2023), on the other hand, believes that the economic consequences of fake news can be crucial, with significant impact on stock markets, business reputations, and last but not least, consumer behavior.

But what are the measures that should be taken to reduce the negative effects of the spread of fake news?

A first step to protect democracies was taken by the European Commission through the Communication on Tackling Online Disinformation, through which it proposed four objectives (Martens et al., 2018):

- increasing transparency about how information is produced or sponsored;
- promoting the diversity of information by encouraging quality journalistic activities, restoring the balance between information creators and distributors;
- promoting information credibility, media education, but also information-friendly online systems, as well as strengthening collective information monitoring capacities.

Although Gupta et al. (2022) consider that, currently, "no viable solution has yet been deployed at a scale," Mwangi, E. (2023) identifies several courses of action meant to reduce the harmful effects of fake news, such as:

- Better digital literacy, educational institutions, and governments should be concerned with developing media literacy programs where individuals are taught to evaluate various sources of information and ultimately manage to distinguish between to be and to seem.
- Increasing the responsibility of technology companies to combat the spread of fake information by investing in technologies, algorithms, and artificial intelligence systems, tools that have the ability to identify fake news.
- Good collaboration between researchers, specialists, and technology companies can lead to finding effective means of detecting false information.
- Developing a set of legislative regulations that, without prejudice to freedom of expression and transparency promotion, take concrete measures against the creators of false information.
- Developing ethical journalistic practices by media organizations through fact-checking and truthful reporting.

In order to stop fake news spreading, Aldwairi and Alwahedi (2018) propose the use of a tool designed to detect and remove web pages containing information that misleads readers.

Combating fake news, most of the time, is unfortunately not a priority for law enforcement bodies or government agencies in the field, nor even for other social arbiters, such as the press, commercial organizations, academics, or activists (Kshetri & Voas, 2017).

# 3 Methodological Framework and Research Results

## *3.1 Study Research Questions and Data Collection*

Following the centralization and presentation of certain theoretical benchmarks regarding the topic addressed, we proposed the following research questions:

RQ1: Do individuals have sufficient access to real economic information through the media?

RQ2: What influences do economic news have on individuals' behavior and what measures can be implemented to reduce their level of confusion regarding real and fake news?

RQ3: What is people's level of concern about economic news in the media and how does it influence their psychological well-being?

The tool used in the study was the questionnaire, applied through the Google forms platform. The questionnaire included a number of 22 questions, 16 being content questions, and 6 respondent identification questions. The content questions offered the respondents predefined options for their answers, or used the 5-point Likert scale for their quantification. The studied sample included a number of 102 respondents, the questionnaire administration period being December 10, 2023–January 20, 2024. The selection of respondents was made randomly, there being no preconditions for completing the questionnaire.

The number of 102 respondents was determined according to the available resources and the time required for data collection and analysis. We also aimed to obtain a sufficiently varied sample in order to get a representative database that would provide relevant insights into the topic. This number was considered adequate for an exploratory study of this nature, allowing meaningful preliminary conclusions to be drawn without requiring a very large study sample size. The process of random selection of respondents did not involve the imposition of specific eligibility criteria. This means that anyone who was available to complete the questionnaire was included in the study without preconditions. This approach aimed at ensuring a diversity of participants and allowing a degree of generalization, but limited to the specifics of exploratory research. The choice of a random selection was appropriate as we did not need to analyze a specific subgroup, but to get a general picture of perceptions among as diverse an audience as possible.

In order to carry out the study and to answer the research questions above, the following objectives were proposed, each having a series of questions associated in the questionnaire:

O1. Assessing the ability of individuals to distinguish between real news and fake news in the media.
O2. Evaluating the influence news has on the behavior of individuals in the economic field.
O3. Analyzing the quality of economic news reporting in terms of accuracy, completeness, and objectivity.
O4. Studying the relevance of economic news in the daily life of the media consumer.
O5. Determining the most frequently used sources of information.
O6. Identifying the psychological effect of economic news on the general public and analyzing the behavior of individuals regarding personal budget management.

## 3.2 The Socio-Demographic Profile of the Respondents

The socio-demographic characteristics of the respondents of the analyzed sample are shown in Table 1.

**Table 1** Demographic profile of study participants ($N = 102$)

| Criterion | $N$ | Variables | $n$ | % |
|---|---|---|---|---|
| Gender | 102 | Male | 49 | 48 |
| | | Female | 53 | 52 |
| Age | 102 | 18–25 years of age | 43 | 42.2 |
| | | 26–35 years of age | 18 | 17.6 |
| | | 36–45 years of age | 29 | 28.4 |
| | | 46–55 years of age | 9 | 8.8 |
| | | Over 55 years of age | 3 | 2.9 |
| The environment of residence | 102 | Urban | 89 | 87.3 |
| | | Rural | 13 | 12.7 |
| Last level of studies | 102 | High-school studies | 54 | 52.9 |
| | | Post-secondary studies | 8 | 7.8 |
| | | Bachelor's level studies | 29 | 28.4 |
| | | Master's level studies | 9 | 8.8 |
| | | PhD level studies | 2 | 2 |
| Monthly net income level | 102 | Under 2500 lei | 46 | 45.1 |
| | | 2500–4000 lei | 2.3 | 22.5 |
| | | Over 4000 lei | 32 | 32.4 |
| Status on the labor market | 102 | Employee | 56 | 55.4 |
| | | Unemployed | 28 | 27.7 |
| | | Entrepreneur | 16 | 15.8 |
| | | Retired | 1 | 1 |

Source: Table elaborated by the authors

Analyzing the structure of the sample, we can see that most respondents are female ($n = 53$, VR = 52%), the share of male respondents being lower (48%). As for the age category of the respondents, most (42.2%) belong to the 18–25 years old group, 17.6% belong to the 26–35 years old category, 28.4% belong to the 36–45 years old group, and the categories of people over 45 years being less represented.

A percentage of 28.4% of the respondents have bachelor's degrees, 8.8% are master's degree. Only 2% finished PhD studies, the rest being people with high school or post-high school education.

The analysis regarding the distribution of respondents according to the area of residence shows us that, of the 102 people who answered the items in the questionnaire, most come from the urban environment (VR = 87.3%), the rural environment being less represented (VR = 12.7%).

Depending on the level of monthly income, most respondents declare that they have an income of up to 2500 lei (45.1%), the percentage of those with an income over 4000 lei being 32.4%. Most of the respondents are employed (55.4%).

The verification of the level of homogeneity and representativeness of the collected responses was achieved by calculating a number of descriptive statistical indicators, namely by determining the average position indicators and the variation indicators.

# 4 Results and Discussion

## 4.1 *Homogeneity and Representativeness of Responses*

The first research question proposed in the study was the following: *Do individuals have sufficient access to real economic information through the media? (RQ1).*

In order to be able to analyze the respondents' perception on this topic, we wanted to first study what is the main source of information that the respondents rely on regarding the information/documentation process.

The overwhelming majority of respondents state that the main source of information is, in their case, the Internet, 94.1% of the interviewed subjects falling into this category, very low percentages being represented, on the other hand, by those who claim they take their information mainly from television, print media, or even radio.

The results obtained from this question confirm the hypotheses resulting from the study of the literature, according to which we are currently living in the Internet era, the online environment being the main source of creation, but also of information consumption. Thus, we are witnessing, especially among the new generations, an "obsolescence" of traditional sources of information, as they select the fastest, most innovative, and, at the same time, interactive way, which confirms, once again, that information is only "one click away." (Fig. 1).

When it comes to the preferred news category, summing up the points 4 and 5 given on the scale by the respondents, we remark that most are keen to economic information and sports information. On the other hand, the subjects participating in the study declare to be less interested in political news and, especially, in mundane news.

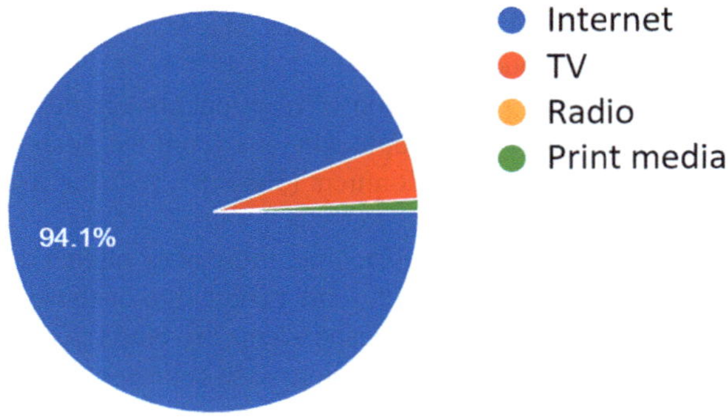


**Fig. 1** Main sources of information

People's relatively high interest in economic information may generally represent a positive signal for a society, knowing that economic and especially financial topics should be a priority for individuals throughout their lives.

In order to analyze the results obtained, for each variable proposed to be analyzed, we determined the descriptive statistical indicators considered relevant for achieving the objectives of this study. They were determined for those questions for which the respondents provided scale answers or the nominal equivalent of the respective scores.

Thus, regarding the type of news that the respondents follow (variable for which the notation TS will be used further on), in general, we find in Table 2 the distribution of responses, respectively, the associated descriptive statistical indicators.

The results obtained, according to Table 2, indicate a relatively large dispersion of the answers given by the surveyed respondents, which can be seen by calculating the dispersion and the variation coefficient, which generally show situations of relative heterogeneity, with $v$ taking values between 49.68% and 63.09% ($f$ = 102). However, the above observation is confirmed, the lowest percentage being recorded in the chapter "Economic News," in this case there is a higher level of representativeness than in the other cases.

The average of the answers ($\overline{x}$) is between 1.90 and 2.45, depending on the category of news to which we are addressing, and the higher level of dispersion may be explained by the fact that the composition of the sample itself is diverse, the respondents being selected in a random way is part of the study. Thus, regarding the type of news they follow, the socio-demographic variables play an important role in giving answers, the level of education, the level of income, even the age, respectively, the gender of the respondents influencing the preferences regarding the type of news followed in general.

The skewness asymmetry coefficient, measuring the asymmetry of the distribution of a variable in relation to its mean, takes a positive value, indicating a relatively asymmetric distribution, with a right skewness, while the degree of flattening of the distribution, measured by the Kurtosis coefficient, is presented in differently for each of the four analyzed variables.

**Table 2** Descriptive statistical indicators

| Variable | $N$ | Minimum | Maximum | Average | Standard deviation | Variation coefficient | Skewness | Kurtosis |
|---|---|---|---|---|---|---|---|---|
| TS_ politics | 102 | 1 | 5 | 2.09 | 1.4333 | 57.33% | 1.0103 | 0.0710 |
| TS_ economics | 102 | 1 | 5 | 2.45 | 1.4829 | 49.68% | 0.4629 | −0.8440 |
| TS_sports | 102 | 1 | 5 | 2.08 | 1.7193 | 63.09% | 0.9917 | −0.2787 |
| TS_ mundane | 102 | 1 | 5 | 1.90 | 1.4217 | 62.69% | 1.2135 | 0.4085 |
| *Scale* | 1—Not at all, 2—Rarely, 3—Neither rarely nor often, 4—Often, 5—Very often | | | | | | | |

Source: Table elaborated by the authors

In general, the "political news" variable has a moderate skew to the right and a near-normal flattening, indicating that most people who answered the questions in the survey follow little political news, but we also find a small number of respondents who follow to a very high extent.

The distribution for the variable "economic news" (with Sk = 0.4629) and "sports news" (with Sk = 0.9917) is flatter than a normal distribution. The majority of respondents therefore follow economic or sports news at a moderate level, but we also find a small percentage of respondents who follow these types of news at a high level. As for the variable "mundane news" (Sk = 1.2135), the skewness is moderately to strongly to the right, being sharper than a normal distribution.

The Kurtosis coefficient of −0.8440 in the case of economic news shows us that the extremes (very little or very much) are less frequent than in the case of a normal distribution, in the respective categories, of those who follow this category of information a lot, being able to include those who are either professionals in the field or carry out a certain type of activity where this information is useful to them to a high extent.

We also aimed to study the extent to which respondents believe that economic news in the media has an impact on them. The results of the calculations made following the centralization of the data are presented in Table 3 and Fig. 2.

People's rather high interest in economic information is also reflected in the way they perceive its effects on their own lives. In this respect, after processing the available information, almost half of the respondents (48%) declare that the economic news in the media does not have much impact on them. Moreover, about 34% of the

**Table 3** Descriptive statistical indicators

| Variable | *N* | Minimum | Maximum | Average | Standard deviation | Variation coefficient | Skewness | Kurtosis |
|---|---|---|---|---|---|---|---|---|
| ec_ news_ impact | 102 | 1 | 5 | 2.73 | 1.4217 | 39.85% | 0.056 | −0.2707 |
| *Scale* | 1—Not at all, 2—Rarely, 3—Neither rarely nor often, 4—Often, 5—Very often | | | | | | | |

Source: Table elaborated by the authors

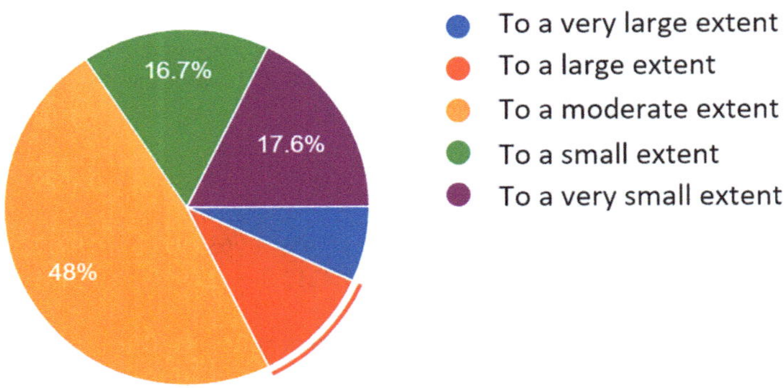


**Fig. 2** The impact of the economic news in the media on the respondents beliefs

subjects interviewed claim that economic news in the media has little or even very little impact on their economic behavior.

With regard to this variable (ec_news_impact), we find that the respondents admit mostly an influence manifested on their own behavior by the economic news in the media. In this sense, we remark an average of $\overline{x} = 2.73$ calculated with regard to this criterion, the degree of dispersion of the answers being lower, fact confirmed by the value of the variation coefficient $v = 39.85\%$, indicating relative homogeneity of the answers. Certainly, also in this case, the impact differs from one person to the next and the representative socio-demographic characteristics. Beyond the elements proposed for the characterization of the sample through the identification questions, we can also complete that list with a series of psychological or socio-cultural characteristics, some of the individuals being in real life much easier to be impressed by "bombastic news" than others, either due to a lower level of knowledge in the field or on the basis of a personal tendency to easily believe media information. With an asymmetry coefficient Skewness = 0.056, therefore very close to 0, we observe that the distribution of the impact of economic news on the respondents is almost symmetrical. So respondents are relatively evenly distributed around the mean in terms of the impact of economic news. The negative value of the Kurtosis coefficient shows a slightly flat distribution of the impact of economic news, with extremes being less frequent than in a normal distribution and most respondents experiencing a moderate, similar, even impact. Therefore, most respondents indicate that economic news is perceived and integrated in a similar way.

As a consequence of this aspect, the extent to which economic news influences the behavior of the respondents may be analyzed through the effects on the personal way of managing one's own budget. From this perspective, the distribution of responses is relatively balanced, with $v = 43.42\%$, the Skewness level being 0.0661, while the calculated value for Kurtosis is −0.7614.

Asked to what extent they believe have sufficient access to actual economic information through the media, more than 40% of subjects believe that, to a high extent, and even to a very high extent, the media provide them with accurate economic information. Only 5.9% of the respondents register a very high level of skepticism, considering that the economic information in the mass media is to a very low extent real. It should be noted that a little more than a third of the subjects (34.3%) believe that they only to some extent have sufficient access to real economic information through the mass media.

As for the extent to which the respondents consider they have sufficient access to real economic information through the media, we find a higher level of representativeness of the answers, expressed through the coefficient of variation $v = 34.46\%$, along with an average $\overline{x} = 3.21$. Most of the scores awarded were therefore above the simple, standard mean of 3, the sample being representative from this point of view (Table 4).

A coefficient of asymmetry equal to −0.0444 shows that the distribution regarding access to real economic information is almost symmetric, the value being very close to zero, with a slight asymmetry to the left. Therefore, respondents who have more or less access to real economic information are relatively evenly distributed

**Table 4** Descriptive statistical indicators

| Variable | *N* | Minimum | Maximum | Average | Standard deviation | Variation coefficient | Skewness | Kurtosis |
|---|---|---|---|---|---|---|---|---|
| real_ec_info_access | 102 | 1 | 5 | 3.21 | 1.1795 | 34.46% | −0.0444 | −0.6860 |
| *Scale* | 1—Not at all, 2—Rarely, 3—Neither rarely nor often, 4—Often, 5—Very often | | | | | | | |

Source: Table elaborated by the authors

around the mean. According to the flattening coefficient, the extremes are less frequent than in a normal distribution, the access distribution being a generally fair one.

However, it is not enough to know that respondents have access to the necessary information. We are still interested to see which sources are used by them to a greater extent for documentation and which of them are recognized for the veracity of the information transmitted.

In the top of Romanian respondents' preferences for the online sources of information, they use to access economic news, the best ranked are Ziarul Financiar (www.zf.ro) and Economica.net with 34.3% and 21.6% of the responses, respectively. On the other hand, the least accessed online sources, in this sense, are *Ceccar Business Magazine* (www.ceccarbusinessmagazine.ro), www.profit.ro, the *Bursa* newspaper (www.bursa.ro), which proves that the respondents are less interested in accounting subjects and, especially, in the stock market, these being to a greater extent dedicated to specialists in the respective fields.

It should be noted that, in addition to Ziarul Financiar (www.zf.ro) and Economica.net, online sources quite accessed by Romanians for reading economic news are also *The Economist* and *Business Magazine*: these sources registering 15.7% and 10.8% respectively of the study participants' preferences.

Out of the desire to outline the most comprehensive framework of the research, the next research question was: *What influences do economic news have on individuals' behavior and what measures can be implemented to reduce their level of confusion regarding real and fake news? (RQ2).*

Although there are also individuals who state that economic news influences them in the process of managing their personal budget to a high extent (16.7%) and even to a very high extent (7.8%), most respondents (37.3%) declare that they are less influenced by this news when it comes to how they manage their financial resources (Fig. 3).

On another note, attention is drawn to the high percentages of those who claim that, in the process of managing their personal budget, economic news influences them to a low extent (17.6%) and even to a very low extent (20.6%).

This can be explained either by the lack of trust in the veracity of certain sources and economic information, or, in other cases, by the relatively poor economic education, which makes them decide not according to the evolution of the markets, but rather according to their habits.

The review of the literature shows that many people do not have the ability or the necessary tools to distinguish between a piece of actual news and one of fake news, especially if it is an information from the economic field.

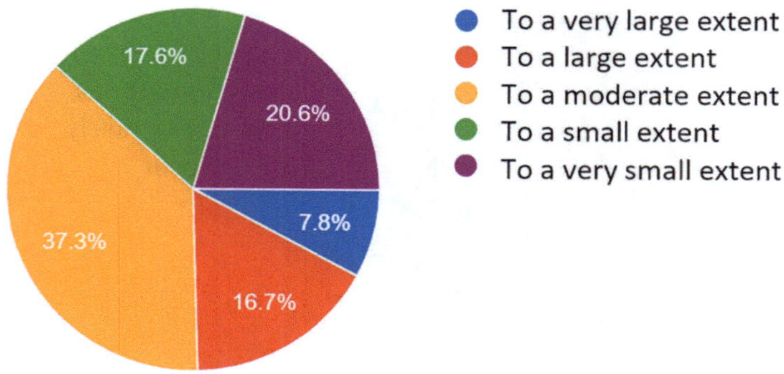


**Fig. 3** The influence of economic news in the process of managing personal budget

In this context, it is very interesting that people generally have a lot of confidence in their ability to distinguish reality from fiction, almost 60% of respondents stating that, in their case, distinguishing between real news and fake news is quite likely. Moreover, 21.6% of subjects are absolutely sure of their ability to differentiate reality from fake news. On the other hand, there are very few who state that they cannot or do not know how to differentiate a real news from a fake news.

The confidence, sometimes too high, in one's own possibilities to differentiate the truth from the lie, correlated with the ability of the creators of fake news to make their message believable, the ease of propagation of these messages, all this means that, as it also appears in the literature, many people are easily duped.

As regards the same aspect of access to truthful economic news, it is interesting to study how often respondents actually check the nature of an economic news in an attempt to determine whether it is true or false, or even if respondents are able to differentiate a true news of one that is not real.

When it comes to how often people check the probability that an economic news story is true or false, the answers are quite mixed. Slightly more than a third of the study participants (36.3%) state that they are concerned about this verification process only in some cases/sometimes. It should be noted that there are quite a large number of those who claim that they check the veracity of economic news quite often (23.5%). Also, a considerable percentage (15.7%) is represented by people who say they always check whether a piece of economic news is true or false (Fig. 4).

Of course, those who admit they check the correctness of economic news quite rarely (17.6%) or even never (6.9%) are not few.

Even if the percentages of those who claim they usually check whether an economic news item is true or not, the question arises, beyond the sincerity of their answers, whether the respondents have the ability to correctly check this item of information, whether they have the power to identify the sources of credible information, which can be a benchmark in this respect.

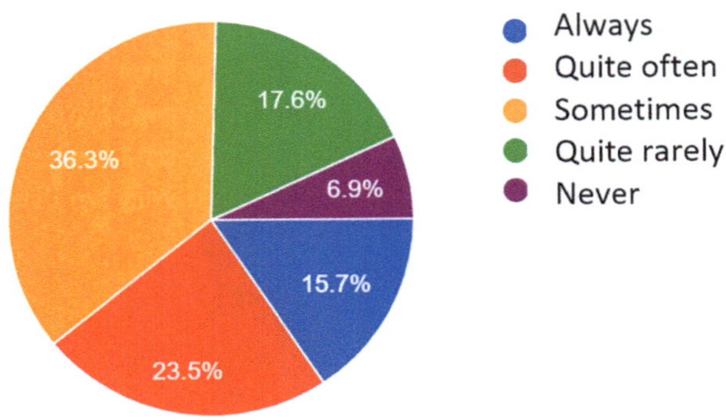


**Fig. 4** How often people check the probability that an economic news story is true or false

**Table 5** Descriptive statistical indicators

| Variable | *N* | Minimum | Maximum | Average | Standard deviation | Variation coefficient | Skewness | Kurtosis |
|---|---|---|---|---|---|---|---|---|
| Real vs. fake | 102 | 1 | 5 | 3.96 | 0.5866 | 19.34% | −0.8625 | 1.7375 |
| News check | 102 | 1 | 5 | 3.21 | 1.1795 | 43.42% | −0.0444 | −0.6860 |
| Multiple sources | 102 | 1 | 5 | 4.16 | 1191 | 26.25% | −1.2827 | 0.9691 |
| *Scale* | 1—Not at all, 2—Rarely, 3—Neither rarely nor often, 4—Often, 5—Very often | | | | | | | |

Source: Table elaborated by the authors

The results of the calculations show us that the answers given by the respondents have a relatively low level of dispersion, and we may speak of the homogeneity and the representativeness of the answers in terms of the ability of the respondents to authenticate a piece of news. The average of 3.96, located toward the upper limit of the scale, shows the fact that the individuals who answered the questions consider they have the ability to differentiate a true item of news from one of the opposite character. In this respect, some of them also verify the reality of news of an economic nature, multiple verification being considered useful to a great extent. In this case we calculated an average $x = 4.16$, therefore most of the answers are positioned very close to the maximum limit of the interval, the sample being in this case also representative, with a calculated $v$ of 26.27% (Table 5).

Slightly more than half of the interviewed subjects (52%) are aware of the fact that it is very important that an economic information is verified from several sources. Also, high percentages are registered by those who consider that the verification has a rather high importance (24.5%) or, at least, average (14.7%).

Correlating the results of this question with those of the previous one, we notice that the percentage of those who are aware of the importance of checking information in several sources is higher than that of people who actually do this check.

Interestingly, most of the participants in this study claim that the economic news broadcast in the mass media has caused them very little or not at all to change their

**Table 6** Descriptive statistical indicators

| Variable | $N$ | Minimum | Maximum | Average | Standard deviation | Variation coefficient | Skewness | Kurtosis |
|---|---|---|---|---|---|---|---|---|
| Investments | 102 | 1 | 5 | 1.90 | 1.0492 | 53.86% | 0.9219 | 0.1361 |
| Purchase of multiple products | 102 | 1 | 5 | 2.35 | 1.4832 | 51.76% | 0.6132 | −0.5618 |
| Purchase of a smaller volume of products | 102 | 1 | 5 | 2.40 | 1.4756 | 50.57% | 0.4821 | −0.7696 |
| Credit contracting | 102 | 1 | 5 | 1.56 | 0.8347 | 58.61% | 1.7802 | 2.9775 |
| *Scale* | 1—Not at all, 2—Rarely, 3—Neither rarely nor often, 4—Often, 5—Very often | | | | | | | |

Source: Table elaborated by the authors

buying and consumption behavior in the sense of purchasing more or fewer products and, moreover, this information influenced them very little or not at all in the process of contracting a loan or investing a certain amount of money.

The explanation for these results may come from the fact that people are generally more receptive to information and advice from their consultants, family, reference groups, opinion leaders, or friends with whom they interact directly than to information that they are presented, often in an impersonal way, in the mass media.

With reference to the previously mentioned behavioral changes, we analyzed to what extent the economic news broadcast in the mass media determined the orientation toward certain specific actions. The calculated values for the descriptive statistical indicators are presented in Table 6.

The distribution of responses is therefore, in the case of all four variants, heterogeneous, with a mean closer to the lower limit of the range and a coefficient of variation exceeding 50%, indicating a greater dispersion of responses.

With regard to how the followed economic news stimulated the making of investments by the respondents, we find that the Skewness value indicates a right-wing asymmetry, meaning that most respondents make fewer investments, generally in small amounts, but we still find in the sample a small number of people making large investments. Also, most respondents purchase fewer products, but the extremes (very large or very small purchases) are less frequent.

As for the contracting of credits, we observe a large positive value in the case of the asymmetry coefficient, indicating that most respondents contract credits in small amounts, only a small number of respondents contracting large amounts.

From the results of this research, it follows that whether it is about choosing a job, information about the saving process, taking out a loan, how people shop, and even investing in a pension fund, the means of information with the greatest impact on consumers of economic news is, currently, the Internet, with this opinion being the overwhelming majority of respondents.

On the other hand, from the results obtained for this question, we understand that, regarding the impact of information sources on individuals, a much smaller role is enjoyed nowadays by television, radio, and, above all, the written press. These results are not at all surprising, the idea being found in most specialized works in this field.

As we have already pointed out, the unprecedented development of the Internet, its widespread use both for the creation and propagation of various information, as well as for its consumption, has made the online environment a fertile ground for the development of fake news, the verification of the veracity of information being much more difficult to carry out online compared to traditional media.

A better education in this regard, from the early grades, could help the population to distinguish more easily between truth and falsehood, which would later impact even the lifestyle, respectively, the quality of life in general.

We thus analyzed, in relation to the third research question, *What is people's level of concern about economic news in the media and how does it influence their psychological well-being? (Q3).*

When asked what is, in general, the impact of economic news in the media on them, the vast majority of the study participants stated that, regardless of whether we are talking about the print media, radio, and even television, the impact of the news delivered is a negative one. On the other hand, opinions are much more divided when we refer to the online environment, with the interviewed subjects answering almost in equal proportions that the news broadcast in this environment has both a negative and a positive impact.

The answers to this question are somehow further confirmation that, in general, news consumers are much more present online than in traditional media.

Regarding the sentiment generated by various categories of news found in the media, we note the following:

When it comes to the economic news regarding pensions, most of the respondents rather have feelings of indifference, this can also be explained by the fact that, in the structure of the sample, most are young, 42.2% representing people between 18 and 25 years old, and 17.6% are people aged between 26 and 35 years old, the prospect of pension being quite distant for them.

It is surprising that the participants in this study were, in general, quite indifferent to the news from the business environment or those related to the way companies are taxed, a fact explained by the fact that, in the structure of the sample, most subjects they have the status of employee.

As for the news regarding the general economic evolution of the country, the respondents are much more concerned and interested in this information, and on the other hand, there are not a few who experience a feeling of concern when they follow the news on this topic.

Also, the results show that people are somewhat more interested and concerned with the news regarding the limitation of the use of cash, the information related to the evolution of the exchange rate, the evolution of utility tariffs, or, even more so, the news related to the increase in food prices. We understand from all these results that people are much more interested in those economic aspects that directly affect their lives.

In general, the average score recorded is above the average score of 3, the values calculated for the coefficient of variation indicating relative homogeneity of the answers ($v$ being between 34.87% and 43.46%). We therefore find a somewhat significant level of responses, the respondents' perception of the topics addressed largely depending on the personal and professional context identified. Thus, a person who is already at the stage of retirement will certainly be much more interested in this topic than young graduates who are at the beginning of their career. On the other hand, the subject of limiting cash will be of less interest to someone who rarely uses this type of payment, compared to other individuals who strongly prefer it. I assigned a score on a 5-point scale to each answer. Thus, the feeling of "fear" was assigned a score of 1, that of "stress" a score of 2, "indifference" was scored with a score of 3, "concern" with a score of 4, and "interest" with a score of 5. Analyzing the averages in Table 7, we find that the feeling generated by these themes varies from indifference to concern, with the highest average score recorded in this regard by the theme regarding the evolution of the exchange rate.

Although it was shown that the online environment is the most suitable place for the creation and dissemination of fake news, most of the respondents (53.9%) consider the Internet to be the most credible means of information, the least credible being in the opinion of the surveyed subjects, print media (13.7%) and, above all, radio (9.8%).

**Table 7** Descriptive statistical indicators

| Variable | *N* | Minimum | Maximum | Average | Standard deviation | Variation coefficient | Skewness | Kurtosis |
|---|---|---|---|---|---|---|---|---|
| The issue of pensions | 102 | 1 | 5 | 3.12 | 1.1822 | 34.87% | −0.377 | 0.1143 |
| Country's general economic development | 102 | 1 | 5 | 3.43 | 1.8531 | 39.67% | −0.5869 | −0.7567 |
| General business environment news | 102 | 1 | 5 | 3.39 | 1.5521 | 36.73% | −0.5063 | −0.4232 |
| Taxation of companies | 102 | 1 | 5 | 3.15 | 1.6548 | 40.87% | −0.2514 | −0.8314 |
| Cash limit | 102 | 1 | 5 | 3.29 | 1.9134 | 41.99% | −0.4316 | −1.0400 |
| Exchange rate evolution | 102 | 1 | 5 | 3.57 | 1.7355 | 36.92% | −0.7256 | −0.4809 |
| Payment of utility bills | 102 | 1 | 5 | 3.28 | 1.7329 | 40.08% | −0.3581 | −0.9093 |
| Food price increase | 102 | 1 | 5 | 3.22 | 1.9535 | 43.46% | −0.2403 | −1.2741 |
| Accessing European funds | 102 | 1 | 5 | 3.15 | 1.4979 | 38.89% | −0.4164 | −0.5006 |
| *Scale* | 1—Not at all, 2—Rarely, 3—Neither rarely nor often, 4—Often, 5—Very often | | | | | | | |

Source: Table elaborated by the authors

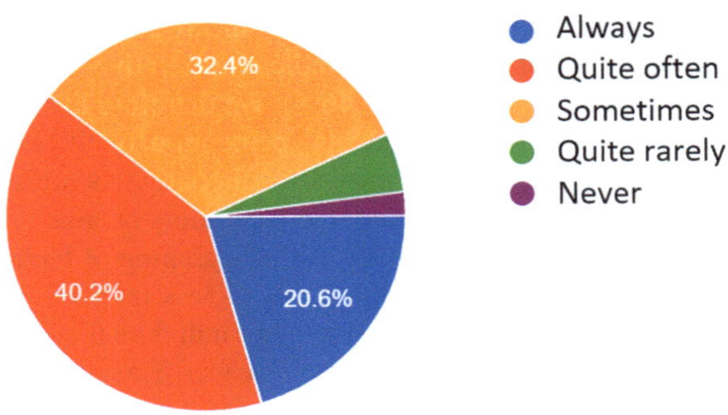


**Fig. 5** Impulsive acting upon learning negative news

Most of the participants in the study (40.2%) state that they very rarely act impulsively upon learning negative news, and 20.6% claim that they never resort to such behavior.

However, around one-third of the respondents (32.4%) state that they sometimes act impulsively upon learning negative news, being quite a few who declare that they often resort to such behavior. Correlating the results obtained with this question, we understand that people are generally more reflexive when learning negative news and do not make immediate decisions (Fig. 5).

When it comes to the quality of economic journalism in Romania, opinions are quite divided. About half of the people surveyed believe that economic journalism in Romania is of average quality, 18.6% perceive a low quality and 16.7% see an extremely low quality in this field. Of course, there are also categories of people, somewhat narrower, who talk about a high and even very high quality of economic journalism in Romania.

## 5 Conclusion

Fake news is not something new, it appeared along with information, it is as old as newspapers or other sources of information, and it spreads quickly, through various media that favor communication.

In the age of the Internet, the speed at which confusing information can be generated and spread has greatly increased. Some create and promote fake information out of a desire to earn more by scaling the number of hits of this "sensational news," others want to gain notoriety, especially if it is in the mundane area or especially in the political area. Last but not least, fake news can lead to substantial changes in the stock market that can translate into considerable economic gains for the creators of that news.

The spread of fake news has far-reaching implications, contributing to the erosion of trust in the media and the erosion of democratic processes. At the same time, it promotes conspiracy theories, manipulates public opinion, leading to polarization and discord in society (Mwangi, 2023).

In general, people cannot or hardly distinguish between fact and fiction, and sometimes they tend to consider information they disagree with as fake news.

There are situations where a certain article contains only a few elements that are not correct, real, and even if the whole news is not fake, it may lead to deception. What's worse is that these fake elements are harder to detect than when we are faced with an entire fake news article.

Whether it is social, political, economic, religious, or otherwise fake news is now more difficult to detect and combat than it is to create and propagate. Identifying and mitigating the effects of fake news is a big challenge due to several human, technological, economic, and, last but not least, legislative factors.

Good collaboration between governmental authorities, legislative authorities, specialists from technology companies, researchers in the field and other stakeholders can lead, if not to stopping the phenomenon of creating and propagating fake news, at least to reducing it, so that the harmful effects of fake news are mitigated. The focus on education, the development of critical thinking of individuals, increasing their ability to analyze and evaluate information, all these can constitute premises of social cohesion and even political and economic stability (Mwangi, 2023).

We therefore observe the high importance of discernment regarding the consumption of especially economic news, which is both numerous and complex.

In the digital age, economic news is ubiquitous, influencing individuals' financial decisions and general perceptions of the state of the economy. However, the complexity of this information, the diversity of media sources, and the direct impact on everyday life pose significant challenges in discerning valuable, useful, and real information.

Economic information is often presented in technical language, full of jargon, and statistical data that risk to discourage people without much knowledge in the field, which can lead to misinterpretations, creating confusion and uncertainty. To address this issue, promoting financial education at all levels, from schools to adult training programs, is essential. By understanding basic economic concepts, individuals can better interpret information and make informed decisions.

In the multitude of media sources, each of them can present its own perspectives. Newspapers, television, online news sites, and social media often provide conflicting information, which can make it difficult to identify reliable sources. This diversity can amplify uncertainty and create a distorted perception of economic reality, in which context it becomes very important for the public to learn to verify the source of information and familiarize themselves with its truthfulness and objectivity.

In this context, the role of the media is amplified, often shaping public economic perceptions. Journalists in the field have a responsibility to provide clear, well-documented, and unbiased information, helping to build trust and educate the public by providing the necessary tools for a correct interpretation of information.

## References


Aldwairi, M., & Alwahedi, A. (2018). Detecting fake news in social media networks. *Procedia Computer Science, 141*, 215–222.

Allcott, H., & Gentzkow, M. (2017). Social media and fake news in the 2016 election. *Journal of Economic Perspectives, 31*(2), 211–236.

Almeida, F., & Mortari, V. (2021). Smartphones, social networks, and fake news: Institutional economics approach to decision making in the twenty-first century. *Journal of Economic Issues, 55*(2), 342–348.

Anderson, C. (2003). *Eyes off the prize: The United Nations and the African American struggle for human rights, 1944-1955*. Cambridge University Press.

Arayankalam, J., & Krishnan, S. (2021). Relating foreign disinformation through social media, domestic online media fractionalization, government's control over cyberspace, and social media-induced offline violence: Insights from the agenda-building theoretical perspective. *Technological Forecasting and Social Change, 166*, 120661. https://doi.org/10.1016/j.techfore.2021.120661

Arencibia, M. G., Corozo, E. H. V., & Cardero, D. M. (2023). Fake news: An analysis from political economy. *Economics, 12*(2), 72–82.

Arnold-Forster, T. (2023). Walter Lippmann and public opinion. *American Journalism, 40*(1), 51–79. https://doi.org/10.1080/08821127.2022.2161665

Bakir, V., & McStay, A. (2018). Fake news and the economy of emotions: Problems, causes, solutions. *Digital Journalism, 6*(2), 154–175.

Baltezarevic, R. (2024). Global economic effects of fake news. In *10th International Black Sea Coastline Countries Scientific Research Conference*, pp. 460–466.

Bezzaoui, I., Fegert, J. & Weinhardt, C. (2022). Distinguishing between truth and fake using explainable AI to understand and combat online disinformation. In ICDS 2022: *The sixteenth international conference on digital society*.

Boomgaarden, H., Van Spanje, J., Vliegenthart, R., & De Vreese, C. (2011). Covering the crisis: Media coverage of the economic crisis and citizens' economic expectations. *Acta Politica, 46*, 353–379. https://doi.org/10.1057/AP.2011.18

Boukes, M., Damstra, A., & Vliegenthart, R. (2021). Media effects across time and subject: How news coverage affects two out of four attributes of consumer confidence. *Communication Research, 48*(3), 454–476.

Buchanan, T. (2020). Why do people spread false information online? The effects of message and viewer characteristics on self-reported likelihood of sharing social media disinformation. *PLoS One, 15*. https://doi.org/10.1371/journal.pone.0239666

Chang, J.-H., Zhu, Y.-Q., Wang, S.-H., & Li, Y.-J. (2018). Would you change your mind? An empirical study of social impact theory on Facebook. *Telematics and Informatics, Elsevier, 35*(1), 282–292. https://doi.org/10.1016/j.tele.2017.11.009

Christov, A. (2019). Economy of the fake news: Business side and effects. *Eastern Academic Journal, 4*, 1–7.

Corbu, N., Bârgăoanu, A., Durach, F., & Udrea, G. (2021). Fake news going viral: The mediating effect of negative emotions. *Media Literacy and Academic Research, 4*(2), 58–87.

Damstra, A., & Boukes, M. (2021). The economy, the news, and the public: A longitudinal study of the impact of economic news on economic evaluations and expectations. *Communication Research, 48*(1), 26–50.

Demartini, G., Mizzaro, S., & Spina, D. (2020). Human-in-the-loop artificial intelligence for fighting online misinformation: Challenges and opportunities. *IEEE Data Engineering Bulletin, 43*, 65–74.

Dentith, M. R. X. (2017). The problem of fake news. *Public Reason, 8*(1–2), 65–79.

Dentith, M. R. X. (2018). What is fake news? *University of Bucharest Review, Truth(s) and Alternative Facts, VIII*(1) (new series).

Di Domenico, G., Sit, J., Ishizaka, A., & Nunan, D. (2021). Fake news, social media and marketing: A systematic review. *Journal of Business Research, 124*, 329–341.

Dimitrakopoulou, D. (2010). *Journalists on the blog: The online discourse of news professionals.*
Eggers, A., & Fouirnaies, A. (2014, July). *The economic impact of economic news.*
Fârte, G.-I., & Obadă, D.-R. (2018). Reactive public relations strategies for managing fake news in the online environment. *Postmodern Openings, 9*(2), 26–44.
Fârte, G. I., & Obadă, D. R. (2021). The effects of fake news on consumers brand trust: An exploratory study in the food security context. *Romanian Journal of Communication and Public Relations, 3*(54), 47–61.
Fraumann, G., & Colavizza, G. (2022). The role of blogs and news sites in science communication during the COVID-19 pandemic. *Frontiers in Research Metrics and Analytics, 7*, 824538.
Gunter, B., Campbell, V., Touri, M., & Gibson, R. (2009, March). Blogs, news and credibility. *Aslib Proceedings, 61*(2), 185–204.
Gupta, A., Kumar, N., Prabhat, P., Gupta, R., Tanwar, S., Sharma, G., et al. (2022). Combating fake news: Stakeholder interventions and potential solutions. *IEEE Access, 10*, 78268–78289.
Hirst, M. (2017). Towards a political economy of fake news. *The Political Economy of Communication, 5*(2).
https://www.bbc.com/news/world-us-canada-21508660
Jaster, R., & Lanius, D. (2018). What is fake news? *Versus, 2*(127), 207–227.
Kelman, H. C. (1961). Processes of opinion change. *Public Opinion Quarterly, 25*, 58.
Konstankevych, I., Kostusiak, N., Shulska, N., Stanislav, O., Yelova, T. & Kauza, I. (n.d.). *Journal of Interdisciplinary Research,* 224–230.
Kshetri, N., & Voas, J. (2017). The economics of "fake news". *IT Professional, 19*(6), 8–12.
Lazer, D. M. J., et al. (2018). The science of fake news. *Science, 359*, 1094–1096. https://doi.org/10.1126/science.aao2998
Lewis, M. (2010). *The big short: Inside the doomsday machine*. Editura W. W. Norton & Company.
Martens, B., Aguiar, L., Gomez-Herrera, E., & Müller-Langer, F. (2018). *The digital transformation of news media and the rise of disinformation and fake news.* JRC Digital Economy Working Paper, No. 2018-02, European Commission, Joint Research Centre (JRC).
McQuail, D. (2010). *McQuail's mass communication theory*. Sage.
Montoro-Montarroso, A., Cantón-Correa, J., Rosso, P., Chulvi, B., Panizo-Lledot, Á., Huertas-Tato, J., Calvo-Figueras, B., Rementeria, M. J., & Gómez-Romero, J. (2023). Fighting disinformation with artificial intelligence: Fundamentals, advances and challenges. *Profesional de la Información, 32*(3), e320322. https://doi.org/10.3145/epi.2023.may.22
Mwangi, E. (2023). *Technology and fake news: Shaping social, political, and economic perspectives.* Authorea Preprints.
Özgöbek, Ö., & Gulla, J. A. (2017, November). Towards an understanding of fake news. In *CEUR workshop proceedings* (Vol. 2041, pp. 35–42).
Parveen, F., Jaafar, N., & Ainin, S. (2016). Social media's impact on organizational performance and entrepreneurial orientation in organizations. *Management Decision, 54*, 2208–2234. https://doi.org/10.1108/MD-08-2015-0336
Prueksuralai, N. (2014). *The role of blogs as news sources: A study of audiences and news professionals in Thailand* (Doctoral dissertation, University of Leicester).
Shu, K., Sliva, A., Wang, S., Tang, J., & Liu, H. (2017). Fake news detection on social media: A data mining perspective. *ACM SIGKDD explorations newsletter, 19*(1), 22–36.
Sterling H. C. (2002). *"The Fireside Chats"—President Franklin D. Roosevelt (1933-1944).* National Registry. https://www.loc.gov/static/programs/national-recording-preservation-board/documents/FiresideChats.pdf
Stiglitz, J. E. (2002). Information and the change in the paradigm in economics. *American Economic Review, 92*(3), 460–501.
Tandoc, E. C. (2019). The facts of fake news: A research review. *Sociology Compass*. https://doi.org/10.1111/soc4.12724
Tandoc, E. C., Lim, Z. W., & Ling, R. (2017). Defining "fake news". A typology of scholarly definitions. *Digital Journalism*. https://www.tandfonline.com/doi/full/10.1080/21670811.2017.1360143?scroll=top&needAccess=true

Vreese, C., Boomgaarden, H., & Semetko, H. (2011). (In)direct framing effects: The effects of news media framing on public support for Turkish membership in the European Union. *Communication Research, 38*, 179–205. https://doi.org/10.1177/0093650210384934
Wardle, C. (2017). *Fake news. It's complicated.* https://medium.com/1st-draft/fake-newsits-complicated-d0f773766c79
ZF, (2015). https://www.zf.ro/zf-25-de-ani/zf-25-de-ani-evenimente-internationale/zf-25-de-ani-2015-dieselgate-scandalul-care-era-sa-distruga-vw-si-22074282

**Suzana Demyen** Permanent Tenure Lecturer within the within the Department of Business Administration of Reşiţa, Faculty of Economics and Business Administration, Babeş-Bolyai University of Cluj – Napoca. The interest in the field of training is directed onto the management of human resources. The research is focused on the following fields: studies on the behaviour of human resources within organisations, analysis of human resources performance, studies of the human value of the enterprise and of the intellectual capital, general management of the organisation.

**Adrian-Costinel Tănase** Permanent Tenure Associate Professor within the Department of Business Administration of Reşiţa, Faculty of Economics and Business Administration, Babeş-Bolyai University of Cluj – Napoca. The focus in the domain of training is directed onto the following fields of interest: analysis of explanatory variables of consumer behaviour, analysis of the entrepreneurial environment, analysis of the determinants of economic on the at micro and macroeconomic level.

**Florenţa-Diana Tănase** Permanent Tenure Associate Professor within the Department of Administration and Public Management, Faculty of Political, Administration and Communication Sciences, Babeş-Bolyai University of Cluj – Napoca. Subjects taught: Economy, Basics of Quality in the Public Sector, Macroeconomic Policies, Strategies of Community and Regional Development, Models of Community Intervention, European Business Environment, Management of Change. The focus in the domain of training is directed onto the following fields of interest: analysis of the entrepreneurial environment, analysis of the determinants of economic competitiveness on the micro and macroeconomic level.

# Holiday Experiences and Media Consumption

**Olga Amarie**

**Abstract** Families, friends, and communities worldwide celebrate shared values during holidays. Strengthening their connections through holidays fosters unity, happiness, gratitude, and a sense of belonging. Many holidays throughout the year provide opportunities to express appreciation, gratitude, and, most importantly, love. This chapter explores different types of holidays and how active participation in holidays and cultural events contributes to happiness, positivity, self-esteem, and improved mental health. The fascinating research on holidays across various cultures is often overlooked, yet it holds much information and benefits. This chapter reflects on this lesser-explored field and contributes valuable insights to the discussion. Unfortunately, from media consumption, we noticed that there are very few public holidays worldwide! Despite their presence in the digital world, many cultural celebrations are unknown and do not receive meaningful attention to make a significant difference in people's lives. In many cases, it is imperative to take appropriate and strategic action to increase the number of public holidays, combat social and digital isolation, and support people with their emotional and social well-being. We also explore how the Internet impacts the way people celebrate holidays. Are our connections as strong and meaningful as they were before the Internet? Celebrating holidays can be challenging, but reflecting on why holidays matter and how much space and time they occupy in the media is essential.

**Keywords** Public holidays · Lesser-known holidays · Eco-friendly holidays · Commercialized holidays · Types of holidays · Online celebrations · Happiness · Digital stress · Holiday stress · Media consumption · Internet usage · Holiday traditions · New Year · Christmas · Paştele Blajinilor · Famadihana · Grand Magal of Touba · National Acadian Day · Juneteenth · Asian holidays · Nicolas Anciaux · Olympics · Celebrating with the dead · Shopping during holidays · Preparing for

---

O. Amarie (✉)
Department of World Languages and Cultures, Georgia Southern University, Statesboro, GA, USA
e-mail: oamarie@georgiasouthern.edu



M. Todorovic et al. (eds.), *Sustainability and Social Responsibility of the Media and in the Media*, CSR, Sustainability, Ethics & Governance,
https://doi.org/10.1007/978-3-032-00086-6_10

holidays · Anticipation of holidays · Holiday-themed course · Teaching and assessing · Survey questions · Social and digital isolation

## 1 Introduction

No matter where people live in the world, they share universal traits, and one of them is celebrating holidays. Hamilton Wright Mabie's quote about the holiday spirit, from his broader body of work and public lectures, captures this universal trait: "Blessed is the season which engages the whole world in a conspiracy of love" (Wright, 1900). Holidays bring love into people's hearts, a universal experience that symbolizes perfect unity shaped by individual and societal factors. In their various forms, holidays act as major facilitators of happiness, feelings of unity and belonging, cross-cultural awareness, and positive convictions. Interestingly, we do not have enough public holidays, and the Internet has impacted the way people understand and celebrate holidays. Furthermore, studies suggest that holidays possess significant biological implications, including enhanced cognitive flexibility, stress management, and overall health: "Although this may seem unbelievable, there is very little scientific literature that explores the direct benefits of holidays on our brain" (Pérez Fernández & de la Torre Martínez, 2023).

Using a mixed methodology approach, quantitative and qualitative analysis, this chapter delves into the influence of holidays on people's health and social well-being and how the Internet has shaped holiday celebrations. From general to specific patterns, the chapter examines a range of famous and lesser-known holidays to give more visibility and meaning to cultural celebrations. The chapter also provides valuable original data from a holiday-themed course, serving as an inspiring resource for other educators. The tools and techniques used were class observations, as well as oral and written surveys, where confidentiality and anonymity are preserved. Planning and anticipation are the key driving forces to picture the exciting things that are to come during the holidays. So, in digital time, with all the stress, anticipation, and expectations during the holidays, can we use the Internet strategically to add more public holidays and stimulate our brains to trigger happier thoughts? The chapter examines the time and effort it takes for cultural celebrations to become public holidays, and the perspective of holiday commercialization, a difficult or pleasant experience, in the technology-enhanced era. The main idea presented in this chapter is that it is vital to maintain a harmony between digital and in-person festivities and leverage modern technology to acknowledge additional public holidays, thereby improving our emotional and social welfare.

With conflicts boiling around the world and an abundance of negative news in the media, holidays serve as a reminder of unity, optimism, love, and worldwide peace. Holidays combat social and digital isolation through solidarity. David Émile Durkheim, a prolific French sociologist, gives us many insights into social rituals, social cohesion, and collective consciousness, helping us interpret holidays. A group of people acts differently compared to its isolated members and a crisis can emerge when isolation trumps solidarity and shared common values (Durkheim, 1960). Holidays help people from different cultures connect and learn about each other. They show us the facts and beliefs of others, helping us see our shared humanity, even if we live far apart or speak different languages (Forgeard, 2023). New Year's Eve is a prime example of a public holiday in many countries, often celebrated with unique cultural traditions and customs. It is the most significant public holiday observed by so many people at once, with government offices, public schools, and many businesses closed.

During the holiday season, people spend much time on the Internet engaging in various repetitive activities: finding events to celebrate, attending live events, communicating with loved ones through messaging apps or video calls, posting photos, broadcasting celebrations, streaming holiday-specific videos (such as Christmas songs or Easter films), sending greetings to family and friends, online shopping for a variety of gifts, booking travel, and accessing many other online services. On New Year's Eve, people from all over the world wish each other "Happy New Year!" in many languages (Faculty, 2024):

Each person sends between 0 and X number of New Year's wishes annually. If we multiply this number by the world's population, there is infinite communication and higher Internet usage, according to Sandvine, a company that tracks web traffic. Web traffic tends to peak significantly during Christmas (Shoemaker, 2014) and New Year's Eve and slows down after January 1 (Sandvine, 2018). What about Happy Birthday wishes? In today's interconnected world, the Internet is flooded with an abundance of daily Happy Birthday wishes in a multitude of languages, catering to individuals from all corners of the globe. The combination of these online activities across nations leads to a noticeable surge in Internet usage. Before the advent of the Internet, from 1970 until 1982, the renowned French writer and filmmaker Georges Perec (1936–1982) shared short prose pieces with his friends and acquaintances to convey New Year's greetings. These pamphlets, titled *Vœux* (*Wishes*), were filled with playful wordings and featured proverbs, common phrases, humor, names of musicians, filmmakers, novelists, book titles, and anecdotes. Each short text was accompanied by a homophonic variation in alphabetical order, challenging to translate into another language. For instance: « 1. Devant les vitrines de Noël d'un grand magasin, un bambin manifeste son scepticisme ou son dégoût pour la plupart des jouets, et montre que ce qu'il a l'intention de se faire offrir, c'est une magnifique bicyclette dont la forme rappelle curieusement celle des premiers véhicules de ce genre. 1. Bah! Beh! Bi beau: but »; « 2. Dans un salon, des dames papotent au sujet de l'adultère 2. Caquet: qui cocu? » (Perec, 1989).

Perec's recognition of the importance of predicting Internet traffic during holidays serves as a prime example of his forward-thinking approach to understanding and managing online communication. Furthermore, his keen observation of people's casual conversations on various topics demonstrates his insightful understanding of human interaction in the digital age.

In addition to Christmas and New Year, are there other events when the world comes together through media and the Internet to celebrate? Absolutely, the Olympic Games. The Olympic Games Opening Ceremony is an internationally acclaimed event that could be regarded as a public holiday worldwide. The Olympics profoundly impact Internet usage due to the considerable demand for connectivity from participants, attendees, and global audiences. Throughout the Olympics, media personnel, athletes, and attendees depend on robust Internet access to communicate, stream, and share their experiences:

> To support the 2024 Olympics' connectivity needs, network providers must prepare to rely on 5G's strength. At an event where champions are sometimes determined in seconds or even nanoseconds, every moment of connectivity will have to count. (Stokel-Walker, 2024)

The Opening Ceremonies of the Olympic Games should be a public holiday because all nations come together to watch, celebrate, and cheer for their best athletes. The 2024 Opening Ceremony in Paris was a great success, featuring Céline Dion performing as a vocal athlete from the Eiffel Tower. The ceremony was very cinematic, giving the impression of watching a live cultural event with special effects and performances along the river Seine. The French extravaganza featured an athlete parade on boats, with Joan of Arc in medieval knight armor riding a white horse and carrying the Olympic torch. It also included famous athletes carrying the Olympic torch, music, dance, art, architecture, fashion shows, projections on the Eiffel Tower, and the lighting of the Olympic cauldron (The Opening Ceremony, 2024). The lighting of the Olympic cauldron during the opening ceremony served as a reminder of the Mongolfier brothers' invention of the hot-air balloon in France. The first piloted ascent, with Jacques-Étienne Mongolfier aboard, occurred in 1783 in the presence of Louis XVI, the King of France. The Olympic Games ceremony should be considered a global holiday, much like New Year's Eve, as it brings people from diverse cultures together through sports and creates shared celebratory memories for all.

## 2 Holiday Experiences in Various Cultures

Widely celebrated and rarely observed holidays benefit us as they emphasize hope, renewal, and new beginnings. They bring value to people's lives in the form of unity, happiness, gratitude, and a sense of belonging. They recharge our energy, free us from high stress levels, improve productivity at work, and reduce depressive behavior. Additionally, they help rebalance the dopaminergic system, ultimately making us feel happy. Although public holidays are not numerous, many political, religious, and social activists worldwide are working to change this fact and include more public holidays in their respective countries.

It is important to note that the number and type of public holidays can vary significantly from one country to another or even from one continent to another. After half a year of holidays in the Roman Empire, we are now down to a handful worldwide. Asia, a continent with a rich diversity of cultures, religions, and traditions, celebrates many public holidays. Nepal has the highest number of public holidays in the world, with 35 holidays. However, employees in Nepal are entitled to only 13 paid public holidays per year. Depending on the region, India observes between 17 and 42 public holidays. Other Asian countries, such as Indonesia (28), Sri Lanka (25), Malaysia (18–24), Myanmar (24), Bangladesh (22), Cambodia (21), Macau (20), Philippines (18) also have a compelling number of public holidays (List of countries, 2024). Curiously, employees in Cambodia are entitled to 27 paid public holidays per year (List of Minimum, 2024), resulting in a fluctuating number of public holidays from year to year or across cultures. The diversity in culture and religion across Asian countries is evident in the multitude of public holidays they observe. These holidays mark significant historical events and honor essential figures.

The Boun Pimay, or Solar New Year, is Laos' most widely celebrated public holiday (Bao, 2024). It lasts for 3 days in April. This holiday coincides with changes in nature and the beginning of the first rains. During this period, life in the country undergoes changes. Laotians clean their houses, host parties, and release birds. People in the streets splash water or use colored water, creating a fun and enjoyable water war experience. Lao Buddhists traditionally take a bath in the Mekong River to purify themselves of past sins. Some people paint their faces with white and red colors. Sand pyramids are built for good fortune. Traditionally, young girls tear men's clothes to symbolize the coming of the New Year.

Curiously, in countries such as Japan, public holidays are designated to express gratitude for the gifts and beauty of nature. Not all cultural events are public holidays, but some exciting festivals celebrate the harvest and the bounties of nature, and the entire community is involved: Shirakawa-go Doburoku and the Cherry Blossom Festival. On the evening of August 31, the residents of Nejapa in El Salvador come together to throw fireballs to commemorate the 1658 El Playon volcanic eruption. During the Las Bolas De Fuego ("Balls of Fire") celebration, teams throw palm-sized fireballs at each other, creating a fiery spectacle. In the Florida Keys in the United States, divers and music enthusiasts gather to listen to underwater tunes during the Underwater Music Festival. This celebration aims to raise awareness for coral preservation. In Statesboro, Georgia (USA), environmentalists and gardeners gather annually for a Community Garden Social. They share their secrets to productive gardens, discuss conservation projects, and work together on preserving the region's farmlands (Smith, 2024). Holidays have become more entertaining and eco-friendly. Drone shows (Bordeaux, 2024) and projections on cathedrals (Saint-Gatien Cathedral in Tours) and famous buildings such as the Submarine Base in Bordeaux, France (Culturespaces, 2020) are replacing traditional fireworks, making celebrations more sustainable. The Internet has played a significant role in modernizing holidays, aligning them with the planet's sustainability goals. These festivals emphasize respect for nature and the diversity of human creativity and traditions.

The process of selecting public holidays involves considering various factors such as culture, religion, history, and the natural environment. These elements are crucial in determining which days are commemorated as public holidays. Significant examples include national holidays like July 4 in the United States and July 14 in France, which hold deep cultural and historical significance for their respective countries. To connect these two public holidays, people think about the Marquis de Lafayette, known as the "Hero of Two Worlds" (Duncan, 2021), for his accomplishments in the service of France and the United States. One would think that Washinton is the common name connecting many cities in the United States, but it is Lafayette. His stories echo throughout the history books, and we think of Lafayette every July 4. But what about other French heroes from the American Revolution? Jennifer Wells said that for quite some time, the history of the American Revolutionary Wars predominantly centered on famous individuals and pivotal battles (Rodriguez, 2021). In the twenty-first century, however, historical accounts become more accurate and detailed, thanks to the Internet, with the emergence of social and political movements. Savannah and Statesboro in Georgia, USA, are home to a forgotten French hero of the American Revolutionary Wars. Nicholas Anciaux, also known as Handshaw or Hand Shaw (potentially a mispronunciation of Anciaux), is believed to be buried on the former Anciaux land in Bulloch County, possibly located on Jones Mill Road (Brannen, 1992). The Brier Creek Chapter of the Daughters of the American Revolution (DAR, 2024) has erected a memorial marker in the field to honor this French and American hero (Swiggart, 2017). In 1807, Anciaux was awarded 1200 acres of land on Mill Creek in Bulloch County in recognition of his military service, which became his last residence (Roger, 2008). From start to finish, Nicolas Anciaux was by Lafayette's side, and he witnessed the surrender of Lord Cornwallis. It is significant for the community to locate his grave and honor his legacy on July 4 at Mill Creek Park in Statesboro, GA, where annual celebrations are held and where Anciaux was the original proprietor. With the assistance of the Internet, drones, modern technology, cadaver dogs, the French community, Anciaux's descendants, and dedicated historians, we hope to fill in the gaps in information about Anciaux and his contributions. It is important to remember the countless military heroes we have read about and also to recognize the lesser-known heroes who sacrificed a lot for a country's independence. They deserve to be celebrated. France is often criticized for its perceived lack of bravery in battles, but during the American Revolution, French soldiers and sailors showed incredible bravery through acts of selflessness and Nicolas Anciaux was one of them.

On August 15, Canada celebrates National Acadian Day, a joyous public holiday on the American continent. This day is dedicated to honoring the Acadian culture and heritage. To commemorate the occasion, Acadians participate in lively street parades where they dress in elaborate costumes and traditional attire adorned with vibrant makeup. The air is filled with sounds of musical instruments, singing, and the clamor of noisemakers, creating a festive and vibrant atmosphere throughout the festivities:

> Une fois la prière terminée, on fera pendant plusieurs minutes, un joyeux tintamarre de tout ce qui peut crier, sonner et faire du bruit: sifflets de moulin, klaxons d'automobiles, clochettes de bicyclettes, criards, jouets, etc. ("Once the prayer is finished, there will be a joyful *tintamarre* lasting for several minutes, featuring anything, everything and everyone that can make noise, shout and ring: mill whistles, car horns, bicycle bells, squawking objects, toys, etc.") (Le Tintamarre, 2010)

They shout, whistle, bang on plates, pots, or anything they find on their way. This tradition is called Tintamarre (Din) and commemorates the Acadians' expulsion from their lands between 1755 and 1764. Historians believe this holiday tradition came from the Middle Ages when people made loud noises to mark sad or joyous events. The first Tintamarre took place on August 10, 1955, in Moncton, New Brunswick, at the instructions of Archbishop Norbert Robichaud on the 200th anniversary of the Deportation of the Acadians. The largest Tintamarre parade, which demonstrates the vitality of the Acadian people, is held in Bouctouche, New Brunswick, Canada (On August 15, 2022).

In Europe, the Principality of Liechtenstein is the only country with no debt and a solid financial sector. It observes between 20 and 22 public holidays annually, showing that citizens can be productive and celebratory at the same time. Other European countries have fewer public holidays, between 7 and 15 per year.

In the Republic of Moldova, there is a unique public holiday called Paştele Blajinilor, or Easter of the Blessed (Cocosila, 2013), which is included in the Labor Code of the country:

Easter of the Blessed is observed on Sunday or Monday, a week after Orthodox Easter. People visit the cemetery to celebrate Easter with their departed loved ones. They color eggs again, bake "pasca" (sweet bread with cheese), share a meal among graves, give away small gifts, listen to a religious ceremony, and talk about their loved ones. It is a beautiful tradition where the dead are included in the holiday celebration. It would be interesting to do future research on emotional associations to see how people feel about this and other holidays depicting sad events. Even a sad celebration is better than none at all.

In Madagascar, the Malagasy people have a cultural ritual called Famadihana, which occurs every 5–7 years during the dry season, from June to September. During Famadihana, a tradition called "zanadrazana" takes place. Family members dig up their deceased ancestors, change their clothes and shrouds, speak to them, ask for blessings, dance with them while carrying their corpses on their heads, circle the tomb, and place them in a new coffin before re-burying them in the family vault. Throughout these ceremonies, they eat "varibemenaka," a rice dish accompanied by beef, drink "toaka gasy," a local rum, and dance to music called "hira gasy." These rituals are a way to celebrate life after death and show respect toward the deceased (Blog, 2019).

Another enjoyable public holiday in Africa is the Grand Magal of Touba, one of the most famous pilgrimages in the world and the largest celebration in Senegal. This holiday is both religious and secular. Villalón wrote that no other religious or secular celebration in Senegal attracts such attention (Villalón, 1995). It is a spiritual celebration, full of entertainment. A religious Mouride Brotherhood celebrates a pilgrimage toward the holy Mouride city of Touba. This holiday commemorates the life, teachings, and departure into exile in 1825 of the founder of this brotherhood, Cheikh Amadou Bamba (1853–1927). Ahmadou Bamba was successively deported to Gabon from 1895 to 1902, to Mauritania from 1902 to 1907, then held under house arrest in Senegal until his passing in 1927. During this holiday, more than three million pilgrims converge worldwide to the city of Touba to meet one of the descendants of Amadou Bamba, also called the Serigne of Touba. It has been anticipated for weeks and is creating general excitement, says De Jong (2010). During this holiday, the faithful pray, touch the walls and door of Amadou Bamba's mausoleum, light lamps, sing poetry by Bamba, and eat meals containing sheep or other meats. The holiday has roots in the colonial past of Senegal, when it was created to defuse tensions and show mutual understanding between the Mourides and the French colonial government (Coulon, 1999). It is clear that culture, religion, history, and nature play essential roles in the selection of public holidays.

It is widely accepted that holidays are crucial for our brain's well-being. However, why are holidays important, and what are the benefits of having different holidays? Surprisingly, there are very few scientific studies that study the positive effects of holidays on our mental health. In 2014, Jessica de Bloom and her colleagues conducted a longitudinal field study on the advantages of recreational travel on cognitive flexibility and originality. They found that holiday travel enhances cognitive flexibility and nurtures creativity:

> Recreational travel may increase creativity by relieving workers from stress, providing diversifying experiences, and increasing positive emotions. Consequently, vacations may boost creativity, apparent in the greater variety (flexibility) and originality of ideas after work resumption. In our longitudinal field study, creativity (measured by Guilford's Alternative Uses Task; independently scored by three raters) was assessed in 46 workers before and after vacation. Potential precursors for creativity changes (i.e., workload, vacation hassles, vacation destination, and positive affect) were also explored. (De Bloom et al., 2014)

The critical point is this: for holidays to be truly beneficial, they must relieve us from work-related stress, including chronic stress, and should not add new stressful situations. What one person finds rewarding during holidays could be stressful or traumatic for others, as experiences and rewards are entirely subjective. In his narrative poem *Childe Harold's Pilgrimage*, Lord Byron, assuming the persona of Childe Harold, describes his disillusionment with some of the Roman Games (*Ludi Apollinares*) in the past. These games were celebrated as an annual festival since 208 BC, taking place in July for 8–9 days. They echoed the broader sentiments of many generations to come, worn by gladiatorial combat, battles, and wars.

> There were his young barbarians all at play,
> There was their Dacian mother—he, their sire,
> Butcher'd to make a Roman holiday—
> All this rush'd with his blood—Shall he expire
> And unavenged?—Arise! ye Goths, and glut your ire! (Lord Byron, 1818)

The above excerpt vividly captures the violence of ancient Roman games, where lives were sacrificed to entertain the masses. In Antiquity, Roman games were very popolar with 175 days dedicated to competitions every year: « Dans l'Antiquité, les jeux du cirque étaient très populaires à travers l'Empire romain: 175 jours étaient consacrés chaque année à l'organisation de ces compétitions! » (Abbott & Haney, 2007). The multitude of celebrations is unreasonable in the today's world, but it is a good reflection on public and private events meant for entertainment and days off work. In other historical holidays, human or animal life was associated with sacrifice and deeply rooted in cultural and religious beliefs. While not widespread, human sacrifice practices varied among different Native American tribes. Many tribes believed that human blood appeased the gods or ensured victory in battles. Captives of the highest status from warring tribes were often decapitated and thrown down temple steps. Other prisoners of war were typically sent into the labor force (Kuroski, 2017). Lupercalia, an ancient festival in Rome, was a violent and sexually charged celebration that involved animal sacrifices and random matchmaking to ward off infertility and evil spirits. It honored the she-wolf that nurtured Romulus and Remus, the legendary founders of Rome, and pleased the Roman fertility god Lupercus. Eid-al-Adha (Feast of Sacrifice) commemorates the willingness of the Prophet Ibrahim to sacrifice his son, Ismail. The sacrifice is symbolized by the ritual slaughter of a sheep, goat, or cow; no human life is sacrificed. In Jewish rituals, lambs were used as sacrifices for sins. Christians believe that Jesus, the Lamb of God, was sacrificed for everyone, making the lamb a powerful symbol. In many countries, lamb is the traditional meal for Easter.

## 3 From Movement to Celebration: The Timeline of Significant Change

How long does it take for a cultural event to become officially recognized as a public holiday? Many enjoyable holidays are currently not public holidays, but they may become so one day. Hopefully, it will not take decades for them to be recognized as essential to people's health and well-being. Some significant holidays for future consideration include, but are not limited to, Mardi Gras (in many countries), Carnival of Binche (Belgium), Toka (alliance between tribes in Vanuatu), Pilou-Pilou (Yam Festival in New Caledonia), Famadihana (Madagascar), Heiva (Acrobats Festival in Tahiti), the FIFA World Cup, International Friendship Day (July 27), and many others.

Juneteenth National Independence Day commemorates the end of slavery in the United States of America (Holidays and Observances, 2024). On June 19, 1865, 2 years after Abraham Lincoln issued the Emancipation Proclamation on January 1, 1863, abolishing slavery, the message arrived in Galveston, Texas, where the last enslaved African Americans were informed of their freedom. In one of her interviews, Opal Lee, "the Grandmother of Juneteenth," said: "I have to tell you that none of us are free until we are all free" (Lee, 2019). The importance of Juneteenth (a combination of June and nineteenth) was acknowledged and celebrated for generations, but it was only recently that, on June 17, 2021, it became a federally recognized public holiday. This resulted from persistent activists, such as Opal Lee, raising awareness of the significance of this day. President Joe Biden signed the Juneteenth National Independence Day Senate Bill S. 475 into law, making it the first new federal holiday since Martin Luther King Jr. Day (MLK Day), observed on the third Monday of January each year and established in 1983. This means that from 1983 until 2021, in the United States, no other public holidays were officially recognized. In 2024, Opal Lee received the Presidential Medal of Freedom, the Nation's highest civilian honor, for her lifetime achievements. She campaigned for decades to make Juneteenth a federal holiday. To promote the idea of a federal holiday, she led 2.5 miles (4 km) of symbolic walks every year, representing the 2.5 years it took for the Emancipation Proclamation news to reach Galveston, Texas. Juneteenth is now a day of reflection, education, and celebration, marked by various traditions and festivities. Some critical traditions associated with Juneteenth include family and community gatherings, church picnics, parades, fireworks, Miss Juneteenth pageants, rodeos, sports, games, music and dance performances, selling clothes and jewelry, fundraising events, the reading of the Emancipation Proclamation, sharing stories about formerly enslaved people, educational events focusing on the struggle for freedom and the achievements of African Americans, and political activism, ongoing pursuits on equality and justice. Sharing a meal plays a central role in Juneteenth celebrations. Famous food traditions include eating red foods (watermelon, strawberries, barbeque, and red velvet cake), symbolizing the resilience and blood the enslaved African Americans shed. The holiday also features Good Fortune Foods (black-eyed peas with red rice, collard greens, and red pepper) and recipes passed down orally through African American generations (Coleman, 2024).

## 4 Catalyst of Change: The Impact of the Internet and the Experience of Anticipation

As we explored in the previous section, holidays are a tapestry of culture, tradition, and personal reflection, offering the opportunity to break the routine of daily life and deepen our connections with one another. Undoubtedly, holidays are a way for people to meet and remember one another. They provide an opportunity to show appreciation, gratitude, and, most importantly, love. By recognizing and celebrating holidays with greater frequency, the world strengthens the bonds that unite it. However, the Internet's invention has significantly changed the way we celebrate. Celebrating has become more frequent and superficial with the rise of social media, messaging apps, video calls, and automated greeting card websites. This has led some to question whether human connections are as strong and meaningful as before the Internet. The impact of the Internet on the strength and meaning of human connections can vary greatly depending on individual experiences and perspectives. Overall, the advent of the Internet has transformed how people celebrate holidays worldwide. The Internet has significantly impacted holiday celebrations, but in what way? There is a paradox here: collectively, we created massive storage clouds and numerous websites dedicated to holidays, but individually, we cannot celebrate any holiday because it feels lonely. The Internet has made it easier to stay connected with people far away, allowing for instant communication and sharing of live events. In many cases, it has strengthened relationships and made holidays more accessible. For instance, social media platforms have made it possible for many people to participate in holidays from their own homes, leading to more inclusive cultural and religious observances. Many sociologists argue that social media plays a pivotal role in the experience of holidays today. Not only do posts foster reciprocity, but social networking also provides us with innovative ways to celebrate collectively. For instance, updating a profile picture with a festive overlay to celebrate a holiday has become a modern shared ritual, much like posting a family photo from a special occasion (Sternheimer, 2015). While the internet allows for more interactions, some argue it does not constantly improve their quality. Online interactions can lack the depth and richness of face-to-face conversations, as they often lack nonverbal cues crucial for effective communication. Additionally, social media platforms tend to encourage broad rather than deep relationships. Having thousands of "friends" or "followers," online does not necessarily equate to having strong and lasting relationships.

In various circumstances, the impact of the Internet on personal relationships and mental health is significant. It is believed that social media can contribute to holiday controversies (Forgeard, 2023) and stress, resulting in increased anxiety, depression, and fatigue. Many people experience the fear of missing out (FOMO) and feel pressured to participate in more celebrations, even if it is beyond their means. The abundance of holiday-related posts on social media can make people feel inadequate and lonely as they compare themselves with others. This constant comparison can lead to feelings of despair, inferiority, financial inadequacy, and unnecessary competitiveness. The Internet has likely altered our perceptions of holidays by making celebrations feel less personal and authentic, especially with the prevalence of automated greeting cards and mass messages replacing handwritten letters or individual meetings. Social media subtly but significantly changes how people experience the holidays, making it challenging to discern genuine joyous moments from overidealized representations of celebrations. However, amid these challenges, there are also opportunities for meaningful connections online. For instance, online communities or forums can provide valuable support and connections for individuals with shared experiences or interests.

Some holidays are the result of marketing campaigns heavily influenced or created by marketing efforts. Black Friday in the United States, celebrated after Thanksgiving on the last Thursday of November, is a notable example that has extended to many countries and different times of the year: Black Friday in July. Black Friday is not a traditional cultural or religious holiday; it is a significant event created by retailers to kick off the Winter Holiday shopping season. Cyber Monday is associated with purchasing technology. Back-to-school block parties help students get back to school. Superhero Day at School is for children to wear their favorite costumes and celebrate their heroes. Singles' Day, popularized by Alibaba, has become the biggest online shopping day in the world. Companies use these occasions to boost sales and capture the holiday spirit. They create campaigns, deals, and ads that generate much holiday excitement. These efforts evoke emotions and promote sharing and generosity, often partnering with charitable organizations and sometimes connecting people worldwide with their loved ones. While these holidays may not have a cultural, historical, or religious origin, Christmas, on the other hand, has spiritual significance but is also heavily shaped by marketing. The increased commercialization of Christmas has allowed several companies to generate 40% of their annual sales during the Christmas holidays. Regardless of their type and nature, these holidays offer businesses an excellent opportunity to increase revenue and connect with like-minded consumers. Through their campaigns, digital or print ads, social media, and deals, businesses worldwide (such as Amazon, Alibaba, Walmart, Samsung, Best Buy, Wayfair, and others) promote consumerism and create pressure to spend more during the holidays.

How much time do people spend preparing for different holidays every year? The critical point is to savor the anticipation or the wait for the holidays. Anticipation plays a crucial role in behavioral economics and cognitive sciences. Advanced information contributes significantly to the experience of anticipation. According to Caplin and others, three specific brain regions, namely the dopaminergic brain regions, show a network of clusters for computing the utility of anticipation and are strongly correlated to preference for advanced information (also known as early resolution for uncertainty (Caplin & Leahy, 2001); information-seeking (Bromberg-Martin & Hikosaka, 2009); or observing (Beierholm & Dayan, 2010)). People prefer discovering what will happen to them and their fate because they derive intrinsic value from information. However, the preference for advanced information is dependent on the reward probability. Human beings are more excited, waiting for actions or rewards. More broadly:

> [...] we identify a novel neural substrate for computing the utility value arising from anticipation. Our results implicate that a functional coupling of three distinctive brain regions links the arrival of advanced information to resolve future uncertainty to the boosted utility of anticipation. We suggest that this boosted anticipatory utility drives a range of behaviors, including information-seeking, addiction, and gambling. Our study may also provide seed for designing individually tailored interventions for psychiatric disorders. (Iigaya et al., 2020)

The time spent preparing for holidays can vary greatly depending on factors such as the holiday itself, personal traditions, gender, and level of involvement in the preparations. For instance, some people may dedicate weeks to getting ready for Christmas, while others may only spend a few minutes. Similarly, the time spent preparing for holidays such as Thanksgiving or Easter can vary widely. It is important to note that holiday preparation often involves multiple activities, including shopping, online shopping, cooking, decorating, and planning events, which can require significant time and effort. Although specific data on holiday preparations is lacking, it is evident that many people invest a substantial amount of time in making holidays unique and memorable. There is great value in preparation and anticipation of holidays.

## 5 Understanding Holidays: My Favorite Holiday and My Least Favorite Holiday

During my teaching career in the United States, I was honored to teach 19 sections of FYE 1410 (First Year Experience): European Holidays with 20–30 students in each section. The European Holidays course focused on helping graduates in all fields face global challenges and understand different cultural perspectives that influence understanding of world issues. This seminar required students to describe factors contributing to their cultural perspective, apply multiple cultural perspectives to global issues, and then use this knowledge through engagement with local communities or problems. Throughout this course, students had the opportunity to explore European holidays by engaging in extensive readings, discussions, and oral presentations. Instead of homework assignments, students shared holiday experiences in class through short oral presentations and visual support such as pictures and videos. When shared, the effort was more significant, and the quality improved. They also watched videos and did oral presentations about their favorite holidays, wrote essays on existing holidays, and even developed new and original holiday ideas. I needed to maintain an interactive, student-centered, communication-based classroom environment that was academically conscientious. I often incorporated research inquiries, comparisons between different cultures, contemporary events in the world, and perspectives from literature, art, and cinema into my teaching style and service-learning projects. Cultural competencies are essential, and students acquire them through holiday traditions. Teaching about religious holidays was challenging every semester because students from different cultures were reluctant to engage in conversations, fearing they might denigrate someone. During class activities, I had international faculty visit my class and do presentations on different holidays. I also invited artists to show students how to make masks and costumes. Using assessment strategies to engage students' interest in online readings or audiobooks, to focus their learning, and to develop communication skills were excellent opportunities to increase cultural competence skills and to give students a view of culture as a way of thinking critically. Reading and listening to books at home allowed students to apply their understanding of the content in class and life situations. In the FYE 1410 Course Evaluations (Amarie, 2018), students wrote: What did you like about the course? "Her knowledge on the subject matter, very direct source." "[The instructor] was very flexible and provided great learning content. This has been my favorite course during the semester and opened my eyes to so much!"

In each of the 19 sections, students were required to complete a comprehensive written Final Exam comprising numerous open-ended questions addressing global issues, specifically focusing on holiday celebrations. Over multiple semesters, I consistently posed two straightforward survey questions to my students:

1 *What is your favorite holiday? Please explain.*

The answer to this question was similar each semester. The vast majority of students, more than 90%, said, "Christmas." Students love Christmas for many reasons. They gather with loved ones, attend parties, share special foods and treats, and create common memories. They cherish traditions, such as decorating the Christmas tree and exchanging gifts. They go in person to see the Nativity Scene and listen to the pipe organ. Many students mentioned the religious significance of the public holiday. For Christians, Christmas celebrates the birth of Jesus Christ, making it a meaningful and spiritual time. They are happy to have time off from school, relax, and enjoy the holiday season.

2 *What is your least favorite holiday? Please explain.*

The answer to this question was similar every semester as well. The vast majority of students, more than 80%, said "Valentine's Day." Some students felt that the holiday has become too commercialized, focusing more on spending money on gifts and less on genuine expressions of love. For young people, in particular, the holiday can create pressure to have a romantic partner or to meet certain expectations, which can be stressful and disappointing. For those who are single on Valentine's Day, the holiday can feel exclusive and highlight their single status, leading to feelings of loneliness and inadequacy.

Teaching this course was an absolute delight, and I am genuinely excited about the prospect of teaching it again in the future in different modalities: face-to-face, hybrid, online synchronous, or online asynchronous. Whether it is in the same format or a different one, I eagerly anticipate the opportunity to share knowledge and inspire students once more.

## 6 Conclusion

Holidays cultivate unity, happiness, gratitude, love, and a profound sense of belonging among individuals and communities. There are many holidays, some widely celebrated and others less known, yet all contribute to people's happiness and combat isolation. These special times throughout the year present many opportunities to express appreciation, celebrate love, and share joy, which are fundamental to our psychological health: emotional, mental, and social well-being. This chapter showed the benefits derived from active engagement online and face-to-face in holidays and cultural festivities, revealing their influence on happiness, self-esteem, social and mental experiences. However, the Internet and social media have not only expanded our ways of celebrating and connecting with others but also brought about new challenges. One of the challenges described was whether our connections can maintain their strength and depth in an increasingly digital world. Another challenge observed was using the Internet strategically to add more public holidays and combat isolation. The conversation about adding more public holidays must be addressed in each country individually and rectified. The limitations of this analysis include the lack of quantitative and qualitative research on human subjects about holidays in different cultures. The chapter identifies a few resources, but more needs to be done to study this interesting universal phenomenon and its benefits on emotional health, productivity, and creativity. Striking a balance between online and offline celebrations is crucial in preserving the essence of holidays. The Internet has revolutionized how we connect and celebrate, and its impact on our relationships is determined by our mindful use of online tools and reinforcement with offline interactions. By examining the significance of holidays and how they are represented in the media, we gained valuable insights into their true worth. Despite the surge in Internet activity during these cherished occasions, holidays remain vital for maintaining human connections, reducing isolation, improving productivity and creativity, and addressing health concerns.

## References

Abbott, S., & Haney, C. (2007). *365 jours pour tester sa culture Générale*. Trivial Pursuit. Play Back.

Amarie, O. (2018). *Online course evaluations. FYE 1410 – European holidays*. Georgia Southern University.

Bao, T. (2024, July 18). *Lao New Year guide in 2024: Sok Di Pi Mai!* Kampá Tour. https://kampatour.com/lao-new-year

Beierholm, U. R., & Dayan, P. (2010). Pavlovian-instrumental interaction in 'observing behavior'. *PLoS Computational Biology, 6*(9), e1000903. https://doi.org/10.1371/journal.pcbi.1000903

Blog. (2019, April 16). *Death around the world: Famadihana in Madagascar.* Funeral Guide. https://www.funeralguide.co.uk/blog/death-around-the-world-famadihana-madagascar

Bordeaux Tourisme & Congrès. (2024). *The Bordeaux Wine festival: Unique drone shows*. https://www.bordeaux-wine-festival.com/drone-show.html

Brannen, D. (1992). *Life in old Bulloch: The story of a Wiregrass County in Georgia, 1796-1940*. Statesboro Regional Library.

Bromberg-Martin, E. S., & Hikosaka, O. (2009). Midbrain dopamine neurons signal a preference for advanced information about upcoming rewards. *Neuron, 63*(1), 119–126. https://doi.org/10.1016/j.neuron.2009.06.009

Caplin, A., & Leahy, J. (2001). Psychological expected utility theory and anticipatory feelings. *The Quarterly Journal of Economics, 116*(1), 55–79. https://doi.org/10.1162/003355301556347

Cocosila, A. (2013, May 13). *Pastele Blajinilor*. CrestinOrtodox. https://www.crestinortodox.ro/datini-obiceiuri-superstitii/pastele-blajinilor-88711.html

Coleman, C. (2024, June 7). *7 Juneteenth foods and traditions*. A&E Television Networks. https://www.history.com/news/juneteenth-foods-traditions

Coulon, C. (1999). The grand Magal in Touba: A religious festival of the Mouride Brotherhood of Senegal. *African Affairs, 98*(391), 195–210. https://doi.org/10.1093/oxfordjournals.afraf.a008007

Culturespaces. (2020). *Bassins des Lumières*. https://www.bassins-lumieres.com/en

DAR. (2024, July 31). *Welcome to the Brier Creek chapter, NSDAR*. Georgia State Society. https://briercreek.georgiastatedar.org/.

De Bloom, J., et al. (2014). Vacation from work: A 'ticket to creativity'?: The effects of recreational travel on cognitive flexibility and originality. *Tourism Management, 44*, 164–171. https://doi.org/10.1016/j.tourman.2014.03.013

De Jong, F. (2010). Remembering the nation. The murid Maggal of Saint-Louis Senegal (Mémoire de la nation. Les Maggal mourides de Saint-Louis, Sénégal). *Cahiers d'Études Africaines, 50*(197), 123–151. https://doi.org/10.4000/etudesafricaines.15810

Duncan, M. (2021). *Hero of two worlds: The Marquis de Lafayette in the age of revolution*. PublicAffairs.

Durkheim, D. E. (1960). *The division of labor in society.* https://archive.org/details/in.ernet.dli.2015.233884/page/n187/mode/2up?q=solidarity

Faculty. (2024). Department of World Languages and Cultures, Georgia Southern University. https://www.georgiasouthern.edu/cah/languages/

Forgeard, V. (2023). *The fabric of society: How holidays shape and define cultures.* https://brilliantio.com/why-are-holidays-important-to-a-culture/#Key_Takeaways

Holidays and Observances in the United States in 2024. (2024). *Time and date*. https://www.timeanddate.com/holidays/us/2024

Iigaya, K., et al. (2020). The value of what's to come: Neural mechanisms coupling prediction error and the utility of anticipation. *Science Advances, 6*(25), eaba3828. https://doi.org/10.1126/sciadv.aba3828

Kuroski, J. (2017). *Human sacrifice in the pre-Columbian Americas: Separating fact from fiction*. ATI. https://allthatsinteresting.com/human-sacrifice.

Le Tintamarre. (2010). *CyberAcadie. L'Histoire acadienne au bout des doigts*. https://web.archive.org/web/20140409125103/http://cyberacadie.com/index.php?/coutumes/Le-Tintamarre.html

Lee, O. (2019). *The real Opal Lee*. https://www.opalswalk2dc.com/.

List of countries by number of public holidays. (2024). Wikipedia. https://en.wikipedia.org/wiki/List_of_countries_by_number_of_public_holidays

List of Minimum Annual Leave by Country. (2024). Wikipedia. https://en.wikipedia.org/wiki/List_of_minimum_annual_leave_by_country

Lord Byron. (1818). *Childe Harold's Pilgrimage*. https://knarf.english.upenn.edu/Byron/charold4.html

On August 15, Let's Celebrate National Acadian Day! (2022). Salut Canada. https://salutcanada.ca/en/listings/on-august-15-lets-celebrate-national-acadian-day/

Perec, G. (1989). *Vœux*. Seuil 11–19.

Pérez Fernández, J., & de la Torre Martínez, R. (2023, July 12). *Even before you take time off, a holiday is good for you*. The Conversation. https://theconversation.com/why-a-holiday-is-good-for-you-even-before-you-take-time-off-209406

Rodriguez, H. (2021). *Unsung heroes of the revolutionary war*. College of Arts & Sciences at Texas A&M University. https://liberalarts.tamu.edu/blog/2021/07/01/unsung-heroes-of-the-revolutionary-war/

Roger, A. (2008). *Remembering Bulloch County: Tales from the Babe to the 'Boro*. The History Press.

Sandvine. (2018). *Sandvine's active network intelligence empowers Telenet to enhance their subscribers' experience*. Sandvine, the App QoE Company. https://www.sandvine.com/press-releases/sandvines-active-network-intelligence-empowers-telenet-to-enhance-their-subscribers-experience

Shoemaker, N. (2014). *What does internet use look like during Christmas?* Big Think. https://bigthink.com/technology-innovation/have-a-connected-christmas-internet-usage-spikes-during-holidays/g Think.

Smith, A. (2024, July 27). *Growing together: Community garden social was a hit*. Grice Connect. https://www.griceconnect.com/local-news/growing-together-community-garden-social-was-a-hit-9263264

Sternheimer, K. (2015). *Sociology and holiday rituals*. https://www.everydaysociologyblog.com/2015/12/sociology-and-holiday-rituals.html
Stokel-Walker, C. (2024). *The Olympics will challenge mobile network providers to satiate 15 million attendees' hunger for internet connectivity*. Business Insider India. https://www.businessinsider.in/tech/news/the-olympics-will-challenge-mobile-network-providers-to-satiate-15-million-attendees-hunger-for-internet-connectivity/amp_articleshow/112023292.cms
Swiggart, C. C. (2017, September 13). *COL Nicolas Anciaux*. Find a Grave. https://www.findagrave.com/memorial/183326122/nicolas-anciaux
The Opening Ceremony. (2024). *Paris 2024*. https://olympics.com/en/paris-2024/the-games/ceremonies/opening-ceremony
Villalón, L. A. (1995). *Islamic society and state power in Senegal: Disciples and citizens in Fatick*. Cambridge University Press.
Wright, M. H. (1900). *My study fire: First series*. Dodd, Mead and Company.

**Dr Olga Amarie** is Professor of French in the Department of World Languages and Cultures at Georgia Southern University. Her current research focuses on nineteenth-century unpublished manuscripts by Juliette Adam, housed by the Lilly Library of Rare Books and Manuscripts at Indiana University-Bloomington. In addition to her individual work on Georges Perec, Hervé Le Tellier, Octave Mirbeau, and Remy de Gourmont, Dr Amarie has contributed to collaborative and interdisciplinary publications: "Amour et séduction dans la Légende de Tristan et Iseut chez Béroul, Thomas, et Albert Cohen" in Entre-Textes: Dialogues littéraires et culturels, Routledge: 2018; "Displacement Vector Analysis on the GO Board in ∈ by Jacques Roubaud" in The Guests of Chance: Hodeporic Poetry, Semicerchio 63, Rivista di Poesia Comparata, Pacini Editore, 2020/2; "Remy de Gourmont 1858-1915" in the Twentieth-Century Literary Criticism (TCLC), Volume 407, Layman Poupard Publishing for Gale/Cengage Learning's Literature Criticism Series, 2021.

# Effective Communication for Sustainable Behavior: Insights from Psychology

**Ana Rita Farias, Bernardo Cruz, Samuel Domingos, Catarina Possidónio, Ana Loureiro, Eliana Portugal, and Sílvia Luís**

**Abstract** As global awareness shifts toward environmental sustainability, understanding how to promote sustainable behavior becomes essential. This chapter examines the relationship between fostering sustainable behavior and implementing effective communication strategies, exploring the psychological factors that influence sustainable practices alongside evidence-based policies and interventions.

The discussion explores the psychological foundations of sustainable behavior, emphasizing how perceptions of environmental risks and emotional responses drive behavioral change. The chapter underscores the importance of tailored communication strategies that resonate with different cultural contexts and audiences, ensuring the effectiveness and impact of messages. An analysis of barriers to sustainable behavior, such as lack of awareness, perceived high costs, and misinformation, is provided, along with strategies to overcome these challenges. The chapter highlights how effective communication can enhance these efforts. It examines the media's influence on public perceptions of sustainability and stresses the need for accurate, truthful, and research-based communication. Evidence-based strategies for promoting sustainable behaviors, managing conflicting information, and fostering informed social debate on sustainability are presented.

By offering insights into customizing interventions and communication strategies, this chapter aims to optimize the effectiveness of these techniques in various domains. It advocates the importance of tailored communication to achieve sustainability goals, emphasizing that effective communication is a key driver in promoting sustainable behavior and fostering a more sustainable future.

A. R. Farias (✉) · B. Cruz · S. Domingos · C. Possidónio · A. Loureiro · E. Portugal · S. Luís
Hei-Lab: Digital Human-environment Interaction Labs, Universidade Lusófona, Lisbon, Portugal
e-mail: ana.rita.farias@ulusofona.pt; bernardo.cruz@ulusofona.pt; samuel.domingos@ulusofona.pt; catarina.possidonio@ulusofona.pt; ana.loureiro@ulusofona.pt; silvia.luis@ulusofona.pt



M. Todorovic et al. (eds.), *Sustainability and Social Responsibility of the Media and in the Media*, CSR, Sustainability, Ethics & Governance,
https://doi.org/10.1007/978-3-032-00086-6_11

**Keywords** Sustainable behavior · Psychological insights · Cultural differences · Environmental action · Behavioral change · Effective communication · Environmental sustainability · Tailored communication · Sustainable Development Goals · Evidence-based recommendations · Climate change · Climate crisis · Eco-friendly practices · Public perception · Risk perception · Risk communication · Attitude–behavior gap · Communication and motivation · Pro-environmental attitudes · Consumer choice · Sustainable communication · Challenges to sustainability · Communication for sustainability · Media communication · Cognitive and emotional responses · Communication innovation · Communication strategies · Communication interventions · Social responsibility · Psychological barriers/psychological obstacles

## 1 Introduction

The growing effects of resource depletion and climate change have made addressing environmental sustainability more urgent in recent years. It is now critical to develop efficient methods to encourage sustainable behavior as societies struggle with these issues. This chapter explores the psychological bases of these behaviors, highlighting the importance of good communication. Through the integration of psychological insights, our goal is to reveal the motivational factors that propel people toward sustainability and pinpoint the obstacles that impede this shift. This research is essential for developing communication plans that work in a variety of cultural contexts and successfully encourage sustainable behavior. We demonstrate the importance of evidence-based interventions and customized communication strategies in accomplishing global sustainability goals through an extensive analysis.

## 2 Psychological Foundations and Sustainable Behaviors

Understanding the psychological foundations driving sustainable behavior is crucial for promoting environmental sustainability effectively. Sustainable or pro-environmental behavior is defined as any action that minimizes harm to the environment or benefits it (Steg & Vlek, 2009).

The current climate crisis results from collective societal decisions and actions over time, leading to severe constraints on life (Intergovernmental Panel on Climate Change [IPCC], 2022). These consequences manifest as extreme temperature variations, more frequent and intense weather events, deteriorating food and water quality, and disease outbreaks (Clayton et al., 2014, 2015, 2021; Corner et al., 2017; IPCC, 2022; Steentjes et al., 2017). In response, sustainability has gained prominence across all societal sectors. European, national, and local institutions have made significant efforts to encourage this transition. Urgent action is required to meet the Paris Agreement commitments, prompting the European Commission to propose reducing net emissions by at least 55% by 2050 (European Commission, 2021). However, despite positive attitudes toward sustainability, actual behavior often lags due to various influencing factors.

Prior research has focused on the "attitude–behavior" gap, examining why pro-environmental attitudes often do not translate into sustainable behavior (Auger & Devinney, 2007; Carrington et al., 2014; Griskevicius et al., 2010; Luchs et al., 2010; Vermeir & Verbeke, 2006). Findings indicate that numerous motivational processes are at play when consumers face environmentally sustainable choices. For instance, those with higher environmental concerns (e.g., Poortinga et al., 2004) and values transcending immediate interests (e.g., Stern & Dietz, 1994) are more likely to engage in green behavior. Cognitively, consumers tend to choose alternatives with higher benefits than costs (Ajzen, 1991; Kaiser et al., 1999). Several factors, such as information quality and quantity (De Pelsmacker & Janssens, 2007) and the context of consumer choices, influence these behavioral intentions (Luchs et al., 2010; van Herpen et al., 2012). Despite these insights, previous research has not fully bridged the gap between intentions and behavior or accelerated the adoption of green practices sufficiently to make a significant change. While the market for environmentally sustainable products is growing, progress remains necessary to achieve the United Nations Sustainable Development Goal 12: Ensure Sustainable Production and Consumption Patterns. Policymakers, non-profits, and companies continue to face challenges in motivating consumers to act on their intentions (Johnstone & Tan, 2015; White et al., 2019).

Given the climate crisis's scope, various scientific fields focus on promoting climate change mitigation, adaptation, and resilience (IPCC, 2021; Global Sustainable Development Report [GSDR], 2023; Steffen et al., 2015). Social sciences and psychology aim to understand what actions can facilitate or hinder sustainable behaviors among different societal groups (Clayton et al., 2021; Newell et al., 2021). The sustainability performance of a group, organization, or country results from a complex network of internal and external factors associated with these entities.

This chapter examines the pivotal role of psychology in fostering these behaviors, emphasizing the need to understand mental processes and social dynamics. Cultural differences significantly shape values, beliefs, and social norms, requiring tailored approaches to encourage sustainability across diverse populations. This chapter explores how psychological insights can enhance the adoption of sustainable behaviors, underscoring the importance of contextualized strategies for effective implementation of the United Nations Agenda for Sustainable Development. Through a comprehensive analysis of public knowledge, attitudes, and perceptions, this chapter highlights the necessity of culturally sensitive and psychologically informed initiatives to overcome resistance and promote widespread environmental action.

## 2.1 Influence of Psychology in Adopting Sustainable Behaviors

In response to the environmental crisis, global governments have focused on collaborative efforts to mitigate its effects. Psychological insights could be key variables in better understanding and promoting the adoption of sustainable behaviors, involving research into mental processes and social dynamics. Cultural differences play a significant role in shaping values, beliefs, and social norms, necessitating customized approaches to foster sustainability across diverse populations effectively.

Successful implementation of the United Nations Agenda for Sustainable Development depends on a contextualized analysis considering cultural particularities and values (Krsnik et al., 2022). Applying a one-size-fits-all approach across different countries can fail and increase resistance to change (Zaval & Cornwell, 2017). Country-specific characteristics, such as attitudes toward conservation, consumption habits, and societal norms regarding environmental responsibility, must be carefully considered and integrated into sustainable development strategies.

Understanding public knowledge and what people still need to know to make informed decisions and change behaviors across different sociocultural contexts is essential (Bostrom et al., 1994; Bruine de Bruin & Bostrom, 2013; Morgan et al., 2002). A nonsensitive approach to these risks can negatively affect the acceptance of environmental initiatives, as cultural factors deeply influence perceptions and behaviors regarding this topic. Understanding public opinion on innovative environmental strategies, such as solar radiation management, is crucial for effective policy implementation. A cross-cultural study (Contzen et al., 2024) on Sustainability Risk Management (SRM) across 20 countries revealed low and conditional acceptance, with higher acceptance in the Global South compared to the Global North. This study underscores the necessity of considering regional attitudes and perceptions in environmental policymaking, as public opinion is significantly influenced by perceived risks and benefits. Integrating these insights can help tailor strategies to specific cultural contexts, enhancing the overall acceptance and success of sustainable development initiatives. Addressing psychological barriers to climate change mitigation and adaptation is essential for fostering sustainable behaviors. A key study identifies seven categories of these "dragons of inaction," including limited cognition, ideological worldviews, social comparisons, sunk costs, discordance, perceived risks, and insufficient behavior change (Gifford, 2011). These barriers highlight the complexity of achieving widespread environmental action and emphasize the need for targeted strategies to overcome them. Understanding these psychological obstacles can inform the development of more effective and culturally sensitive sustainability initiatives, enhancing global efforts toward environmental conservation.

Psychology plays a crucial role in this context, supporting behavior change through analyses tailored to different groups. Tailoring information channels, message formats, and communication styles to each population appears to facilitate change (Ardoin et al., 2023). Additionally, the effectiveness of these strategies may depend on other factors beyond content and information channels such as radio or television. Some studies highlight the importance of evoking emotions (e.g., fear-inducing messages to raise awareness of consequences and urgency), cognitive factors (e.g., message complexity and literacy levels), and cultural characteristics (De Vries, 2020; Kolandai-Matchett & Armoudian, 2020; Zaval & Cornwell, 2017). Various approaches have been explored to improve sustainable communication, such as using simpler figures (e.g., climate change annual reports) and adapting communication to the target audience's educational level (Bruine de Bruin et al., 2024; Wege et al., 2024). Furthermore, climate change impacts are more severe in economically disadvantaged regions due to their limited adaptive capacity (Leichenko & Silva, 2014). Effective communication strategies should prioritize low-cost interventions that align with local practices and community dynamics. Encouraging community collaboration in resource management can inspire collective action and strengthen resilience, while incorporating traditional knowledge and local languages increases the relevance and impact of the message.

## 2.2 Psychological Factors that Encourage or Hinder Eco-friendly Practices

Among the various concepts studied in the field of sustainable behaviors, emotions significantly influence sustainable behavior, as raising awareness of the negative consequences of actions and the need for change can motivate individuals to adopt new behaviors (Abu Bakar et al., 2021). Several studies in recent years, conducted in different environments and contexts, have tried to understand the relationship between a person's emotional activation and their resulting behavior (Brosch, 2021; Hui et al., 2023; Kals & Maes, 2002; Zelenski & Desrochers, 2021). These studies have explored various contexts, such as exposure to futuristic scenarios (e.g., images of advanced societies, videos of climate consequences), physical environments (e.g., hotels and restaurants), and other forms (e.g., bottle labels, posters; Rohling et al., 2016; Taff et al., 2022). The results generally indicate that both positive emotions (e.g., enthusiasm for change) and negative emotions (e.g., fear of future consequences) can trigger new behaviors.

Classic studies by Melissa Finucane and Paul Slovic, alongside their colleagues, explore the intricate relationship between risk perception and affect, fundamentally transforming our understanding of how people evaluate risks and benefits. Slovic (1987) highlighted how risk perception is influenced by psychological factors, challenging the assumption of rational decision-making in risk assessment. Building on this, Finucane et al. (2000) introduced the "affect heuristic," demonstrating that people's judgments of risks and benefits are often guided by their emotions rather than logical analysis. This concept was further explored by Slovic et al. (2004, 2007), who examined how affective responses can overshadow analytical reasoning in risk evaluations. Emotions play a pivotal role in shaping public responses to risk, often leading to biases impacting decision-making processes. Slovic and Peters (2006) emphasized the interplay between risk perception and affect, underscoring the importance of affective processes in understanding how people perceive and react to potential hazards. These studies underscore the significant role of emotions in risk perception and decision-making, challenging traditional notions of rationality in these domains.

Mass mobilization relies on constructing clear messages, properly adapted to the target audience (Bruine de Bruin & Bostrom, 2013; Morgan et al., 2002). The way the message is perceived by the public can trigger different orientations, including favorable and aligned actions or resistance (De Vries, 2020). Studies have shown that using complex technical terms (i.e., jargon) with lay audiences can make it difficult to understand the message and reduce engagement (Krsnik et al., 2022). Even if the message is clear and well-constructed from an expert's point of view, if it does not respect the characteristics of the target audience, it can trigger disinterest, indifference, detachment, or diffusion of responsibility, as seen with the bystander effect (Zaval & Cornwell, 2017). Therefore, targeting the audience and using effective messages in sustainable communication is crucial to ensure that information resonates with the audience, leading to greater engagement. For example, generational differences significantly impact communication strategies. Younger audiences, tech-savvy and environmentally conscious, engage more with digital platforms such as social media, while older generations prefer traditional media such as newspapers or television. Tailoring messages to these preferences, using familiar channels and relatable content, enhances engagement and facilitates behavior change across age groups.

While emotions' role in behavioral change is undeniable, people's knowledge about the problem is equally important when adopting new habits (Hui et al., 2023). Campaigns, advertisements, and other sustainable initiatives promoting environmental literacy have also been studied (Hu & Chen, 2016; Rohling et al., 2016). Encouraging actions through messages based on the ease and simplicity of the same action seems effective in changing behaviors (Taff et al., 2022). Conversely, while also effective, messages focused on morality seem less influential in guiding behavior compared with ease-centered messages. These initiatives have significantly increased people's engagement and literacy levels (Chang et al., 2019). When applied within a multigenerational group, the involvement of the parties is greater (Hu & Chen, 2016).

Educational programs may fail to yield desired effects if not suitably tailored to the target audience, such as considering their literacy level on the subject (Hui et al., 2023; Kolandai-Matchett, 2009). Whether disseminated through advertisements, programs, or other forms, it is imperative that these initiatives incorporate a clear and feasible action plan for everyone (De Vries, 2020). By doing so, the risk of bias related to comprehension is minimized both spatially and temporally (De Vries, 2020).

Realizing how emotions and tailored messages shape sustainable actions can highlight the need for educational programs (Brosch & Steg, 2021; White et al., 2019). When these programs fit the audience and have clear action plans, they not only enhance people's knowledge but also encourage them to engage in more initiatives. The role of social responsibility within organizations should not be devalued; it must be properly aligned with these initiatives. Exploring this can show how organizations play a key role in encouraging positive change at a societal level (Abbas & Dogan, 2022).

### 2.3 Overcoming Barriers and Embracing Success in Sustainable Behavior

To promote sustainability effectively, it is crucial to acknowledge and tackle the barriers hindering the widespread adoption of sustainable behaviors. Understanding psychological barriers is key to shaping how we feel, think, and act about sustainability (Milfont & Schultz, 2016). Despite efforts to boost environmental awareness, many remain unaware of the importance of sustainable practices (Leiserowitz et al., 2020a, b). Additionally, the perception that sustainable options are more expensive and industries' hesitation to invest in sustainability exacerbate the challenge (Lorenzoni et al., 2007). We must address these psychological barriers and explore strategies that inspire and enable individuals to make sustainable choices.

Behavioral theories, such as the Theory of Behavioral Change, offer relevant insights into the underlying factors influencing individual decisions and actions. Research suggests that intensifying the promotion of environmental literacy could be a key lever to foster behavioral change (Hu & Chen, 2016; Hui et al., 2023; Rohling et al., 2016). This involves enhancing individuals' trust, knowledge, awareness, and understanding of environmental issues and their perceived efficacy in contributing to positive change. For example, Lorenzoni et al. (2007) found that a lack of awareness and understanding about climate change can act as a significant barrier to sustainable behavior. Similarly, Oreg and Katz-Gerro (2006) emphasize the importance of values and beliefs in shaping pro-environmental behavior, underscoring the need for targeted educational interventions to address misconceptions and promote positive attitudes.

Communication plays a crucial role in promoting sustainable behaviors. While news channels are a primary source of information about environmental issues (Eurobarometer, 2019), traditional media may not be sufficient to effectively communicate complex environmental issues to the public (Maibach et al., 2010). Innovative communication strategies, such as social media campaigns and interactive educational platforms using digital influencers and community representatives, could be more effective in engaging diverse audiences and fostering meaningful behavioral change. Krsnik et al. (2022) highlight the need for targeted awareness-raising campaigns, as a significant portion of the Slovenian population was unaware of the Agenda for Sustainable Development. Promoting sustainable behaviors within social networks and communities can create a supportive and reinforcing environment for behavior change (Bamberg & Möser, 2007).

The perceived high costs of sustainable products remain a significant barrier for many consumers (Asim et al., 2022). Schultz and Kaiser (2012) suggest that price sensitivity is often influenced by perceptions of product quality and durability. Initiatives aimed at enhancing consumer trust in sustainable products, such as certification programs and product labeling schemes, could help address this barrier and stimulate demand for sustainable alternatives. Strategies to lower production and marketing costs are essential to make these products more accessible. Increasing investments in sectors such as packaging and the food industry could lead to economies of scale and lower costs for sustainable products, making them more accessible to consumers and creating opportunities for small organizations to drive further innovation (Agarwal et al., 2023; Capodaglio, 2023; Cruz et al., 2022; Findrik & Meixner, 2023).

Recognizing the barriers is essential, but drawing inspiration from success stories and best practices is equally important. Exemplary cases in Japan and the United States offer valuable insights into the transformative potential of sustainable initiatives (Shenyoputro & Jones, 2023; Zaman, 2022). These success stories underscore the pivotal role of psychology in driving positive change and highlight the need for collective action across various societal levels (Zaman & Newman, 2021).

Despite the complexity of this behavioral change, exemplary cases have emerged in the transition to a green economy. In Japan, the city of Kamikatsu has made notable progress toward carbon neutrality through waste reuse policies and incentives for the circular economy. A study by Shenyoputro and Jones (2023) highlighted Kamikatsu's remarkable progress, including innovative recycling capacity, tax relief policies, and incentives for companies producing sustainable products. These measures have not only reduced carbon emissions but also fostered a thriving circular economy, where waste is viewed as a valuable resource.

San Francisco in the United States is another success story. Zaman (2022) underscores the city's commitment to zero waste policies, which have significantly enhanced resource reuse and management. By aligning public policies with the principles of the circular economy, San Francisco has implemented incentives to promote repair, reuse, and the purchase of sustainable products. These initiatives have demonstrated the feasibility of managing waste streams more efficiently and sustainably.

These cases show that change is possible and highlight the importance of a collective movement sensitive to the cause. For significant results, it is important to involve all stakeholders, including political, organizational, and community levels (Zaman & Newman, 2021). Studies emphasize promoting shared knowledge toward sustainability (Topal et al., 2021). By fostering awareness and establishing platforms for dialogue and cooperation involving government, businesses, nongovernmental organizations (NGOs), and communities, we can stimulate collective action toward sustainable behaviors. The media plays a key role in this, as it is a primary source of information on environmental and sustainability matters in many countries (Eurobarometer, 2019). Clear and consistent messages from the media can stimulate action at different societal levels and encourage the emergence of more success stories worldwide (Culliford et al., 2023; Shahzalal & Hassan, 2019). By understanding and addressing these barriers through targeted education, innovative communication, and strategic investment, we can create an environment that encourages and supports sustainable behaviors across diverse populations. Additionally, drawing inspiration from successful examples can provide practical insights and foster a collective movement toward a more sustainable future.

# 3 Effective Communication for Sustainability

Effective communication is pivotal in changing behavior, particularly in the context of sustainability. Implementing measures without adequately informing and contextualizing the recipients can fail to achieve the desired effect and foster resistance due to a lack of trust and transparency. In the current climate crisis, all information on this topic has the potential to significantly influence attitudes and behaviors.

Genç (2017) identifies three major approaches to sustainability-oriented communication: Communication about Sustainability (CaS), Communication of Sustainability (CoS), and Communication for Sustainability (CfS). CaS focuses on discussing emerging problems and concepts related to sustainability. CoS involves operationalizing measures, often through vertical, unidirectional communication (e.g., political measures, European directives). CfS combines both approaches, aiming for a holistic transformation of society.

Effective communication for sustainability requires a nuanced approach that considers the complexity of human behavior, the power of media, and the importance of credibility and transparency. By leveraging behavioral insights and utilizing various communication strategies, we can foster a more sustainable future. Specific examples, such as the successful implementation of green nudges, demonstrate the practical application of these principles. Through tailored messages, credible communicators, and the strategic use of media, we can significantly influence public perceptions and behaviors toward sustainability, driving societal transformation and environmental protection.

## *3.1 The Media's Impact on Shaping Public Perceptions of Sustainability*

The media plays a central role in shaping public understanding and responses to sustainability. Numerous studies highlight the media's significant impact on public perceptions and behaviors regarding environmental issues (Carmichael & Brulle, 2017; Wu et al., 2023). Through various channels such as news coverage, newspapers, and social media, media organizations influence how we perceive and address sustainability challenges (Harcourt et al., 2020; Sampei & Aoyagi-Usui, 2009).

Research shows that the way environmental stories are presented can profoundly influence audience attitudes and behaviors toward sustainability. Buschini and Guillou (2022) describe mechanisms such as diffusion, effusion, propaganda, and propagation, illustrating how environmental messages are disseminated, amplified, and communicated to create a lasting impact on public perceptions.

Social media exposure to environmental information can shape individuals' perceptions and beliefs about climate change, motivating them to engage in pro-environmental causes (Schäfer, 2012). Media coverage often aligns with broader societal trends and movements, such as activism led by younger generations influenced by social media (Alam et al., 2024). For instance, the #FridaysForFuture movement, led by Greta Thunberg, gained significant traction through social media, mobilizing millions of young people worldwide to demand action on climate change. By amplifying activist voices and providing platforms for environmental movements, media organizations shape public discourse and policy agendas related to sustainability.

### *3.2 Interventions and Communication Strategies: Principles of Crafting Messages that Resonate with Diverse Audiences*

News channels are the primary source of climate information for Europeans, making it essential to understand how these sources can enhance or hinder message dissemination. Media entities hold significant social responsibility as their information influences public behavior and attitudes (Jha & Verma, 2023). They can stimulate new knowledge, promote organizational progress, and regulate practices through public opinion.

Effective communication involves integrating credible speakers, designing tailored messages that meet citizens' informational needs, and ensuring clarity and transparency (Riley et al., 2021). In today's media landscape filled with information and disinformation, the credibility of the communicator and their organization is crucial (Jain et al., 2022). Studies indicate that people exhibit higher levels of pro-environmental behavior when exposed to influencers rather than scientists (Smith et al., 2023). Aligning the message with the audience, facilitated by appropriate communicators, enhances public engagement (Abu Bakar et al., 2021). Emotional involvement further boosts engagement (Culliford et al., 2023; Richter et al., 2021). For example, campaigns featuring celebrities known for their environmental advocacy, such as Leonardo DiCaprio or Emma Watson, can significantly boost public interest and engagement in sustainability issues. Tailoring messages to specific audiences, such as using local languages and cultural references, can also increase the effectiveness of communication strategies.

All things considered, a successful communication strategy for advancing sustainability should incorporate a mix of informational and emotional appeals, audience segmentation, and well-defined goals. Culturally relevant messaging, reliable and transparent sources, and feedback systems to improve tactics are essential elements. Broader reach is ensured by combining digital and traditional media, and collaborations with neighborhood organizations can improve community involvement and message credibility.

### *3.3 Research-Based Strategies for Influencing Environmental Attitudes*

Recent studies explore how communication shapes sustainable behaviors (e.g., Shahzalal & Hassan, 2019; Kolandai-Matchett & Armoudian, 2020). A hybrid approach combining face-to-face and online communication seems most effective, extending campaign reach and deepening engagement (Kolandai-Matchett et al., 2021; Nayeem et al., 2023). Textual information can sometimes have a more significant impact than visual stimuli, especially when using compelling narratives and emotion-based language (Shen, 2023; Skoric et al., 2022).

Nudges, subtle encouragements based on behavioral economics, are central tools in promoting sustainable choices (Ngo et al., 2022; Vandenbroele et al., 2020). For instance, framing messages to emphasize gains rather than losses can elicit more positive responses (Kause et al., 2021). Highlighting the benefits of sustainable behaviors encourages adoption. The concept of "green nudges," interventions that encourage eco-friendly behaviors without restricting choice, is particularly relevant. These nudges leverage human tendencies such as loss aversion and inertia to make sustainable choices more accessible and appealing (Santos Silva, 2022; Thaler & Sunstein, 2008).

Successful green nudges include default options for renewable energy, significantly increasing participation rates. For example, a study in Germany found that when households were automatically enrolled in green energy programs with an option to opt-out, participation rates soared compared to opt-in programs (Pichert & Katsikopoulos, 2008). Eco-labels on products can nudge consumers toward more sustainable choices by providing clear and accessible information about the environmental impact of their purchases (Andor & Fels, 2018).

Multisensory experiences that trigger emotional responses can shape decision-making processes, making sustainability messages more impactful. Media organizations, such as radio stations, can leverage the intimate quality of broadcasts to deeply connect with listeners, fostering empathy and understanding (Kolandai-Matchett & Armoudian, 2020). Showcasing local success stories can motivate action and cultivate a sense of collective efficacy.

### 3.4 *Strategies for Customizing Communication to Different Contexts*

As we transition into the digital age, traditional news presentation methods such as television, radio, and newspapers are increasingly moving online, incorporating more user-generated content. This shift enables news to be constantly updated and introduces a new dynamic between readers and journalists through the ability to comment on publications. In an era marked by growing public skepticism toward the news, online news can shorten the interaction between the public and journalists, helping counteract the rise in misinformation perceptions (Allen et al., 2020; Altay et al., 2023). Given the significant visibility and influence that news organizations have on public attitudes and behavior, ensuring the reliability of information while maintaining trust is crucial.

Emerging technologies such as artificial intelligence (AI) and virtual reality (VR) offer groundbreaking opportunities for improving sustainability communication. AI can dynamically analyze audience behavior, tailor messages to different demographics, and optimize timing for message delivery to enhance engagement. Meanwhile, VR enables immersive experiences, helping individuals visualize the consequences of their choices, such as the environmental impact of packaging or dietary habits. For example, Farias et al. (2024) employed VR restaurant scenarios to explore the interplay of dietary choices, packaging preferences, and environmental attitudes, demonstrating that immersive simulations can effectively foster pro-environmental behaviors. By integrating these tools, communication strategies can create more personalized and impactful experiences, empowering audiences to make sustainable decisions grounded in emotional and cognitive engagement.

To address the public's perceived risk regarding news credibility, Gunster (2017) suggests several strategies. These include focusing on populations more likely to embrace change, adapting the discourse to align with their values and norms, and illustrating points with real-life examples. Additionally, Dernbach (2003) advocates for a new orientation in news reporting on climate change, emphasizing that it should not solely focus on negative events and catastrophes. Instead, it is essential to clearly present success stories and action plans to engage and motivate people toward positive change (Wege et al., 2024).

Presenting communication approaches in various formats simultaneously can significantly promote sustainable behavior. However, using multiple information channels without proper monitoring can increase the risk of spreading misinformation (Guenther et al., 2024). Gathering feedback from the target audience helps improve content suitability and ensures accurate transmission of information (Kolandai-Matchett & Armoudian, 2020; Kolandai-Matchett et al., 2021; Nayeem et al., 2023). Encouraging debates can effectively promote reflection and critical thinking on sustainability issues. Similar to political debates, public reactions on social media can stimulate discussions (Fontenla-Pedreira et al., 2023). Promoting debates on sustainability can enhance public knowledge and critical thinking, leading to the desired social change (Rodriguez-Dono & Hernández-Fernández, 2021).

## 4 Conclusion

This chapter has explored the critical interplay between psychological insights and effective communication strategies to foster sustainable behavior. By examining the psychological foundations of sustainable actions, we have highlighted how perceptions of climate and other environmental risks, emotional responses, and cultural contexts drive or hinder behavioral change. The analysis has underscored the necessity of tailored communication strategies that resonate with diverse audiences to overcome barriers such as lack of awareness, perceived high costs, and misinformation.

Furthermore, the chapter has elucidated the role of media influence in shaping public perceptions and promoting eco-friendly practices. Effective communication, grounded in psychological understanding and evidence-based strategies, emerges as a pivotal tool in managing conflicting information and fostering informed social debate on sustainability.

Through comprehensive exploration, this chapter has emphasized the importance of customizing interventions to various cultural and contextual landscapes. It advocates for the strategic use of emotional engagement, clarity, and transparency in messages, leveraging the credibility of communicators and the power of media to drive societal transformation. The integration of psychological insights with innovative communication techniques holds the potential to optimize the adoption of sustainable behaviors, thereby contributing to a more sustainable future. Effective communication is not just a supplementary tool but a fundamental driver in achieving global sustainability goals.

## References

Abbas, J., & Dogan, E. (2022). The impacts of organizational green culture and corporate social responsibility on employees' responsible behaviour towards the society. *Environmental Science and Pollution Research, 29*(40), 60024–60034. https://doi.org/10.1007/s11356-022-20072-w

Abu Bakar, M. F., Wu, W., Proverbs, D., & Mavritsaki, E. (2021). Effective communication for water resilient communities: A conceptual framework. *Water, 13*(20), 2880. https://doi.org/10.3390/w13202880

Agarwal, A., Shaida, B., Rastogi, M., & Singh, N. B. (2023). Food packaging materials with special reference to biopolymers-properties and applications. *Chemistry Africa, 6*(1), 117–144. https://doi.org/10.1007/s42250-022-00446-w

Ajzen, I. (1991). The theory of planned behavior. *Organizational Behavior and Human Decision Processes, 50*(2), 179–211. https://doi.org/10.1016/0749-5978(91)90020-T

Alam, F., Tao, M., Rastogi, R., Mendiratta, A., & Attri, R. (2024). Do social media influencers influence the vaccination drive? An application of source credibility theory and uses and gratification theory. *Technological Forecasting and Social Change, 198*, 122973. https://doi.org/10.1016/j.techfore.2023.122973

Allen, J., Howland, B., Mobius, M., Rothschild, D., & Watts, D. J. (2020). Evaluating the fake news problem at the scale of the information ecosystem. *Science Advances, 6*(14), eaay3539. https://doi.org/10.1126/sciadv.aay3539

Altay, S., Berriche, M., & Acerbi, A. (2023). Misinformation on misinformation: Conceptual and methodological challenges. *Social Media + Society, 9*(1). https://doi.org/10.1177/20563051221150412

Andor, M. A., & Fels, K. M. (2018). Behavioral Economics and Energy Conservation – A Systematic Review of Non-price Interventions and Their Causal Effects. *Ecological Economics, 148*, 178–210. https://doi.org/10.1016/j.ecolecon.2018.01.018

Ardoin, N. M., Bowers, A. W., & Wheaton, M. (2023). Leveraging collective action and environmental literacy to address complex sustainability challenges. *Ambio, 52*(1), 30–44. https://doi.org/10.1007/s13280-022-01764-6

Asim, N., Su'ait, M. S., Badiei, M., Mohammad, M., Akhtaruzzaman, M. D., Rajabi, A., Amin, N., & Ghazali, M. J. (2022). Perspectives in biopolymer/graphene-based composite application: Advances, challenges, and recommendations. *Nanotechnology Reviews, 11*(1), 1525–1554. https://doi.org/10.1515/ntrev-2022-0087

Auger, P., & Devinney, T. M. (2007). Do what consumers say matter? The misalignment of preferences with unconstrained ethical intentions. *Journal of Business Ethics, 76*(4), 361–383. https://doi.org/10.1007/s10551-006-9287-y

Bamberg, S., & Möser, G. (2007). Twenty years after Hines, Hungerford, and Tomera: A new meta-analysis of psycho-social determinants of pro-environmental behaviour. *Journal of Environmental Psychology, 27*(1), 14–25. https://doi.org/10.1016/j.jenvp.2006.12.002

Bostrom, A., Morgan, M. G., Fischhoff, B., & Read, D. (1994). What do people know about global climate change? 1. Mental models. *Risk Analysis, 14*, 959–970. https://doi.org/10.1111/j.1539-6924.1994.tb00065.x

Brosch, T. (2021). Affect and emotions as drivers of climate change perception and action: A review. *Current Opinion in Behavioral Sciences, 42*, 15–21.

Brosch, T., & Steg, L. (2021). Leveraging emotion for sustainable action. *One Earth, 4*(12), 1693–1703. https://doi.org/10.1016/j.oneear.2021.11.006

Bruine de Bruin, W., & Bostrom, A. (2013). Assessing what to address in science communication. *Proceedings of the National Academy of Sciences of the United States of America, 110*, 14062–14068. https://doi.org/10.1073/pnas.1212729110

Bruine de Bruin, W., Rabinovich, L., Weber, K., Babboni, M., Ignon, L., Wald, R., Dean, M., Kashdan, A., & Luz, S. (2024). Improving figures for climate change communications: Insights from interviews with international policymakers and practitioners. *Climatic Change, 177*, 57. https://doi.org/10.1007/s10584-024-03704-7

Buschini, F., & Guillou, E. (2022). Diffusion, propagation, propaganda: And then came effusion. A new mode of communication for social representations. *Papers on Social Representations, 31*(2), 11.1–11.21.

Capodaglio, A. G. (2023). Biorefinery of sewage sludge: Overview of possible value-added products and applicable process technologies. *Water, 15*(6), 1195. https://doi.org/10.3390/w15061195

Carmichael, J. T., & Brulle, R. J. (2017). Elite cues, media coverage, and public concern: An integrated path analysis of public opinion on climate change, 2001–2013. *Environmental Politics, 26*(2), 232–252. https://doi.org/10.1080/09644016.2016.1263433

Carrington, M. J., Neville, B. A., & Whitwell, G. J. (2014). Lost in translation: Exploring the ethical consumer intention–behavior gap. *Journal of Business Research, 67*(1), 2759–2767. https://doi.org/10.1016/j.jbusres.2012.09.022

Chang, H. H., Tsai, S. H., & Huang, C. C. (2019). Sustainable development: The effects of environmental policy disclosure in advertising. *Business Strategy and the Environment, 28*(8), 1497–1506. https://doi.org/10.1002/bse.2325

Clayton, S., Manning, C., & Hodge, C. (2014). Beyond storms & droughts: The psychological impacts of climate change. *American Psychological Association and ecoAmerica.*. https://ecoamerica.org/wp-content/uploads/2014/06/eA_Beyond_Storms_and_Droughts_Psych_Impacts_of_Climate_Change.pdf

Clayton, S., Devine-Wright, P., Stern, P. C., Whitmarsh, L., Carrico, A., Steg, L., Swim, J., & Bonnes, M. (2015). Psychological research and global climate change. *Nature Climate Change, 5*, 640–646. https://doi.org/10.1038/nclimate2622

Clayton, S., Manning, C. M., Speiser, M., & Hill, A. N. (2021). *Mental health and our changing climate: Impacts, inequities, responses*. American Psychological Association and ecoAmerica. https://www.apa.org/news/press/releases/mental-health-climate-change.pdf

Contzen, N., Perlaviciute, G., Steg, L., Reckels, S. C., Alves, S., Bidwell, D., et al. (2024). Public opinion about solar radiation management: A cross-cultural study in 20 countries around the world. *Climatic Change, 65*. https://doi.org/10.1007/s10584-024-03708-3

Corner, A., Arnold, A., Böhm, G., Mays, C., Pidgeon, N., Poortinga, W., Poumadère, M., Scheer, D., Sonnberger, M., Steentjes, K., & Tvinnereim, E. (2017). *European perceptions of climate change – Six Recommendations for Public Engagement.* Climate Outreach. https://orca.cardiff.ac.uk/id/eprint/99014/1/EPCC%20Recommendations_final.pdf

Cruz, R. M., Krauter, V., Krauter, S., Agriopoulou, S., Weinrich, R., Herbes, C., Scholten, P. B., Uysal-Unalan, I., Sogut, E., Kopacic, S., Lahti, J., Rutkaite, R., & Varzakas, T. (2022). Bioplastics for food packaging: Environmental impact, trends and regulatory aspects. *Food, 11*, 3087. https://doi.org/10.3390/foods11193087

Culliford, A. E., Bradbury, J., & Medici, E. B. (2023). Improving communication of the UK sustainable healthy dietary guidelines the eatwell guide: A rapid review. *Sustainability, 15*(7), 6149. https://doi.org/10.3390/su15076149

De Pelsmacker, P., & Janssens, W. A. (2007). Model for fair trade buying behaviour: The role of perceived quantity and quality of information and of product-specific attitudes. *Journal of Business Ethics, 75*, 361–380. https://doi.org/10.1007/s10551-006-9259-2

De Vries, G. (2020). Public communication as a tool to implement environmental policies. *Social Issues and Policy Review, 14*(1), 244–272. https://doi.org/10.1111/sipr.12061

Dernbach, J. C. (2003). Achieving sustainable development: The centrality and multiple facets of integrated decision making. *Indiana Journal of Global Legal Studies, 10*(1), 247–285. https://doi.org/10.1353/gls.2003.0006

Eurobarometer. (2019). *Attitudes of Europeans towards the Environment - Summary*. European Commission. https://ec.europa.eu/clima/sites/clima/files/eurobarometer-2019-climate-change-report_en.pdf

European Commission (14 July, 2021). *European Green Deal: Commission proposes transformation of EU economy and society to meet climate ambitions*. European Commission - Press release. https://ec.europa.eu/commission/presscorner/detail/en/IP_21_3541

Farias, A. R., Lane, H., Killingsworth, J., Warden, J. M., & Wais, S. (2024). Exploring the interplay of pro-environmental attitudes, dietary choices, and packaging preferences: A virtual reality restaurant scenario study. *Challenges, 15*(1), 1. https://doi.org/10.3390/challe15010001

Findrik, E., & Meixner, O. (2023). Drivers and barriers for consumers purchasing bioplastics – A systematic literature review. *Journal of Cleaner Production, 410*, 137311. https://doi.org/10.1016/j.jclepro.2023.137311

Finucane, M. L., Alhakami, A., Slovic, P., & Johnson, S. M. (2000). The affect heuristic in judgments of risks and benefits. *Journal of Behavioral Decision Making, 13*(1), 1–17. https://doi.org/10.1002/(SICI)1099-0771(200001/03)13:1<1::AID-BDM333>3.0.CO;2-S

Fontenla-Pedreira, J., Maiz-Bar, C., & Rodríguez-Martelo, T. (2023). Use of Twitter during televised election debates: Spanish general election (28 April 2019) vs. French general election (24 April 2022). *Societies, 13*(3), 70. https://doi.org/10.3390/soc13030070

Genç, R. (2017). The importance of communication in sustainability & sustainable strategies. *Procedia Manufacturing, 8*, 511–516. https://doi.org/10.1016/j.promfg.2017.02.065

Gifford, R. (2011). The dragons of inaction: Psychological barriers that limit climate change mitigation and adaptation. *American Psychologist, 66*(4), 290–302. https://doi.org/10.1037/a0023566

Global Sustainable Development Report [GSDR]. (2023). *Times of Crisis, Times of Change. Science for accelerating transformations to sustainable development.* United Nations. https://sdgs.un.org/sites/default/files/2023-09/FINAL%20GSDR%202023-Digital%20-110923_1.pdf
Griskevicius, V., Tybur, J. M., & Van den Bergh, B. (2010). Going green to be seen: Status, reputation, and conspicuous conservation. *Journal of Personality and Social Psychology, 98*(3), 392–404. https://doi.org/10.1037/a0017346
Guenther, L., Ibrahim, Y., Lüdemann, J., O'Brian, C. I., Pavenstädt, C. N., Sievert, I. J., Brüggemann, M., Rödder, S., & Schnegg, M. (2024). Social constructions of climate futures: Reframing science's harmful impact frame across news media, social movements, and local communities. *Environmental Communication, 18*(3), 1–17. https://doi.org/10.1080/17524032.2024.2305827
Gunster, S. (2017). Engaging climate communication: Audiences, frames, values and norms. In R. A. Hackett, S. Forde, S. Gunster, & K. Foxwell-Norton (Eds.), *Journalism and climate crisis: Public engagement, media alternatives* (pp. 49–76). Routledge.
Harcourt, R., Bruine de Bruin, W., Dessai, S., & Taylor, A. (2020). What adaptation stories are UK newspapers telling? A Narrative Analysis. *Environmental Communication, 14*(8), 1061–1078. https://doi.org/10.1080/17524032.2020.1767672
Hu, S., & Chen, J. (2016). Place-based inter-generational communication on local climate improves adolescents' perceptions and willingness to mitigate climate change. *Climatic Change, 138*(3-4), 425–438. https://doi.org/10.1007/s10584-016-1746-6
Hui, X., Raza, S. H., Khan, S. W., Zaman, U., & Ogadimma, E. C. (2023). Exploring regenerative tourism using media richness theory: Emerging role of immersive journalism, metaverse-based promotion, eco-literacy, and pro-environmental behavior. *Sustainability, 15*(6), 5046. https://doi.org/10.3390/su15065046
Intergovernmental Panel on Climate Change [IPCC]. (2021). *Climate change 2021: The physical science basis. Contribution of working group I to the sixth assessment report of the intergovernmental panel on climate change.* Cambridge University Press. https://doi.org/10.1017/9781009157896
Intergovernmental Panel on Climate Change [IPCC]. (2022). *Climate change 2022: Impacts, adaptation and vulnerability. Contribution of working group II to the sixth assessment report of the intergovernmental panel on climate change.* Cambridge University Press. https://doi.org/10.1017/9781009325844
Jain, R., Luck, E., Mathews, S., & Schuster, L. (2022). Creating persuasive environmental communicators: Spokescharacters as endorsers in promoting sustainable behaviors. *Sustainability, 15*(1), 335. https://doi.org/10.3390/su15010335
Jha, A. K., & Verma, N. K. (2023). Social media sustainability communication: An analysis of firm behaviour and stakeholder responses. *Information Systems Frontiers, 25*(2), 723–742. https://doi.org/10.1007/s10796-022-10257-6
Johnstone, M. L., & Tan, L. P. (2015). Exploring the gap between consumers' green rhetoric and purchasing behaviour. *Journal of Business Ethics, 132*(2), 311–328. https://doi.org/10.1007/s10551-014-2316-3
Kaiser, F. G., Wölfing, S., & Fuhrer, U. (1999). Environmental attitude and ecological behaviour. *Journal of Environmental Psychology, 19*(1), 1–19. https://doi.org/10.1006/jevp.1998.0107
Kals, E., & Maes, J. (2002). Sustainable development and emotions. In P. Schmuck & W. P. Schultz (Eds.), *Psychology of sustainable development* (pp. 97–122). Kluwer Academic Publishers. https://doi.org/10.1007/978-1-4615-0995-0_6
Kause, A., Bruine de Bruin, W., Domingos, S., Mittal, N., Lowe, J., & Fung, F. (2021). Communications about uncertainty in scientific climate-related findings: A qualitative systematic review. *Environmental Research Letters, 16*(5), 053005. https://doi.org/10.1088/1748-9326/abb265
Kolandai-Matchett, K. (2009). Mediated communication of 'sustainable consumption' in the alternative media: a case study exploring a message framing strategy. *International Journal of Consumer Studies, 33*(2), 113–125. https://doi.org/10.1111/j.1470-6431.2009.00754.x

Kolandai-Matchett, K., & Armoudian, M. (2020). Message framing strategies for effective marine conservation communication. *Aquatic Conservation: Marine and Freshwater Ecosystems, 30*(12), 2441–2463. https://doi.org/10.1002/aqc.3349

Kolandai-Matchett, K., Armoudian, M., & Thrush, S. (2021). Communicating complex marine science: Does media format matter? *Aquatic Conservation: Marine and Freshwater Ecosystems, 31*(7), 1772–1790. https://doi.org/10.1002/aqc.3560

Krsnik, S., Erjavec, K., & Klopčič, M. (2022). Impact of citizens' personal values, knowledge, awareness, informing, advertising, and truth of environmental and climate challenges in support of the sustainable development goals. *Sustainability, 14*(12), 7333. https://doi.org/10.3390/su14127333

Leichenko, R., & Silva, J. A. (2014). Climate change and poverty: Vulnerability, impacts, and alleviation strategies. *Wiley Interdisciplinary Reviews: Climate Change, 5*(4), 539–556.

Leiserowitz, A. A., Maibach, E., Roser-Renouf, C., Feinberg, G., & Rosenthal, S. (2020a). *Climate change in the American mind.* University of Washington.

Leiserowitz, A., Maibach, E., Rosenthal, S., Kotcher, J., Bergquist, P., Ballew, M., Goldberg, M., Gustafson, A., & Wang, X. (2020b). *Climate change in the American mind: April 2020.* Yale University and George Mason University.

Lorenzoni, I., Nicholson-Cole, S., & Whitmarsh, L. (2007). Barriers perceived to engaging with climate change among the UK public and their policy implications. *Global Environmental Change, 17*(3–4), 445–459. https://doi.org/10.1016/j.gloenvcha.2007.01.004

Luchs, M. G., Naylor, R. W., Irwin, J. R., & Raghunathan, R. (2010). The sustainability liability: Potential negative effects of ethicality on product preference. *Journal of Marketing, 74*(5), 18–31. https://doi.org/10.1509/jmkg.74.5.018

Maibach, E. W., Nisbet, M., Baldwin, P., Akerlof, K., & Diao, G. (2010). Reframing climate change as a public health issue: An exploratory study of public reactions. *BMC Public Health, 10*, 1–11. https://doi.org/10.1186/1471-2458-10-299

Milfont, T. L., & Schultz, P. W. (2016). Culture and the natural environment. *Current Opinion in Psychology, 8*, 194–199. https://doi.org/10.1016/j.copsyc.2015.09.009

Morgan, M. G., Fischhoff, B., Bostrom, A., & Atman, C. J. (2002). *Risk communication: A mental models approach.* Cambridge University Press. isbn:978-0-521-00256-1.

Nayeem, T., Pawsey, N., Murshed, F., Baumgartner, L., Boys, C., & Rayner, T. (2023). Modern sustainable fish screens: A study on developing effective communication with water users. *Sustainability, 15*(9), 7694. https://doi.org/10.3390/su15097694

Newell, P., Srivastava, S., Naess, L. O., Torres Contreras, G. A., & Price, R. (2021). Toward transformative climate justice: An emerging research agenda. *Wiley Interdisciplinary Reviews: Climate Change, 12*(6), e733. https://doi.org/10.1002/wcc.733

Ngo, C. C., Poortvliet, P. M., & Klerkx, L. (2022). The persuasiveness of gain vs. loss framed messages on farmers' perceptions and decisions to climate change: A case study in coastal communities of Vietnam. *Climate Risk Management, 35*, 100409. https://doi.org/10.1016/j.crm.2022.100409

Oreg, S., & Katz-Gerro, T. (2006). Predicting proenvironmental behavior cross-nationally: Values, the theory of planned behavior, and value-belief-norm theory. *Environment and Behavior, 38*(4), 462–483. https://doi.org/10.1177/0013916505286012

Pichert, D., & Katsikopoulos, K. V. (2008). Green defaults: Information presentation and pro-environmental behaviour. *Journal of Environmental Psychology, 28*(1), 63–73. https://doi.org/10.1016/j.jenvp.2007.09.004

Poortinga, W., Steg, L., & Vlek, C. (2004). Values, environmental concern, and environmental behavior: A study into household energy use. *Environment and Behavior, 36*(1), 70–93. https://doi.org/10.1177/0013916503251466

Richter, I., Sumeldan, J., Avillanosa, A., Gabe-Thomas, E., Creencia, L., & Pahl, S. (2021). Co-created future scenarios as a tool to communicate sustainable development in coastal communities in Palawan, Philippines. *Frontiers in Psychology, 12*, 627972. https://doi.org/10.3389/fpsyg.2021.627972

Riley, R., de Preux, L., Capella, P., Mejia, C., Kajikawa, Y., & de Nazelle, A. (2021). How do we effectively communicate air pollution to change public attitudes and behaviours? A review. *Sustainability Science, 16*, 1–21. https://doi.org/10.1007/s11625-021-01038-2

Rodriguez-Dono, A., & Hernández-Fernández, A. (2021). Fostering sustainability and critical thinking through debate - A case study. *Sustainability, 13*(11), 6397. https://doi.org/10.3390/su13116397

Rohling, K., Wandersee, C., Baker, L. M., & Tomlinson, P. (2016). Communicating climate change: A qualitative study exploring how communicators and educators are approaching climate-change discussions. *Journal of Applied Communications, 100*(3), 9. https://doi.org/10.4148/1051-0834.1232

Sampei, Y., & Aoyagi-Usui, M. (2009). Mass-media coverage, its influence on public awareness of climate-change issues, and implications for Japan's national campaign to reduce greenhouse gas emissions. *Global Environmental Change, 19*(2), 203–212. https://doi.org/10.1016/j.gloenvcha.2008.10.005

Santos Silva, M. (2022). Nudging and other behaviourally based policies as enablers for environmental sustainability. *Laws, 11*(1), 9. https://doi.org/10.3390/laws11010009

Schäfer, M. S. (2012). Online communication on climate change and climate politics: A literature review. *Wiley Interdisciplinary Reviews: Climate Change, 3*(6), 527–543. https://doi.org/10.1002/wcc.191

Schultz, P. W., & Kaiser, F. G. (2012). Promoting pro-environmental behavior. In S. D. Clayton (Ed.), *The Oxford handbook of environmental and conservation psychology* (pp. 556–580). Oxford University Press. https://doi.org/10.1093/oxfordhb/9780199733026.013.0029

Shahzalal, M. D., & Hassan, A. (2019). Communicating sustainability: Using community media to influence rural people's intention to adopt sustainable behaviour. *Sustainability, 11*(3), 812. https://doi.org/10.3390/su11030812

Shen, Z. (2023). Mining sustainable fashion e-commerce: social media texts and consumer behaviors. *Electronic Commerce Research, 23*(2), 949–971. https://doi.org/10.1007/s10660-021-09498-5

Shenyoputro, K., & Jones, T. E. (2023). Reflections on a two-decade journey toward zero waste: A case study of Kamikatsu town, Japan. *Frontiers in Environmental Science, 11*, 443. https://doi.org/10.3389/fenvs.2023.1171379

Skoric, M. M., Zhang, N., Kasadha, J., Tse, C. H., & Liu, J. (2022). Reducing the use of disposable plastics through public engagement campaigns: An experimental study of the effectiveness of message appeals, modalities, and sources. *International Journal of Environmental Research and Public Health, 19*(14), 8273. https://doi.org/10.3390/ijerph19148273

Slovic, P. (1987). Perception of risk. *Science, 236*(4799), 280–285. https://doi.org/10.1126/science.3563507

Slovic, P., & Peters, E. (2006). Risk perception and affect. *Current Directions in Psychological Science, 15*(6), 322–325. https://doi.org/10.1111/j.1467-8721.2006.00461.x

Slovic, P., Finucane, M. L., Peters, E., & MacGregor, D. G. (2004). Risk as analysis and risk as feelings: Some thoughts about affect, reason, risk, and rationality. *Risk Analysis, 24*(2), 311–3220. https://doi.org/10.1111/j.0272-4332.2004.00433.x

Slovic, P., Finucane, M. L., Peters, E., & MacGregor, D. G. (2007). The affect heuristic. *European Journal of Operational Research, 177*(3), 1333–1352. https://doi.org/10.1016/j.ejor.2005.04.006

Smith, M., Finkler, W., & Aitken, R. (2023). Connecting people with science: A proof-of-concept study to evaluate action-based storytelling for science communication. *Sustainability, 15*(15), 11655. https://doi.org/10.3390/su151511655

Steentjes, K., Pidgeon, N., Poortinga, W., Corner, A., Arnold, A., Böhm, G., et al. (2017). *European Perceptions of Climate Change (EPCC): Topline findings of a survey conducted in four European countries in 2016*. Cardiff University. https://orca.cardiff.ac.uk/id/eprint/98660/7/EPCC.pdf

Steffen, W., Richardson, K., Rockstrom, J., Cornell, S. E., Fetzer, I., Bennett, E. M., et al. (2015). Planetary boundaries: Guiding human development on a changing planet. *Science, 347*(6223), 1259855–1259855. https://doi.org/10.1126/science.1259855

Steg, L., & Vlek, C. (2009). Encouraging pro-environmental behaviour: An integrative review and research agenda. *Journal of Environmental Psychology, 29*(3), 309–317. https://doi.org/10.1016/j.jenvp.2008.10.004

Stern, P. C., & Dietz, T. (1994). The value basis of environmental concern. *Journal of Social Issues, 50*(3), 65–84. https://doi.org/10.1111/j.1540-4560.1994.tb02420.x

Taff, B. D., Miller, Z., Lawhon, B., Freeman, S., & Newman, P. (2022). Effective communication and campground recycling: Lessons learned from Yosemite, Grand Teton, and Denali National Parks. *Land, 11*(10), 1872. https://doi.org/10.3390/land11101872

Thaler, R. H., & Sunstein, C. R. (2008). *Nudge: Improving decisions about health, wealth, and happiness*. Yale University Press. https://doi.org/10.15695/hmltc.v33i2.3307

Topal, H. F., Hunt, D. V. L., & Rogers, C. D. F. (2021). Sustainability understanding and behaviors across urban areas: A case study on Istanbul City. *Sustainability, 13*, 7711. https://doi.org/10.3390/su13147711

van Herpen, E., van Nierop, E., & Sloot, L. (2012). The relationship between in-store marketing and observed sales for organic versus fair trade products. *Marketing Letters, 23*, 293–308. https://doi.org/10.1007/s11002-011-9154-1

Vandenbroele, J., Vermeir, I., Geuens, M., Slabbinck, H., & Van Kerckhove, A. (2020). Nudging to get our food choices on a sustainable track. *Proceedings of the Nutrition Society, 79*(1), 133–146. https://doi.org/10.1017/S0029665119000971

Vermeir, I., & Verbeke, W. (2006). Sustainable food consumption: Exploring the consumer "Attitude – behavioral intention" gap. *Journal of Agricultural and Environmental Ethics, 19*, 169–194. https://doi.org/10.1007/s10806-005-5485-3

Wege, L., Bruine de Bruin, W., & Kause, A. (2024). Public understanding of climate change terminology in Germany. *Climatic Change, 177*, 81. https://doi.org/10.1007/s10584-024-03725-2

White, K., Habib, R., & Hardisty, D. J. (2019). How to SHIFT consumer behaviors to be more sustainable: A literature review and guiding framework. *Journal of Marketing, 83*(3), 22–49. https://doi.org/10.1177/0022242919825649

World Health Organization. (2021). *Climate change and health.* https://www.who.int/news-room/fact-sheets/detail/climate-change-and-health

Wu, M., Long, R., Chen, H., & Wang, J. (2023). Is social media more conducive to climate change communication behavior? The mediating role of risk perception and environmental values. *Environment, Development and Sustainability, 1–27.* https://doi.org/10.1007/s10668-023-03870-1

Zaman, A. (2022). Zero-waste: A new sustainability paradigm for addressing the global waste problem. In K. E. Björnberg, S. O. Hansson, M. Belin, & C. Tingvall (Eds.), *The vision zero handbook: Theory, technology and management for a zero casualty policy* (pp. 1–24). Springer International Publishing. https://doi.org/10.1007/978-3-030-23176-7_46-1

Zaman, A., & Newman, P. (2021). Plastics: Are they part of the zero-waste agenda or the toxic-waste agenda? *Sustainable Earth, 4*, 1–16. https://doi.org/10.1186/s42055-021-00043-8

Zaval, L., & Cornwell, J. F. (2017). Effective education and communication strategies to promote environmental engagement. *European Journal of Education, 52*(4), 477–486. https://doi.org/10.1111/ejed.12252

Zelenski, J. M., & Desrochers, J. E. (2021). Can positive and self-transcendent emotions promote pro-environmental behavior? *Current Opinion in Psychology, 42*, 31–35. https://doi.org/10.1016/j.copsyc.2021.02.009

Zhang, Y., Zhou, Y., Liu, Y., & Xiao, Z. (2023). Research on the influencing mechanism of the effect of brands' sustainable behaviors on consumer attitudes: An empirical study on clothing brands. *Sustainability, 15*(3), 2351. https://doi.org/10.3390/su15032351

**Ana Rita Farias** (PhD in Psychology) is Senior Researcher and Professor at Hei-Lab: Digital Human-Environment Interaction Lab. She has held research positions and been a guest researcher in psychology, business, and economics departments at institutions such as Utrecht University (NL), University of Wisconsin (USA), and Católica Lisbon Research Unit in Business and Economics (CUBE). In recent years, she has shifted her focus from pure Psychology to Behavioural Economics, exploring how digital environments influence consumer decisions and promoting sustainable practices through behavioural insights. She has led and collaborated on numerous research projects funded by national (FCT) and international agencies (laCaixa Foundation, NIMH, EC, FCT-EC).

**Bernardo Cruz** has a master's degree in Social and Organizational Psychology by Universidade Lusófona de Humanidades e Tecnologias. He is an assistant researcher at Hei-Lab: Digital Human-Environment Interaction Labs. In addition to being a committee member for the Society for Risks Analysis—Iberian chapter for the second year, he is currently working in BioLaMer, a European project.

**Samuel Domingos** has a PhD in Psychology, specialization in Social Psychology by ISPA—Instituto Universitário. He is a researcher at Hei-Lab: Digital Human-Environment Interaction Labs, and assistant professor at Universidade Lusófona and Universidade de Évora in Portugal. He is also a board member of the Society for Risk Analysis Europe (SRA-E) and of the Society for Risk Analysis Europe Iberian Chapter (SRA—Europe Iberian Chapter). His research interests focus on understanding public perceptions in different domains, such as natural and man-made hazards, new and emerging technologies, and current societal challenges, with a particular interest in risk and crisis communication.

**Catarina Possidónio** holds a PhD in Psychology with a specialization in Social Psychology. Her research has the primary goal of promoting more sustainable and healthier consumer behaviours. She is particularly dedicated to designing and implementing studies that provide valuable insights and actionable recommendations for effectively communicating and promoting sustainability. Catarina is also a lecturer at several institutions, including Lusófona University and the Piaget Institute, where she teaches at both the Bachelor's and Master's levels. Additionally, she serves as the Scientific Editor of the peer-reviewed journal *Psicologia*, published by the Portuguese Association of Psychology.

**Ana Loureiro** is Associate Professor at Universidade Lusófona and a researcher at HEI-Lab. She holds a PhD in Social and Environmental Psychology by ISCTE-IUL. She collaborates in projects aimed to intervene in the community, using different methodologies such as co-creation and stakeholder's engagement, and is currently a board member of PSICAMB—Asociación de Psicología Ambiental. Her research is focused on human experience of natural and urban environments, and environmental behaviour sociopsychological determinants.

**Eliana Portugal** is finalizing her Master's degree in Social and Organizational Psychology at Universidade Lusófona de Humanidades e Tecnologias. She has a research scholarship from HEI-Lab: Digital Human Environment Interaction Labs where she is working as a research assistant. There she collaborates in the TechMap project focusing on the assessment of minimum level of proficiency in digital literacy, and in the BioLaMer project, focusing on industry and consumer perceptions of new biopolymer technologies.

**Sílvia Luís** has a PhD in Social Psychology by Universidade de Lisboa, Faculdade de Psicologia. She is an assistant professor at Universidade Lusófona, where she is co-coordinator of the master's in Organizational and Social Psychology. She is a board member of the Iberian Society for Environmental Psychology (PsicAmb) and of the Institute of Behavioural Technology (INTEC). Her research interests focus on how psychosocial variables shape the person–environment interaction and on the implications of this interaction for people and for the environment.

# "They Don't Promote It, in the Capitalist Society that Promotes the Meat Industry": Echo Chambers and Media Mistrust Among Women Vegans in the UK and Croatia: A Contribution to the Debate on the Social Responsibility of the Media

**Martina Topić-Rutherford, Marija Geiger Zeman, and Gabrielle Round**

**Abstract** This chapter presents findings from an original study conducted in the UK and Croatia on vegan women and their media consumption. The authors conducted 20 interviews in Croatia and 28 in the UK and explored media consumption, media trust, and the role of social media in news consumption. A two-tier thematic analysis has been used to analyze the findings. The authors first analyzed each country's dataset and then conducted a joint thematic analysis for both countries. The findings show commonalities of views and experiences of women vegans in the two countries. In both cases, women mention hostility, bias, and misinformation in the mainstream media, leading them to express a lack of trust in the media and turn toward vegan news sources and echo chambers, increasing also social media news consumption. The findings are analyzed in the context of wider media representation of veganism and the social responsibility of the media framework. The authors conclude that based on the previous research, as well as these two datasets, there is a problem with a profit-driven mainstream media system as well as a lack of diversity in newsrooms that influences a situation in which only journalists who are friendly toward veganism write accurate and detailed news reports.

---

M. Topić-Rutherford (✉)
Department of Advertising and PR, College of Communication and Information Sciences, The University of Alabama, Tuscaloosa, AL, USA
e-mail: mtopic@ua.edu

M. G. Zeman
Ivo Pilar Institute of Social Sciences, Zagreb, Croatia
e-mail: Marija.GeigerZeman@pilar.hr

G. Round
University of Salford, Salford, UK
e-mail: G.A.L.Round1@salford.ac.uk



M. Todorovic et al. (eds.), *Sustainability and Social Responsibility of the Media and in the Media*, CSR, Sustainability, Ethics & Governance,
https://doi.org/10.1007/978-3-032-00086-6_12

**Keywords** The UK · Croatia · Vegans · Women · Mainstream media · Media trust · Misinformation · Media bias · Stereotypes · Prejudice · Trivialization · Echo chambers · Media consumption · Social media · Social media news · Trolling · Newspapers · Plant-based diet · Vegan lifestyle · Vegan groups · Alternative media · Media social responsibility · Social responsibility · Hutchins Commission · Qualitative interviews · Thematic analysis

## 1 Introduction

Public polls are continually showing a lack of trust in the mainstream media. For example, the UK has one of the lowest levels of trust in the world. The 2023 Edelman study shows that only 37% of British citizens trust the media (Majid, 2023), and the Ipsos Veracity Index has shown that only 29% of Britons trust journalists to tell the truth (Majid, 2022). While in both cases, this low level of trust presents an increase from previous polls, the reality remains that the mainstream media lost trust, and this, as recent political events revealed, has led to the proliferation of fake news and polarization in many countries in the world. In the UK context, citizens report they trust their neighbors, coworkers, and people in the local community to a greater extent than journalists, thus opening a question of the role of the media in society and whether the media are socially responsible and fulfilling their role. In Croatia, "research on public perception of media and media content" conducted in 2020 showed that 24.9% of respondents trust journalists, 48.4% neither trust nor do not trust them, and 34.8% do not trust journalists (Jourlab, n.d.) although most respondents (60.3%) are informed about social and political events via television and 48.1% via internet portals. Also, the results showed that the younger population (18–22) prefers information from Internet portals (52.5%) and social networks (50.1%). The Eurobarometer Media & News Survey 2022 also confirmed the great popularity of television (75%), especially among 55+ respondents (84%), and Internet portals in the general population (48%) in Croatia. In the flood of fake news and misinformation, public television and radio stations are recognized as the most reliable sources of information (Eurobarometer/Media and News Survey 2022/ Country Factsheets Croatia).

At the same time, research studying media coverage continually warns of poor editorial standards and bias in media coverage, with some authors calling for a change in the way the media operates. For example, in a monograph exploring the environmental coverage of the British press, Topić (2021) argued that the press promotes economic growth and environmental preservation, with the two not going together, and thus calling not just for a change in how we treat the planet but also for the change in the media system. Topić (2021) called for a media system that will be socially responsible, arguing this can only happen if media become nongovernmental organizations (NGOs) that obtain public funding but remain independent, with the market orientation of the media not working due to commercialism and interests of owners impeding the work of journalists and editors and preventing the media to fulfil their purpose of informing citizens and acting as guardians of the democracy.

## 2 Veganism and the Mainstream Media

Voluntary abstention from meat consumption is a practice that has existed since prehistorical times (Spencer, 1996). According to the Vegetarian Society (n.d.), 4.5% of people in the UK are either vegetarians or vegans and in 2019, people in the UK ate 17% less meat than in 2008. The Equality Act also protects vegans (Jamieson, 2010) and veganism thus counts increasingly as a human right. According to Veganuary data, the number of vegans increased from 400,000 to 700,000 participants from 2020 to 2023, and 1.8 million in 2024 (Bumbridge Sutton, 2024). However, despite the rise of this lifestyle and its legal protections, MacInnis and Hodson (2017) argued that vegans still experience discrimination like those experienced by other minority groups, thus often being a target of bias and discrimination including having to cease contact with some family members and friends upon embracing veganism and that they sometimes feel anxious about revealing their lifestyle in public (ibid).

Ecofeminist theory has been arguing for decades that historically, women were more likely to engage with environmental affairs and join environmentalist protests (Holy, 2007; Giacomini, 2014; Topić, 2020, Topić, 2021). However, environmentalist affairs and food choices came to the media agenda in the past decade and while these topics were traditionally topics from women's pages and historically covered by women journalists, with the rise of prominence of these topics, women journalists did not rise in relevance and moved to general news sections; instead, these topics now get frequently covered by male journalists who remain dominant in national politics and business, as the two most prestigious beats in journalism (Christmas, 1997; Burkhart & Sigelman, 1990; Franks, 2013; Topić, 2018). Topić (2021) argued that the majority of environmental topics are now covered by men who write about them from a neoliberal point of view and enforce the notion that environmentalism and economic growth go together, with which they can be seen as saturating news coverage with very little changing. In addition, this same analysis has shown that the *Daily Mail* as the most-read newspaper in the UK, responsible for setting the public and the political agenda, writes about environmentalists negatively calling them middle class and privileged, thus also attempting to pity the working-class Britons against those who support the environmentalist cause (ibid).

Some authors argued that the media created a *vegaphobia* (Cole & Morgan, 2011) by negatively writing about veganism and using characterizations such as "ridiculing veganism, characterizing veganism as asceticism, describing veganism as difficult or impossible to sustain, describing veganism as a fad, characterizing vegans as oversensitive, characterizing vegans as hostile" (p. 139). These findings are from the UK press; however, an Australian study confirmed the findings (Ragusa et al., 2014). In addition, BBC, which prides itself on impartiality and is the most trusted media in the UK (Smith, 2023, Majid, 2023a), in 2020, published a sensationalist article arguing that veganism could affect one's brain despite the author admitting there is no evidence for such claim (Lockwood, n.d.). Other famous UK examples of denouncing vegan and vegetarian lifestyles include the infamous incident with Piers Morgan almost vomiting on live TV after trying Gregg's vegan sausage roll (Bergman, 2019). Lockwood (n.d.) thus argues that apart from boycotting some forms of media, people need to stay vigilant because the mainstream media want to see vegans as the enemy and advises vegetarians and vegans to get their news from trusted sources such as Plant Based News, Tenderly, Live Kindly, Vegconomist, Sentient Media, and Dodo. This is accompanied by the advice that for non-vegan but less unfriendly toward vegans ("better than the mainstream media," ibid., n.p.), one should read Guardian's Animal Farmed series, ByLine Times, Double Down News, Open Democracy, DeSmogUK, and the Conversation.

In Croatia, this topic has not been covered in research and there is no knowledge of vegan representation in the media. Thus, this chapter approaches this topic for the first time.

## 3 Social Responsibility of the Media

The social responsibility of the media is a relevant topic because of the decreasing trust of the public in the mainstream media and the rise of fake news and disinformation (Lee, 2019; Watson, 2020). Media are often seen differently than corporations because of media influence over attitudes and behaviors, and the fact media work in the open market is sometimes criticized because media being reliant on external income also means they are facing market competition (Ingenhoff & Koelling, 2012; Owen & Wildman, 1992; Bardoel & d'Haenens, 2004; Meier & Perrin, 2007; Hou & Reber, 2011). Therefore, media have been criticized for placing their commercial interest ahead of the interests of the public and for a stakeholder approach to satisfying the needs of their audiences with what they cover and how they report on affairs (Plaisance, 2000). This criticism also came because media has a more important role in society than traditional corporations since, apart from already-mentioned attitudes and behaviors, media also have an impact on social norms and values (Van Leidekerke, 2004).

Marx ([1849]1959) argued that the press needs to be a watchdog and "the tireless denouncer of those in power, the omnipresent eye, the omnipresent mouthpiece of the people's spirit that jealously guards its freedom" (p. 231, cited from Sandoval, 2013, p. 44–45). In this view, the media should not be a trade (ibid), and this argument was supported by authors who argued that media needs to be studied within the context of political economy and capitalism (Sandoval, 2013; Mosco, 2009; Garnham, 1998). This means studying the media from the point of its dual role centered on the economic aspect and social and cultural aspects (Sandoval, 2013) with the economic aspect being centered on profit-generating focus of the mainstream media, commercial interest that impairs media's social work and inevitable exploitation of journalists and other media employees, working in the free market and facing market pressures, which then results in uniformity and conformism (Garnham, [1986]2006; Smythe, [1977]1997; Fuchs, 2010, 2011; Schiller, 1997; McChesney, 2004). Sandoval (2013) argued that the corporate social responsibility concept is flawed because corporations will not go the extra mile if it affects profit and they absolutely have to; applying this to the media, this means that media are also corporations and would not engage in protecting social interests if it impairs profit generation. However, there has never been an agreement on how media should operate and what the media business goal should be.

The debate on the role of media in society has existed since the 1940s when the Commission on the Freedom of the Press, popularly known as the Hutchins Commission was formed at the University of Chicago. This commission tackled the social responsibility of the press and released a report in 1947, "A Free and Responsible Press." According to the Hutchins Commission, the press has the following obligations:

- To provide a truthful, comprehensive, and intelligent account of the day's events in a context that gives meaning.
- To serve as a forum for the exchange of comment and criticism.
- To develop a representative picture of the constituent groups in society.
- To be responsible for the presentation and clarification of the goals and values of society.
- To provide full access to the day's intelligence (cited from Bivins, 2004, p. 42).

Siebert et al. (1963) developed the theory of social responsibility of the media and argued that media must use their powerful position to deliver information to citizens, and if the media fail in this, then there is a case for media regulation. Owens-Ibie (1994) argued that media are responsible to citizens, the government, media owners, and also to themselves, and failing to fulfil this four-tier responsibility decreases trust and accountability. The supporters of the social responsibility of the media argue that media cannot just report on issues but need to open a debate. Ward (2008) argued that there was a "disillusionment with the liberal hope that an unregulated press would be a responsible educator of citizens on matters of public interests" (p. 298). Topić (2021) argued that media form a part of the wheel of neoliberalism and contribute toward perpetuating the status quo on the environment and corporate social responsibility; the latter is also seen as part of the wheel of neoliberalism and called for independent media that would be publicly funded and formed as some sort of nongovernmental organization but without market competition so that the media can again serve society freed from advertising and any other revenue pressure.

## 4 Method

Drawing from the literature presented above, particularly respective of calls to boycott mainstream media and engage in consuming alternative media sources (Lockwood, n.d.), we conducted qualitative interviews to explore media consumption and media trust of women vegans. The reason for a vegan study being centered on women lies in the fact that, according to some research studies, women are more likely to be vegans (Scholzl & Lenhart, 2023, Gorvett, 2020, Khalidi, 2020). In addition, while studies are looking at media trust, these are usually conducted generally and do not differentiate between genders. This is the case even after environmentalism and food, two traditional areas of women's interest (Topić, 2018), came to the news agenda. Therefore, for this study, 28 qualitative interviews were conducted in the UK with women vegans. In Croatia, 20 women were interviewed. Saturation was reached in both samples. The research was conducted between January and April 2023 in Croatia and January and May 2023 in the UK. Interviews in the UK were conducted in English language, whereas interviews in Croatia were first conducted in Croatian language and then transcripts were translated to English by the Croatian author of this chapter.

In the UK, interviews were recruited by using personal contacts of the two authors and snowballing as well as posting a call to vegan social media groups. The recruitment process ran smoothly for most of England; however, the researchers failed to find interviewees from London or other parts of the UK. Thus, the sample consists of women in the north and south of England only, but without London. The interviewee demographic is outlined in Table 1 and interviewees are labeled with an initial from a researcher who interviewed them, followed by a number.

In Croatia, 17 interviews were conducted face-to-face, and three were conducted online, whereas in the UK, all interviews were conducted online. Interviewees in Croatia were recruited in two ways: (1) using personal contacts and a snowball method and (2) by contacting two of the most well-known Croatian nongovernmental organizations for the promotion of animal rights and veganism (Prijatelji životinja from Zagreb and Udruga Pobjede from Osijek). Participant demographics are present in Table 2, using pseudonyms selected by interviewees. Due to anonymization, the names of the villages where the research participants live are not listed, but a larger place near the place of residence is indicated.

The questions were derived from the literature review outlined above and thus, we asked women which newspapers they read regularly, which TV stations they watch regularly, whether they listen to the radio, and if so, which station. We also asked questions about social media use (whether they use social media and if so, which channels). After this general consumption inquiry, we also asked women to what extent they get news from social media, and if they do get news from social media, which channels. These questions were relevant to explore to what extent social media news is used, following some reports that the public in both countries uses social media for news (Maher, 2024, Jourlab, n.d.), which further exacerbates the issue of media trust. In addition, we asked women about their views on how

**Table 1** The UK sample

| Interviewee no. | Age | Urban vs rural | Qualifications | City |
|---|---|---|---|---|
| G1 | 65 | Urban | City and guilds | Manchester |
| G2 | 37 | Urban | PhD | Manchester |
| G3 | 29 | Urban | Professional doctorate | Salford |
| G4 | 27 | Urban | UG degree | Manchester |
| G5 | 64 | Suburban | Professional doctorate | Manchester |
| G6 | 29 | Suburban | Masters | Oldham |
| G7 | 37 | urban | Masters | Salford |
| G8 | 60 | Suburban | Masters | Rutland |
| G9 | 45 | Rural | Masters | Glossop |
| G10 | 27 | Urban | bachelors | Manchester |
| G11 | 62 | Urban | Masters | Chorlton (Manchester) |
| G12 | 35 | Urban | A Levels | Luton |
| G13 | 30 | Urban | Level 3 PT, or A Levels | Manchester |
| G14 | 53 | Urban | Level 4 | Bradford |
| G15 | 29 | Suburban | Masters | Southampton |
| M1 | 42 | Suburban | PGDIP | Leeds |
| M2 | 61 | Urban | PhD | Doncaster |
| M3 | 56 | Urban | Masters | West Linsey |
| M4 | 32 | Urban | Masters | Manchester |
| M5 | 44 | Urban | PhD | Leeds |
| M6 | 61 | Suburban | Bachelors | Ilkley |
| M7 | 24 | Rural | Bachelors | Dover |
| M8 | 24 | Rural town | Bachelors | Skipton |
| M9 | 25 | Urban | Masters | Leeds |
| M10 | 20 | Urban | A Levels | Bristol |
| M11 | 31 | Urban | Bachelors | Leeds |
| M12 | 32 | Urban | LPC | Leeds |
| M13 | 51 | Rural village | PhD | Hull |

vegan affairs are presented in the mainstream media, whether they consider any mainstream media as unreliable or sensationalist about veganism, and then how they see the representation of veganism on social media. Finally, we asked interviewees how they get news about veganism and veganism-related affairs.

The questions were copied from a questionnaire, and all answers from interviewee responses to media questions were outlined, question by question. After the creation of a transcript as per the questions outlined above, we coded our data by following the same coding process for both datasets, which was implemented by both research teams:

Step 1: Familiarization with the data,
Step 2: Generating initial codes,
Step 3: Searching for themes,
Step 4: Reviewing themes,

**Table 2** Croatian sample

| Pseudonym | Age | Urban vs rural | Qualifications | City |
|---|---|---|---|---|
| Hepburn | 37 | Urban | PhD | Zagreb |
| Sneky | 50 | Urban | MSc | Zagreb |
| Vanja | 50 | Rural | Masters | Near Sveti Ivan Zelina |
| Sara | 42 | Urban | Masters | Zagreb |
| Ruža | 38 | Urban | Masters | Zagreb |
| Klara | 23 | Urban | BA | Zagreb |
| Paula | 57 | Urban | Secondary school | Zagreb |
| Alborada | 53 | Urban | Masters | Zagreb |
| Maja | 37 | Urban | Masters | Zagreb |
| Tara | 70 | Urban | Secondary school | Zagreb |
| Mara | 51 | Urban | Secondary school | Zagreb |
| Vega | 30 | Rural | Secondary school | Near Sveti Ivan Zelina |
| Monika | 18 | Urban | Student | Zagreb |
| Zobena kaša | 22 | Urban | Student | Zagreb |
| Mirela | 36 | Urban | BA | Zagreb |
| Zaira | 28 | Urban | Masters | Zagreb |
| Ana | 24 | Urban | BA | Rijeka |
| Anja | 30 | Urban | Masters | Split |
| Lavanda | 28 | Urban | Secondary school | Split |
| Crazy Cat Lady | 46 | Urban | Masters | Samobor |

Step 5: Defining themes,
Step 6: Write-up (Maguire & Delahunt, 2017).

A thematic analysis was carried out according to a guide offered by Braun and Clark (2006). Thematic analysis is "a systematic approach to the analysis of qualitative data that involves identifying themes or patterns of cultural meaning; coding and classifying data, usually textual, according to themes; and interpreting the resulting thematic structures by seeking commonalities, 15 relationships, overarching patterns, theoretical constructs, or explanatory principles" (Lapadat, 2010, p. 926). In the presentation of findings, we first present thematic analysis findings for each country, the UK and Croatia, and then we provide a joint analysis of both countries in the final section of this chapter. Therefore, a two-tier thematic analysis was conducted, first on each country's sample and then on two samples together to provide for a more meaningful and in-depth analysis. We visualized our findings and then offered an analysis of each country using direct quotes from interviewees, as per the usual thematic analysis approach proposed by Braun and Clarke (2006).

Thematic analysis is not linked to any specific theory. The method presents a sensemaking approach to data analysis and is helpful when working with a large dataset, such as in this case where we analyze transcripts from a total of 48 interviews. The thematic analysis was deemed relevant for this study because it enabled us to identify trends in data and common sentiments about the representation of veganism in the mainstream media, as well as inquire about media consumption and trust among women vegans in two countries.

Institutional research ethics approvals were granted by the Ivo Pilar Institute of Social Sciences, Zagreb, Croatia and Leeds Beckett University where the first and third authors were based at the time the study was conducted. The study was conducted without formal funding and in the author's free time. The analysis and all writing have been done since both UK authors changed institutions.

# 5 Findings

## *5.1 The UK*

In the UK context, interviewees had almost nothing positive to say about the mainstream media, and some interviewees no longer consume their content because of negative media coverage of veganism. The main thematic analysis for the UK data centers on three main themes: *mainstream media as unreliable and biased*, *stereotypical representation of vegans in the mainstream media*, and *the rise of social media consumption* (Fig. 1).

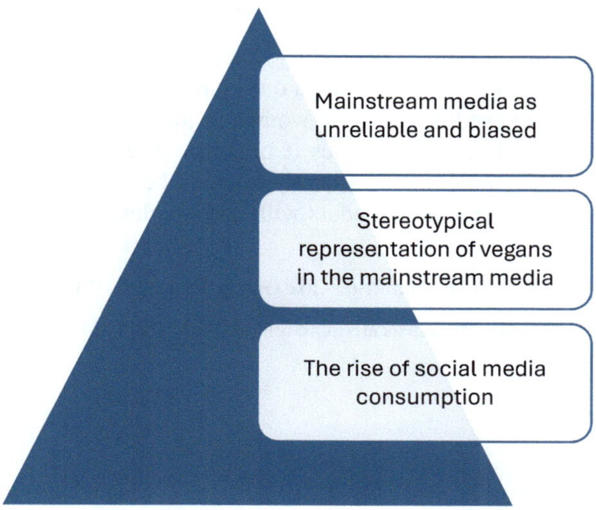


**Fig. 1** Thematic findings from the UK sample

## 5.2 Theme 1: Mainstream Media as Unreliable and Biased

The majority of interviewees see the mainstream media as unreliable generally, but two exceptions emerged to an extent: the Guardian and the BBC. These two news outlets are also the most consumed among participants, BBC both as an online portal where it came as second most consumed after the Guardian and as the most consumed TV channel. Radio 4 and 6, which are part of the BBC radio network, are also mentioned as the most listened to, albeit several interviewees mentioned they only listen to the radio when driving. The Guardian has the most mentions when it comes to newspaper consumption and reliability generally, while tabloids and the *Daily Mail* (which is often accused of being a tabloid due to sensationalist writing but is not formally classified as such) as the least reliable. However, an interesting discrepancy emerged because when interviewees were asked which media they consume, they answered mainly the Guardian, then secondarily BBC for news reading and BBC for TV media consumption, but when asked which mainstream media they would consider reliable, the majority replied none of them. This signals a large distrust of the mainstream media, which goes in line with general data on the low media trust in the UK (Majid, 2022, 2023). Many interviewees also replied that only vegan sources are reliable, and this is largely linked to the hostility of the mainstream media about veganism, where the sentiment is that vegans are picked on.

> I would say particularly in things like the Daily Mail. There is a lot of anti-vegan, almost propaganda. It's always linked to health, as well. And there's almost a sort of 'we told you so' about it. So, articles saying things like, actually veganism isn't healthy. There's always that. It's always very negative about veganism and the people that they'll pick to talk about … (M19)

> None, they are all slanderous and biased. (M28)

> Only vegan ones, like plant-based news and vegan papers. (G13)

Some participants also made interesting statements that the tone of the coverage depends on the journalist, thus opening a question of diversity in newsrooms. This question has been opened in works on women in journalism, arguing that women historically brought diversity in topics (Christmas, 1997; Burkhart & Sigelman, 1990; Franks, 2013; Topić, 2018). Equally, it could be said that more diversity in newsrooms would engage more readers with news outlets and potentially increase media trust.

> The Guardian, particularly certain writers like George Montbello. (M21)

> None specifically; some journalists are more reliable than others. (M24)

## 5.3 *Theme 2: Stereotypical Representation of Vegans in the Mainstream Media*

The reason for the lack of trust in the media, expressed in theme 1, can be likely found in theme 2, where interviewees focused on the stereotypical representation of vegans in the news media, The interviewees particularly emphasized the way vegans are portrayed, citing examples of hippies, crackpots, false representation of veganism as deadly, changing the name of veganism due to negativity when media do want to say something positive, and also they emphasized the profit-orientation of the media, which links with the literature cited earlier in this chapter with many authors calling for independent and socially responsible media freed from profit orientation and challenging the market orientation of the media (Sandoval, 2013; Garnham, [1986]2006; Smythe, [1977]1997; Fuchs, 2010, 2011; Schiller, 1997; McChesney, 2004). For example, interviewees described how they are represented in the media and recalled some experiences and memories of media articles:

> ... we're mainly described as crackpots and that we don't know what we're doing, and we have no strength, our energy and, you know, we're going to be falling over soon because we can't do anything. (G1)

> They use every opportunity to say something bad about it; they don't promote it, in the capitalist society that promotes the meat industry. (G5)

> I remember one about someone who was vegan, and they died on mount just forgotten the name of the mountain in Scotland (...) but that was also negative because that was focused on 'Look, they're vegan and they died. Look at that.' It's not so much the news article, but the comments that it causes. I think sometimes it can be very negatively based; sometimes it's neutral, but I'd never say it was positive. But again, that could be impacted because of the comments that have been on it. (G6)

These media articles signify bias and stereotypes, and this portrayal does not cause just harm to the vegan community but makes it legitimate that people cannot trust the mainstream media when the portrayal is perceived as inaccurate or biased. Inaccuracy and bias are seen as a particular problem, and most interviewees emphasized this issue and cited examples. For example, some interviewees noticed that the mainstream media write about veganism as a fad or excessively mention only certain types of food as well as portray vegans as rich, lefty lunatics and similar, which echoes findings from Topić's (2021) book on environmentalism in the British press where she emphasized that some newspapers, in particular, the *Daily Mail*, portray environmentalists negatively and attempt to pity working class people against the middle class:

> ... lot that are very like, oh, it's a fad and vegans spend loads of money and eat avocado, toast and all of that stuff. And it's very like modern thing and there's also that debate of kind of, it's a rich white privilege as well. Which I don't think it is because like it's like the basics. (G10)

> I think really badly. Yeah, I think that it's like, it's something that people can have a go at. And I think that that's fuelled well. Obviously, it'll be fuelled by the amount of money that is in the meat and dairy industry and the people who own the media will probably have

> some fingers in those pies. And so, they want that to continue. Then of course, you've got the sort of the cruelty to animals' type thing where you know the fox hunting and the shooting and all the rest of it, which is all you rich people anyway that and they're in charge of the media. So, I think it's in their interest to keep veganism as looking like something that's really weird, something that's, you know, these lefty lunatics and blah blah and I think they do it really well. I think you know, when you look at some of the comments and articles it's like, these bloody vegans, what they need is a good bit beef burger and yeah it seriously annoys me with the way that it's portrayed in the media. I think they're those with the biggest pockets who have got the loudest voices. (G14)

> … doesn't feel it's presented much unless it's very extreme like a big protest and then it's talked about dismissively. I've never seen veganism treated seriously in a newspaper. (M24)

## 5.4 *Theme 3: The Rise of Social Media News Consumption*

Interviewees use social media, where Facebook is the most popular, closely followed by Instagram, and Twitter is the third most popular network. In line with wider trends in media trust (Majid, 2022, 2023), interviewees said they get news from social media; the majority of interviewees largely obtain news in this way, while ten interviewees said they do not at all use social media for news. Therefore, there seems to be a division about social media news, and since there are no previous studies on vegan media consumption, it is not clear whether this data goes in line with the trend of decreased media trust, so more vegans consume media in this way or whether the trust in social media is decreasing. However, social media is important in news consumption, which poses a challenge to the mainstream media but also democratic processes due to the proliferation of fake news and misinformation on social media (Lee, 2019, Watson, 2020).

Interviewees stated they consume social media by following vegan news channels and vegan individuals, and admit they live in their circle or a bubble. Vegan individuals and vegan news are seen as the most reliable and are mentioned by interviewees as those that can be trusted. The channels also have a practical use. For example, interviewees mentioned they obtain recipes from social media and obtain information, but they also spoke of the issue of trolling and aggressive behavior from non-vegans, thus signaling that a safe space for vegans does not exist; they face stereotypes and bias from the mainstream media and trolling and abuse on social media:

> … not monolithic because social media is the sum of its users. So, in my bubble, people are really sympathetic to veganism, and I get a lot of recipes from social media. Every once in a while, I am confronted with anti-vegan posts on social media, sometimes deliberately, sometimes not. But for example, apparently after the Veganuary every year comes February dairy and that was something that the dairy industry started in order to counter pro-vegan views in January. So, but then, you know, what happened was that vegans started (…) hashtagging heavy dairy and pointing out the cruelty of the dairy industry. And so you know, one day last year I posted a recipe for oat milk and then just said, you know, hashtag February and there was a bit of a pile on it. It was kind of organised because I looked at the, accounts that were piling on and they were just commenting on every vegan post that was

> hashtag of the day with the same things about how if you don't consume dairy, then it'll destroy your immune system and you'll get COVID and you'll die. (G2)

> I try to follow just vegan things, so I probably see a lot of the good and on Instagram, for example, where you create your followers, whereas on Facebook, I'm a member of a lot of vegan groups, and occasionally, you will get people who come in and post pictures of bacon or make jokes about vegans being weak or protein, lacking protein or things like that so it can be presented quite badly in those situations. Or if a paper like the Metro runs an article about veganism being better. Comments will often go against that. (M19)

In addition, some interviewees said they like to follow NHS doctors because they share reliable information, thus showing that some interviewees actively seek reliability when consuming information.

> A lot of you know, like I follow a lot of doctors. And as in like NHS doctors on social media who share a lot of what I believe is really reliable and like useful content for people who are vegan or want to be vegan. And like me, like health conscious', the sorts of things that come up on my feed are derogatory to veganism; snowflake type culture, saying everyone who is vegan is that. (M27)

## 5.5 *CROATIA*

In the Croatian sample, four main themes were identified: *media landscape*, *the power of social networks*, *misrepresentations*, and *new approaches*. Each of the main themes also had a subtheme, which is outlined in Fig. 2.

## 5.6 *Theme 1: Media Landscape*

Interviewees emphasized that they follow the media in different intensities; a part said they are engaging with abstention from any media consumption and thus do not read the newspapers, watch TV, or listen to the radio, while the others said they follow the mainstream media to get information on daily events, and for this, mainly consume the most popular web portals and newspapers such as *Index.hr*, *Jutarnji list*, *Večernji list*, and *Slobodna Dalmacija*. The analysis has shown that there is a distinction between mainstream media available to everyone and those that tend to be hostile or indifferent toward veganism as opposed to alternative media. This distinction is particularly based on a wider view of the media because alternative media does not only include newspapers and news portals but also fanzine, music, and blogs of anarchist and radical left background, activist documentaries, and activist web pages that promote animal rights.

The positive view of alternative media is accompanied by a mainly critical view of the mainstream media, which many participants labeled as "the extended hand of profit and the industry" (Crazy Cat Lady), and participants emphasized their negative influence and unwillingness to report important things. In addition,

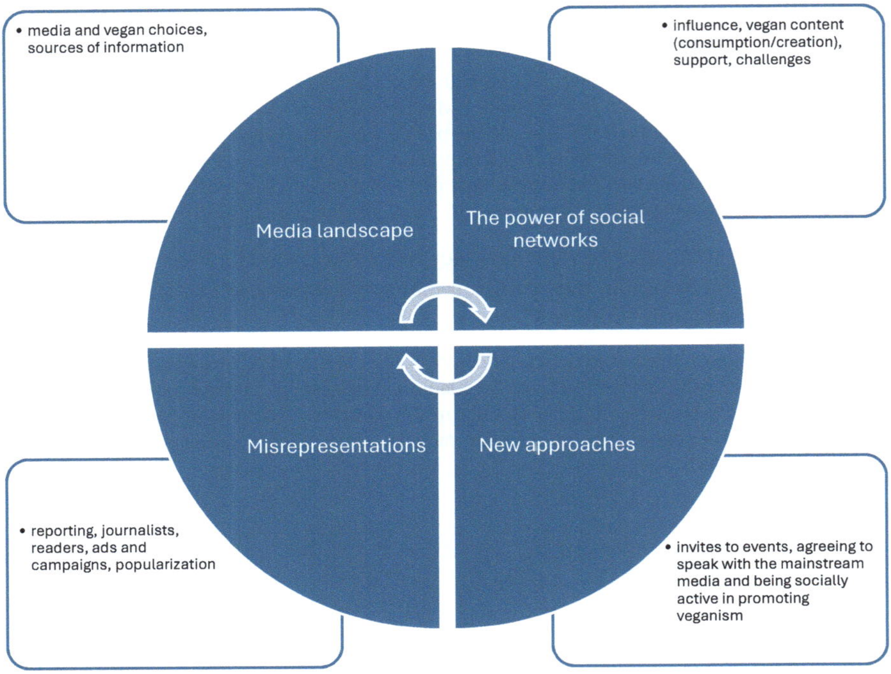


**Fig. 2** Thematic findings from a Croatian sample

interviewees said that news portals have a visible advantage in providing information about veganism over newspapers and this is mainly because the mainstream media show resistance toward veganism, which then pushes vegans toward independent information searching and they continually try to independently learn and find independent information, including some interviewees mentioning books and research studies. The interviewees particularly recognize, as a reliable source, the NGO Animal Friends (Prijatelji životinja) and see them as an important factor in promoting veganism and publicly communicating about animal protection and defending animal rights. One part of the interviewees follow web pages and public information announcements of this NGO and emphasizes the quality of articles and visual materials about animal suffrage. These findings echo UK findings where participants also emphasized vegan sources as reliable, and some mentioned they follow NHS doctors; thus findings in both cases signaling a lack of trust in the media and turning toward findings reliable sources for themselves.

> I mean, the source is very important to me. It is not like I will read anything (…) If I am looking for something related to veganism, then I look as to whether the source is propaganda or if the source is vegan, or something scientific. (Vega)

## 5.7 Theme 2: The Power of Social Networks

Interviewees emphasized the influence of social media in disseminating information about veganism since the topic of veganism is insufficiently present in the mainstream media. These alternative sources of information include social media (Facebook, Instagram, YouTube) and collaborative communication channels (WhatsApp). The majority of interviewees emphasized the positive influence of social media when changing their lifestyle to veganism and in continuous education and information about veganism and the vegan movement. Moreover, some interviewees think browsing vegan profiles can be an encouragement and support for a (permanent) switch to veganism.

Similarly to UK findings, Croatian interviewees also emphasized there is negativity on social media, but they presented positive views of social media networks because these platforms allow them to change mindsets about veganism, so they share information about meals they are preparing on their Facebook profiles, share photos and video materials linked to animal suffrage and animal protection, support homing domestic animals, share information on sustainable and zero-waste lifestyles, and so on. However, some interviewees said they are being careful, so their posts do not sound as if they are attacking or insulting anyone because they do not want to cause counter-effects:

> I am posting photos along with a thoughtful statement, maybe some people resent this, some do not. I am continuing this way and will not give up. I am posting photos of food, occasionally also photos of suffrage because, had I not seen those photos, I would not have started on the path, that I think is the right one so I do not hold back. I like posting about little animals I or my social circle care for; I like to talk about some aspects of ecology, food, and animal protection, post photos where we are happy in the company of those animals, where these animals realize what is comfortable and what is not; I have lots of photos with farm animals which instigates an emotional reaction in people, and photos of food are also very important to me, I think that people like them. (Paula)

The majority of interviewees consider Facebook important because of information available on the site (recipes, vegan products, vegan restaurants, etc.), articles, and people, and some also emphasize the importance of vegan groups on social networks. These groups are larger, more anonymous, and heterogenous, where people who participate are engaged with veganism, and veganism is the only topic that is shared. Apart from being places to connect vegans, these groups are an excellent source of information, empowerment, support, and this is particularly relevant in the period of transition; and groups also serve as a platform for debate, sharing advice, and exchanging recipes:

> I think that social media, generally, are an excellent tool one can easily use with a view to share something that is important to them. So, if someone wants to follow vegan profiles and then think about veganism, then this is excellent (...) it is not like I know a lot of vegans and if they are in some group, that serves as an Internet community, it is possible to identify with those people who can tell you something you didn't know. It is great because you have a community that shares recipes and talks about it, there are different people of different ages, views, and opinions but this is not actually important because we only talk about

veganism and this is great for me (…) I get recipes, new information when new products become available when something is on sale. I think I received a lot of information on the Internet, some things I would never expect to be vegan, this is where I find out about it. (Zobena kaša).

## 5.8 Theme 3: Misrepresentations

According to the estimate of the interviewees, veganism themes such as animal rights, food, events, the announcement of activist actions, and so on are not regular and not visible in Croatian mainstream media. But occasionally, there are articles about veganism, for example, the most famous daily newspapers online (24 sata/24sata.hr., Jutarnji list/jutarnji.hr., Večernji list/večernji.hr) publish holiday cookbooks with vegan recipes. Mainstream media are seen as promoting the culture of eating meat (e.g., adverts for dairy and meat products that are based on misinformation, incorrect reporting, journalism bias, etc.), which echoes UK findings where some participants labeled British mainstream media as profit-oriented and capitalist. Several interviewees emphasized that there is a little bit of space that is opening up toward veganism in the Croatian mainstream media. Still, in this first phase, there is no adequate debate about veganism, and interviewees argue that it is important that this topic is covered in the public sphere. As with the UK findings, Croatian findings also show the relevance of a journalist and the importance of increasing diversity in newsrooms because some journalists are more interested and friendlier toward veganism than others.

These main media (…) I cannot say that I have ever seen or noticed something related to veganism. (Zaira)

OK, this is changing now to an extent because we can see trends that are entering the country from abroad, so it is not such a taboo, and the writing is no longer patronising but it is not supportive either. On the other hand, *Večernji list* every once in a while, publishes their own cookbooks so these articles can be, before holidays usually, here, try some traditional stuff without animal products. Then there is vegan Easter, so try something different, so it basically depends on the journalist. (Crazy Cat Lady).

The interviewees argued that the mainstream media promote the culture of eating meat and meat traditions, and when they do publish an article about veganism, these are usually negative in tone in which they write about veganism in an unclear way or in a way that gives partial or incorrect information. For example, veganism is mixed with vegetarianism, while veganism is often, in a sensationalist way, shown as a problem, trivialized, and the good sides of veganism are mocked. The mainstream media also do not emphasize the advantages and importance of vegan food, particularly for the planet's preservation, and they also promote ambivalent and contradictory messages such as that after a report on veganism, the TV can often also have a report on meat tradition. Activist actions are incorrectly reported with censorship practice, and topics linked to veganism are presented more softly, particularly when animal suffrage is reported because these topics are treated as unpleasant and extreme.

> I read mainstream media. I (op. online portal) annoys me because they are writing in a bombastic way such as someone died because they were vegan. They constantly write about that topic in a stupid way. And also, G.hr writes positively and then craps up generally. When you give plus/minus information, this perpetuates the status quo. People say, what now, then I won't try anything, and the comments underneath articles are horrible. (Sara)

Interviewees from an activist background also shared experiences of being interviewed by journalists who are not vegans or vegetarians. Interviewees said that aside from these journalists not having any knowledge of vegan topics and the vegan movement, they also behave aggressively when communicating with vegans and defend the dominant meat-based position. Journalists are often biased; they mock grease-free food, glorify meat consumption, and promote the meat culture in the public sphere. When they deal with vegan topics, what transpires, according to interviewees, is their lack of experience and knowledge, accompanied by a lack of information and the lack of understanding of the topic, which results in errors and wrong interpretations when reporting about the topic. In this way, the mainstream media negatively affects the vegan movement. Some interviewees also emphasized negative, aggressive, and poorly informed comments from uninformed readers who see articles about veganism as a form of imposition, thus also signaling hostility toward vegans generally, similar to the UK findings.

> Aha, well, these people I follow, who are vegans, their attitudes are great, but on the other hand, random people who accidentally found some article, their views are critical and horrible. (Monika)

The popularisation of veganism, according to interviewees is visible in articles and reports that present veganism as a healthy eating without ethical component and the ideology. Popularization and reducing vegans to food only mask the idea and the goal of vegan politics but a part of the interviewees emphasized that this is the way for veganism to enter the public sphere and dominate media discourse, despite somewhat stereotypical portrayals of vegans as boho hippies, a finding that echoes UK findings where interviewees also spoke of stereotypes of vegans as rich, privileged hippies and crackpots.

> There is a bit of space nowadays and some popularisation. I think this is positive. Of course, the debate is not very deep, but this is the first phase – we can at least hear about the problem, and the problem is at last becoming visible in public discourse. (Anja)

## *5.9 Theme 4: New Approaches*

The interviewees argued that there are different approaches to representing veganism, which are currently in the minority. Crazy Cat Lady emphasized an example of her guest talk on a local TV station. There are also radio shows about the nutritional aspects of veganism, which are useful, but the main point of difference seems to be the approach of the journalist because some journalists are more interested and friendlier than others.

> Here, the fact I was a guest on a TV station last week, that journalist is vegan, actually vegetarian but going more towards veganism, and she was completely open towards the debate, she gave me prompts about what to talk about which otherwise I could not say, and that topic would never be a topic with someone else because another journalist would immediately need to defend himself and throw counter-arguments, you know, and she was really, give me more, more, more, which was great. If there were more journalists like her, things would be great. (Crazy Cat Lady)

These findings open an important question on the influence of journalists and the diversity of newsrooms and whether hiring more vegans would change things, a question already asked about women in newsrooms and their positive influence on environmental affairs' writing (Topić, 2021).

## 6 Instead of Conclusion: Echo Chambers as a Way of Escaping Mainstream Media Bias

The above analysis was conducted in two different national contexts; however, there are similarities in findings that frequently echo each other. The final thematic analysis, taking into consideration both cases, shows that the issue with the mainstream media, based on interviewee responses, lies in the fact they cannot trust the media, so they live in their own echo chambers and follow vegan sources, consuming news via social media. Still, they recognize the importance of a journalist and the fact media work for profit (Fig. 3).

Interviewees in both countries mentioned the promotion of meat culture because of which veganism is portrayed negatively, and there is often misinformation because of it. This is often seen as due to the profit-driven business model of the contemporary mainstream media.

> Negatively. I think that the minute you see a positive article about and I've noticed that they've swung away from saying vegan, the saying plant-based, so I think there is that sort of stigma around kind of extinction rebellion, you know, all that sort of stuff that negativity, they're like, they don't want that link to veganism. They're saying plant-based. But for every article, you see, praising a plant-based diet, all of a sudden, there'll be 10 articles saying scientists have found that you know, not eating red meat does this or something about having too much leafy green? I mean, it's absolute nonsense, and I can see the way it's moving. And you can see what they're doing to counteract anybody who's trying to adopt this lifestyle. (G9 -UK)

> … the extended hand of the profit and the industry, which is so strong. (Crazy Cat Lady-Croatia)

The fact that media work for profit is visible in findings even when not directly referenced by the interviewees because it is well known that the meat industry promotes its products via the mainstream media and thus affects media revenues. In addition, the issue remains the diversity in newsrooms, which has been recognized

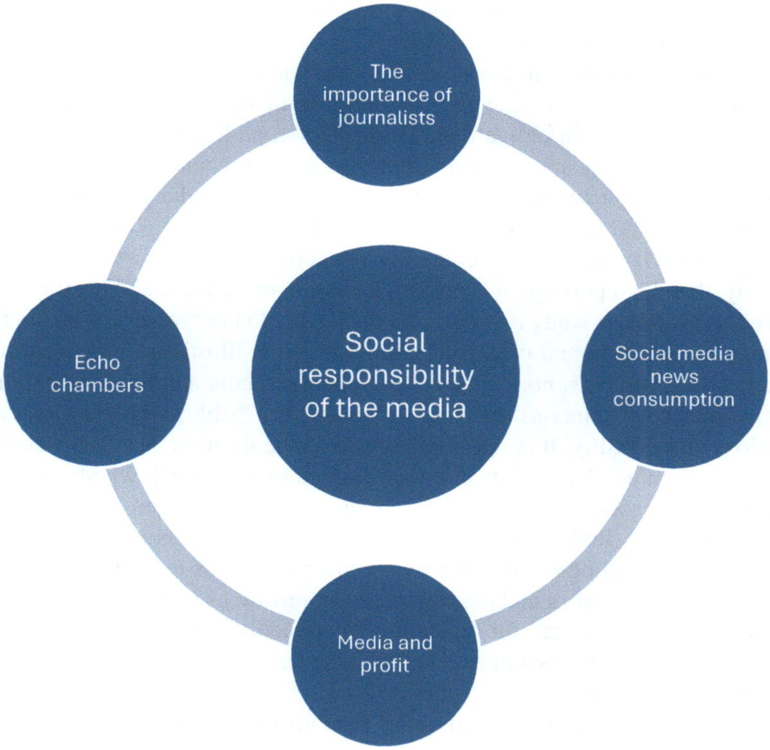


**Fig. 3** A comparative thematic analysis

by researchers, and because of this, only friendly or informative articles or reports one can find are those where the journalist is understanding and friendly. The fact that many interviewees say that media are hostile, biased, and disseminate incorrect information, indeed opens up a question of the social responsibility of the media.

If we are to look at media social responsibility theory, it is clear that interviewees—and other literature we cited—criticize the mainstream media because of their profit orientation and the fact they operate as a trade in the open market (Sandoval, 2013), which to an extent impairs their ability to do social work due to commercial pressures resulting in uniformity and conformism (Garnham, [1986]2006; Smythe, [1977]1997; Fuchs, 2010, 2011; Schiller, 1997; McChesney, 2004) where it is too easy to mock and criticize veganism as the "other." Vegans in this study said they feel criticized and treated with hostility by both the media and the general public on social media. Thus, it comes as no surprise they are turning toward their own echo chambers and consuming mainly vegan sources for news. The question remains to what extent will this worsen and whether this is a wider problem, as these two studies were qualitative and cannot be generalized. Further research, on a larger scale, is needed to confirm these findings and warn about a (further) potential issue with media representation and bias.

The findings from this study also open a question of media regulation and social responsibility of the media. The already mentioned Hutchins Commission outlined that the press has certain obligations (Bivins, 2004, p. 42), and based on our findings, it seems that they are failing in doing so, at least when this study and vegans are at stake. In other words, it cannot be said that the media provide "a truthful, comprehensive, and intelligent account of the day's events in a context that gives meaning" (ibid) because vegans continually report, in literature we cited as well as this study, that the reporting on veganism is biased, inaccurate, and hostile. In addition, research and this study revealed that the media do not "serve as a forum for the exchange of comment and criticism" (ibid), which is illustrated with examples of media hostility and bias, nor can it be said that the media are developing "a representative picture of the constituent groups in society" (ibid) when one group face exclusions and hostility. It is not possible to comment on the last two elements of Hutchins' recommendations because this study did not research whether the media generally act "responsible for the presentation and clarification of the goals and values of society" (ibid) or whether they "provide full access to the day's intelligence" (ibid) but failing three out of five recommendations indicates a wider problem with the media's work and the way news organizations in two countries operate. Therefore, authors who call for regulation when media fail in fulfilling their social role (Siebert et al., 1963) seem as relevant today as they were back when they first wrote their recommendations.

The media should strive toward social responsibility and serve the public they inform about issues of the day. At present, the media lack accountability and are losing trust despite nominally being responsible to citizens and governments, not just media owners (Owens-Ibie, 1994). It comes as no surprise that many are again debating regulation of the press or further regulation (see Media Reform Coalition, 2021, UK Parliament, n.d.).

## References

Bardoel, J., & d'Haenens, L. (2004). Media meet the citizen: Beyond market mechanisms and government regulations. *European Journal of Communication, 19*(2), 165–194.

Bergman, S. (2019). Piers Morgan almost vomits after tasting Greggs' vegan sausage roll on Good Morning Britain. *The Independent,* January 7. Retrieved from https://www.independent.co.uk/life-style/piers-morgan-greggs-vegan-sausage-roll-good-morning-britain-snowflakes-a8716071.html

Bivins, T. (2004). *Mixed media: Moral distinctions in advertising, public relations, and journalism.* Lawrence Erlbaum Associates Publishers.

Braun, V., & Clarke, V. (2006). Using thematic analysis in psychology. *Qualitative Research in Psychology, 3*(2), 77–101.

Bumbridge Sutton, A. (2024). Veganuary 2024 – Success or Failure? *Food Navigator Europe,* March, 1. Retrieved from https://www.foodnavigator.com/Article/2024/03/01/veganuary-2024-success-or-failure#

Burkhart, F. N., & Sigelman, C. K. (1990). Byline bias? Effects of gender on news article evaluations. *Journalism Quarterly, 67*(3), 492–500.

Christmas, L. (1997). Women in journalism: Chaps of both sexes? – Women decision-makers in newspapers: Do they make a difference? *Report for Women in Journalism Organisation*. Retrieved from http://womeninjournalism.co.uk/wp-content/uploads/2012/10/Chaps-of-both-Sexes.pdf

Cole, M., & Morgan, K. (2011). Vegaphobia: Derogatory discourses of veganism and the reproduction of speciesism in UK national newspapers. *The British Journal of Sociology, 62*(1), 134–153.

Eurobarometer. (2022). *Eurobarometer/Media and News Survey 2022/Country Factsheets Croatia*. Retrieved from https://europa.eu/eurobarometer/surveys/detail/2832

Franks, S. (2013). *Women and journalism*. I.B. Tauris.

Fuchs, C. (2010). Labor in informational capitalism and on the Internet. *The Information Society, 26*(3), 179–196.

Fuchs, C. (2011). *Foundations of critical media and information studies*. Routledge.

Garnham, N. (1998). Political economy and cultural studies: Reconciliation or divorce. In J. Sorey (Ed.), *Cultural theory and popular culture: A reader* (pp. 600–612). Pearson.

Garnham, N. ([1986]2006). Contribution to a political economy of mass communication. In Durham, M.G., & Kellner, D. (eds.), *Media and cultural studies: KeyWorks* (pp. 201–229). Blackwell.

Giacomini, T. (2014). Ecofeminism and system change. Women on the frontlines of the struggle against fossil capitalism and for the solar commons. *Canadian Woman Studies, 31*(1–2), 95–100.

Gorvett, Z. (2020). *The mystery of why there are more women vegans*. BBC, February 18. Retrieved from https://www.bbc.com/future/article/20200214-the-mystery-of-why-there-are-more-women-vegans

Holy, M. (2007). *Mitski aspekti ekofeminizma*. TIM Press.

Hou, J. H., & Reber, B. H. (2011). Dimensions of disclosures: Corporate social responsibility (CSR) reporting by media companies. *Public Relations Review, 37*, 166–168.

Ingenhoff, D., & Koelling, A. M. (2012). Media governance and corporate social responsibility of media organizations: An international comparison. *Business Ethics: A European Review, 21*(2), 154–167.

Jamieson, A. (2010). Vegans should be protected from discrimination, says equality watchdog. *The Telegraph,* March 7. Retrieved from http://www.telegraph.co.uk/news/religion/7392222/Vegans-should-be-protected-from-discrimination-says-equality-watchdog.html

Jourlab. (n.d.). *Istraživanje javnosti o medijima i medijskom sadržaju (presentation)*. Retrieved from https://min-kulture.gov.hr/UserDocsImages/dokumenti/Doc/Istra%C5%BEivanje%20percepcije%20javnosti%20o%20medijima%20i%20medijskom%20sadr%C5%BEaju%20-%20%C5%A0to%20publika%20%C5%BEeli.pdf

Khalidi, F. (2020). *Why are more women vegan than men?* Women Media Center, January 13. Retrieved from https://womensmediacenter.com/fbomb/why-are-more-women-vegan-than-men

Lapadat, J. (2010). Thematic analysis. In A. J. Mills, G. Durepos, & E. Wiebe (Eds.), *Encyclopedia of case study research* (pp. 926–928). Sage.

Lee, T. (2019). The global rise of "fake news" and the threat to democratic elections in the USA. *Public Administration and Policy: An Asia-Pacific Journal, 22*(1), 15–24.

Lockwood, A. (n.d.). Why the mainstream media continues to go after vegans. *Surge,* October 8. Retrieved from https://www.surgeactivism.org/articles/why-the-mainstream-media-continues-to-go-after-vegans

MacInnis, C. C., & Hodson, G. (2017). It ain't easy eating greens: Evidence of bias toward vegetarians and vegans from both source and target. *Group Processes and Intergroup Relations, 20*(6), 721–744.

Maguire, M., & Delahunt, B. (2017). *Doing a thematic analysis: A practical, step-by-step guide*. Retrieved from https://www.researchgate.net/publication/349506918_Doing_a_Thematic_Analysis_A_Practical_Step-by-Step_Guide

Maher, B. (2024). Two-thirds of Brits follow local news, mostly via social media and TV. *The Press Gazette,* January 12. Retrieved from https://pressgazette.co.uk/publishers/regional-newspapers/ofcom-local-news-research/#:~:text=Almost%20two%2Dthirds%20(65%25),as%20through%20the%20tech%20platforms

Majid, A. (2022). Trust me, I'm a journalist: New survey shows growing faith in profession. *Press Gazette,* November 25. Retrieved from https://pressgazette.co.uk/media-audience-and-business-data/trust-in-journalists/

Majid, A. (2023). Trust in media up in the UK but remains among lowest in the world, Edelman survey finds, *Press Gazette,* January 18. Retrieved from https://pressgazette.co.uk/media-audience-and-business-data/trust-media-increase-uk-us-edelman/

Majid, A. (2023a). UK newsbrand trust rankings: The Sun is least-trusted and BBC and FT lead the way. *The Press Gazette,* May 31. Retrieved from https://pressgazette.co.uk/media-audience-and-business-data/uk-newsbrand-trust-rankings-sun-least-trusted-bbc-ft-most/

McChesney, R. W. (2004). *The problem of the media. U.S. communication politics in the 21st century*. Monthly Review Press.

Media Reform Coalition. (2021). *Media influence matrix: United Kingdom.* Retrieved from https://cmds.ceu.edu/sites/cmcs.ceu.hu/files/attachment/basicpage/1926/mimuk0327.pdf

Meier, W. A., & Perrin, I. (2007). Media concentration and media governance. *Communications, 32*(3), 336–343.

Mosco, V. (2009). *The political economy of communication*. Sage.

Owen, B. M., & Wildman, S. S. (1992). *Video economics*. Harvard University Press.

Owens-Ibie, N. (1994). Press responsibility and public opinion in political transition. *Africa Media Review, 8*(1).

Plaisance, P. L. (2000). The concept of media accountability reconsidered. *Journal of Mass Media Ethics, 15*(4), 257–268.

Ragusa, A. T., Crampton, A., & Masterman-Smith, H. (2014). *Reproducing speciesism: A content analysis of Australian media representations of veganism. A conference paper.* Retrieved from https://www.researchgate.net/publication/274253326_Reproducing_speciesism_a_content_analysis_of_Australian_media_representations_of_veganism.

Sandoval, M. (2013). Corporate Social (Ir)Responsibility in Media and Communication Industries. *Javnost – The Public: Journal of the European Institute for Communication and Culture, 20*(3), 39–57.

Schiller, H. (1997). Manipulation and the packaged consciousness. In P. Golding & G. Murdock (Eds.), *The political economy of the media* (Vol. I, pp. 423–437). Elgar.

Scholzl, A. E., & Lenhart, J. (2023). Masculinity and veganism: The effect of linking vegan dishes with masculinity on men's attitudes toward vegan food. *Frontiers in Communication, 8*. https://doi.org/10.3389/fcomm.2023.1244471

Siebert, F., Peterson, T. B., & Schramm, W. (1963). *Four theories of the press: The authoritarian, libertarian, social responsibility, and Soviet communist concepts of what the press should be and do*. University of Illinois Press.

Smith, M. (2023). Which media outlets do Britons trust in 2023? *YouGov*, May 25. Retrieved from https://yougov.co.uk/politics/articles/45744-which-media-outlets-do-britons-trust-2023

Smythe, D. W. ([1977]1997). Communications. blindspots of Western Marxism. In P. Golding and G. Murdock (eds.), The political economy of the media, I, 438–464. : Elgar.

Spencer, C. (1996). *The heretic's feast: A history of vegetarianism.* University Press of New England.

The Vegetarian Society. (n.d.). *Facts and figures.* Retrieved from https://vegsoc.org/facts-and-figures/

Topić, M. (2018). Not Bloke-ified enough? Women Journalists, Supermarket Industry and the Debate on Sugar in the British Press (2010-2015). *Newspaper Research Journal, 39*(4), 433–442.

Topić, M. (2020). Ekofeministička analiza vrijednosti i filozofija globalnih eko-selâ: Jesu li žene sklonije kolektivizmu i anti-hijerarhiji nego muškaraci? In G. Đurđević & S. Marjanić (Eds.), *Ekofeminizam: Izmedju zelenih i ženskih studija*. Durieux.
Topić, M. (2021). *Corporate social responsibility and the environmental affairs: An ecofeminist critique of neoliberalism*. Routledge.
Topić, M., Diers Lawson, A., & Kelsey, S. (2021). Women and the squander cycle in food waste in the United Kingdom: An ecofeminist and feminist economic analysis. *Social Ecology/ Socijalna Ekologija, 30*(2), 219–253.
UK Parliament. (n.d.). *Chapter 2: History of press regulation in the UK*. Retrieved from https://publications.parliament.uk/pa/ld201415/ldselect/ldcomuni/135/13505.htm
Van Liedekerke, L. (2004). Media ethics: From corporate governance to governance, to corporate social responsibility. Communications, 29(1),27–42.
Ward, S. (2008). Journalism ethics. In K. Wahl-Jorgensen & T. Hanitzsch (Eds.), *The handbook of journalism studies*. Routledge.
Watson, A. (2020). *Fake news - Statistics & facts*. Statista. Retrieved from https://www.statista.com/topics/3251/fake-news/#dossierSummary__chapter4

**Martina Topić-Rutherford** is a behavioural sociologist and mass communication scholar. Currently, she is an associate professor in public relations leadership at the University of Alabama, College of Communication and Information Sciences, Department of Advertising and PR, USA. Before that, she worked at Leeds Beckett University where she held roles such as graduate teaching assistant, lecturer in public relations, senior lecturer in public relations, and reader at Leeds Business School (2014–2023). She is editor-in-chief of the *Corporate Communications: An International Journal* and *Northern Lights* journals and editor-in-chief of the book series, *Women, Economics and the Labour Relations* (Emerald). Since 2021, she has been leading the EUPERA research network, Women in PR (previously a research project, 2018–2021). She was a research lead for the #WECAN project (Women Empowered through Coaching and Networking) funded by the UK's Department of Work and Pensions and the European Social Fund (2020–2023), and she was an author and lead of the British Academy-funded project on Women in Advertising (2018–2019). She is an author of Corporate Social Responsibility and Environmental Affairs in the British Press: An Ecofeminist Critique of Neoliberalism (Routledge, 2021) and Workplace Culture in Mass Communication Industries (Routledge, 2023). She is a member of the American Sociological Association, British Sociological Association, and EUPRERA. She is the author and lead of the Comms Women advocacy initiative (https://www.commswomen.uk)

**Marija Geiger Zeman** (PhD) is Senior Research Scientist in Sociology at the Institute of Social Sciences Ivo Pilar in Zagreb (Croatia). Geiger Zeman is the author of many papers and conference presentations focused on gender issues, media, body, and ageing issues. Her empirical work is based on qualitative methodology. In 2010, Geiger Zeman received the Annual Science Award by the Croatian Parliament in the field of social sciences.

**Gabrielle Round** , BA (Hons), MSc, PgCAP, FHEA, is Lecturer in Enterprise at the University of Salford, UK. This work supports students in starting their own businesses and involves close work with internal investors, growth hubs, and charities such as Young Enterprise and JA Europe. Gabi is nearing the completion of her PhD with Leeds Beckett University, supported by the #WECAN project. The focus of this research is the exploration of solutions for gender disparity in small-medium enterprise (SME) succession processes. Gabi's research interests focus predominantly on business topics, with a feminist lens, with a spotlight on gender disparity and progressing solutions for equity in the business environment. Passionate about environmental issues and animal rights, Gabi has added veganism and eco-feminism to her research areas in recent years and hopes to continue this trajectory.

# The Interference Between STEM Education and Media to Maximize the Potential of Future Generations

**Mihaela Tinca Udriştioiu**

**Abstract** The media plays an important role in shaping scientific literacy and increasing public trust in science and technology, preparing future generations for a rapidly evolving technological world. The next generation should be trained not for today's jobs but for tomorrow's jobs. It is up to each society to support the future workforce's education. At the same time, it is up to the skill of educators to choose active and interactive teaching methods and techniques that keep students' attention in an environment that stimulates them to use technology, developing their creativity and capacity for innovation.

Given the features of future professions, which are increasingly based on modern technologies such as artificial intelligence, robots, big data, the Internet of Things, and others, the media has a moral obligation and an ethical commitment to promote the latest scientific discoveries and cutting-edge technologies. Through the prism of journalistic responsibility, mass media should provide accurate information in an accessible way without misleading the public.

This chapter aims to analyze the role of STEM education in training future generations of citizens. It presents how mass media can help STEM education reach the public. This discussion is important because STEM education leads to valuable technical skills and develops a system of thinking through which to approach any problem from multiple points of view, leaving no room for interpretation and error. On the shoulders of today's educators lies the responsibility of forming technical skills in tomorrow's generations so that humanity can progress in a correct, pacifist, and sustainable direction and in which to enjoy the benefits of technology. By promoting scientific and technological progress, the media can inspire new generations to focus on careers in high-difficulty STEM fields.

**Keywords** STEM education · Artificial intelligence · Robots · Digitization · Humanity progress · Science popularization · Sustainable society · Scientific

---

M. T. Udriştioiu (✉)
Department of Physics, Faculty of Sciences, University of Craiova, Craiova, Romania
e-mail: mtudristioiu@central.ucv.ro



M. Todorovic et al. (eds.), *Sustainability and Social Responsibility of the Media and in the Media*, CSR, Sustainability, Ethics & Governance,
https://doi.org/10.1007/978-3-032-00086-6_13

literacy · Media coverage · Media responsibility · Science journalism · Awareness campaigns · Green energy · Climate change · Pollution · Threats · Human health · Biodiversity · Future generations · Technical abilities · Scientific progress · Public perceptions · Communication · Education image · STEM students · STEM careers · Future professions · Online education · Social media · Streaming platforms · Science magazines · Radio · Television · Internet

## 1 Introduction

During the COVID-19 pandemic, the media shaped the public understanding of the pandemic crises. At the beginning of this crisis, the media informed people about the virus, safety measures, and government responses. Also, the media influenced public behavior, helping them understand the seriousness of this situation. The media promoted mask-wearing and social distancing. It also disseminated experts' interviews and showed images from the overwhelmed hospitals (Tsoy et al., 2021). At the same time, media platforms become a source of misinformation .Theories of conspiration spread in the context of the lack of official information about the origins of the virus and the role of the vaccine in stopping the pandemic (Dixon&Clarke, 2013; Vai et al., 2020). Regarding policy and government action, policymakers often adjusted their communication strategies based on how the media portrayed the pandemic. This example highlights the power of media in shaping public perception, influencing behavior, and impacting government decisions (Malecki et al., 2021) and justifies why this power should be used in education.

The COVID-19 pandemic has just shown us how fragile humanity is in the face of such viruses. From this perspective, humanity needs to rethink its stance toward education and prioritize science and technology so that progress can continue on its natural path. The pandemic has shown the need of each society for a scientific explanation of the evolution of the virus so that people understand what they will go through, what they will be exposed to, what the risks are, when the vaccine can be made, and the need for vaccination to come back to everyday life. In each country affected by the pandemic, the media presented the communication strategy of the decision-makers in the public health field to the large public. In countries such as those in Northern Europe (Irfan et al., 2022) and Spain (de Sola, 2021), the scientific community was trusted to communicate to the public what would happen in the following days, weeks, and months, why vaccination was necessary and what should be done to minimize the effects of the pandemic. This communication approach proved winning and saved many lives in the countries that used it. Other countries, such as the former communist countries of South-Eastern Europe, used constraints (such as self-declarations) and fines to make people stay home. They cultivated a sense of fear by presenting daily statistics on deaths (Cernicova-Buca & Palea, 2021; Nikolova et al., 2022; Demirbilek et al., 2020).

During the COVID-19 pandemic, online platforms made education possible, allowing for social distancing and isolation. Educators quickly learned how to use online platforms and create engaging content for their students (Udristioiu & Amarie, 2024; Raykova et al., 2023, 2024). Digitization occurred unprecedentedly during and after the pandemic (Amankwah-Amoah et al., 2021).

The complexity of the interference between STEM education and media has increased and refined over the last 20 years (Robertson et al., 2023). It appears that this complexity has its duality (Weingart, 1998). On the one hand, media channels bring complex scientific research and innovation closer to the public. Information has become accessible because of innovation in digital media (Pérez-Seijo & Silva-Rodríguez, 2024). On the other hand, media affects the quality of communication and science coverage (Reinman et al., 2019; Rosenthal, 2020; Schäfer, 2017; Chen et al., 2023a, 2023b). Mass media brings current scientific results to the public and promotes researchers' image and work. It can influence the scientific credibility of image vectors, implicitly of researchers (Boothby et al., 2021). Often, the media attracts public interest by presenting events regarding crises, minimizing or ignoring essential issues due to political and economic influences (Cassandra syndrome) (Weingart, 1998).

Younger generations are overwhelmed by information coming from all directions, and their capacity for synthesis has decreased significantly (Bradley & Bradley, 2006). It is difficult for many young people to read, understand a text, and extract the main ideas (Frankel et al., 2021). An important percentage of these generations grew up with a screen in front of them, and to avoid getting bored, information must come at the pace of touch (Kardaras, 2016). Moreover, students' concentration has decreased, and educators should be well-trained and work hard to keep their attention high (Le, 2021). In this context, it is important to understand and evaluate the factors that lead to real learning, the criteria based on which students select their sources of information and how they assess the degree of trustworthiness of the sources (Takahashi & Tandoc, 2016).

In most cases, scientific communities have learned to appreciate the benefits of visibility in the media. Communication departments have been created in research institutes or universities evaluating the effects of institutional communication and regulating the contact of members of the respective community with the media through strict rules (Peters, 2013). Like companies, universities and research institutes have developed departments of communication and academic image through which they communicate the most important research results, disseminate information about events organized in the scientific community, and present projects of involvement in the development of the community and the economic environment.

Finally, the budget of many universities depends on the number of students enrolled (Korfmann et al., 2021). From this perspective, the importance of the academic image department increases over time. How well these institutions communicate at the institutional level depends on the professionalism and involvement of the employees in these departments. It has become standard for academic institutions and research institutes to have institutional accounts on social media platforms (Facebook, YouTube, Twitter, Instagram) and communicate with their target audience, present results and events, express their mission and values, share information, and answer questions.

The following sections will present the benefits of the interference between science, technology, education, and media, which topics focused on technology, innovation, and education are covered by the media, how media can support STEM education, and a few conclusions on this topic.

## 2 Science, Technology, and Education Need the Support of Mass Media

### *2.1 Modern Society Is Technology-Hungry, and the Media Knows This*

The media knows that technology attracts public attention and does everything possible to meet this need. Although stakeholders are involved and want to explain technological progress, their desire to shape public perceptions and understanding (O'Connell et al., 2011) makes the unaltered transmission of the scientific message challenging to achieve (Landrum et al., 2019). The media first presents crises, local, national, and international events, and scientific discoveries. It appeals to the emotions of the target audience.

Journalists need to objectively present the events and the reasons behind important decisions by decision-makers for local communities in crises, promoting a sense of trust in the media and increasing public confidence in the ability of authorities to solve problems. Science journalists publish daily articles in newspapers, magazines, and websites that popularize science and explain scientific discoveries, theories, and technologies in a way everyone can understand. While science journalism is in decline, blogging and podcasting are growing in importance (Brumfiel, 2009). Science experts blog and podcast about scientific news and explain complex abstract concepts.

Social networks have significantly changed the way people communicate and use technology. Companies pay to obtain information about user preferences or to assess the public's reaction to their products (Shchepina & Surikov, 2022). Social network users see a text, video, or image, analyze it, express their opinion (like, dislike, and all variations), and eventually pass on the information. Based on machine learning algorithms, the information provided by social networks about user's reactions is automatically analyzed (Gasco et al., 2019) and can be exploited for a fee. It can be used by companies and stakeholders that manage elections, crises, wars, and natural cataclysms. Based on the location data collected from social networks, it is possible to evaluate how crowds move during huge city events (Lorente-Riverola & Ruiz-Sánchez, 2018).

The interference between modern science, technology, and mass media helps people understand global issues, cooperate in crises, help each other, and promote peace worldwide. Mass media contributes to increasing the public's scientific literacy, stimulating curiosity, and understanding of the importance of science in everyday life (Slovic, 2013; Barashkova et al., 2019). Documentaries about science and technology reach homes via electromagnetic waves (radio, TV, streaming platforms), addressing large audiences. Newspapers, magazines, and websites publish articles about the latest discoveries and innovations, broadcast interviews with scientists, or explain complex phenomena and technologies. An impressive number of documentaries are broadcast on TV or streaming platforms. Everyone finds information about public events dedicated to science and technology through social networks, blogs, and podcasts.

## 2.2 *How Can Media Be Involved in Promoting STEM Education?*

Formal education shapes future citizens' minds, helping students understand what a healthy ecosystem means, why the interactions between species are important, what happens when a species disappears, and how it affects the environment in which they live (Boni & Calabuig, 2017). Education guides young generations in unravelling the mysteries of biodiversity, helping them to understand why biodiversity is vital to the beautiful blue planet, how this biodiversity can be conserved, what natural resources are available (fossil and renewable), and which are the threats to the environment (air, water, and soil pollution, climate change, deforestation, loss of habitats and species) (Bermudez et al., 2022). Furthermore, education helps students think in terms of cause and effect. Later, this feature will help them make responsible environmental decisions locally and globally as future citizens. With educators' help in organizing curricular and extracurricular activities for their students, the education system contributes to developing feelings of responsibility and empathy toward the environment, as well as an appreciation of the beauty and complexity of nature in future citizens (Wang et al., 2023).

These goals are achieved over time and repeated at various levels of child development through nature exploration and observation, active participation in laboratory activities, involving students in campaigns to raise awareness of environmental issues, and carrying out homework and projects. Civic involvement is encouraged by stimulating students to volunteer in environmental protection activities (planting trees, cleaning rivers of plastic bottles, monitoring air quality, cleaning parks of household waste), helping the elderly or disadvantaged children, and helping organize events for the general public. Environmental education contributes to an awareness of the importance of a clean environment for human health. It also helps citizens become aware of problems in their community and want to get involved in solving them. Environmental education sensitizes future generations, helps them move toward sustainable and clean energy sources, and encourages them to choose environmentally friendly technology. It has been shown that through education with an environmental component, citizens are more likely to support green economies and make sustainable consumption choices (Zsóka et al., 2013).

Another direction in which the media is involved, along with education, is that of raising public awareness on issues related to public health, climate change and environmental protection (monitoring environmental quality, presenting innovative solutions in solving environmental challenges, preserving biodiversity), safety and prevention (road traffic safety, combating domestic violence, cybersecurity), human rights and social inclusion (e.g., rights related to work, education and access to opportunities, women's rights, ethnic minorities), public health of the population (vaccination and disease prevention, mental health, healthy lifestyle), and education and culture (protecting cultural heritage, promoting reading). Public figures such as the actor Leonardo DiCaprio and Prince William use media platforms to support environmental causes and encourage innovative solutions to environmental problems. Developing awareness campaigns with the media's support is an opportunity to educate the public. Triple R is known worldwide through awareness campaigns promoting waste reduction, reuse and recycling practices through TV spots, posters, and online advertisements (Anua et al., 2022). World Wildlife Fund (WWH) runs a well-known event, "Earth Hour," an annual campaign encouraging people to turn off the lights for an hour to raise awareness of climate change (Jechow, 2019). Greenpeace and the WWH share information and mobilize youth in environmental action to Generation Z. They use new digital technologies such as generative artificial intelligence to promote their targeted messages through TikTok, Instagram, and Slack channels (Muncie, 2024).

Mass media has an important role in presenting current STEM topics such as innovations and scientific discoveries, the rise in sea and ocean levels due to the melting of glacier caps, the effects of extreme events induced by climate change, and environmental pollution. It makes citizens aware of how rising global temperatures affect human health and biodiversity and how human activities might influence global warming. Media also helps popularize solutions and showcase available technologies. Live streaming of high-impact events, such as conferences, webinars, and workshops, brings together political leaders and environmental activists to discuss finding solutions to mitigate the effects of climate change and slow climate warming.

Another way of interacting education with the media is through educational games and shows dedicated to children. Eco-Schools and National Geographic Kids have been developed to teach children about the importance of a clean environment for their health and why everyone should respect "Mother Nature." Minecraft education edition includes lessons and activities on sustainability and environmental protection, which educators can use to teach children through play. WWF's Footprint Calculator is an interactive online tool that helps users calculate their carbon footprint and learn how to reduce it. Conferences, workshops, exhibitions, and science festivals are organized worldwide, allowing young generations to see live experiments, listen to well-known scientists who give talks on hot topics, and participate in interactive discussions with researchers (Fox et al., 2014). Science museums offer education through exhibitions and activities that encourage learning through direct interaction (Sampson & Turgo, 2018). In addition, during events such as Night at the Museum, people who cannot afford to visit those museums have access to information and can see museum collections.

## 3 Media Coverage of STEM Subjects

TV channels, radio, written and online press, and streaming platforms present hot topics in science and technology topics, from the macroscopic to the quantum level and from the cosmos to the depths of the oceans. Green and renewable energy sources, energy storage technologies, environmental impacts, climate change awareness and media usage, energy-saving behavior in the framework that everyone pays more for green energy, and the space industry are only a few such topics (Arlt et al., 2011; Barkemeyer et al., 2017; Liobikienė & Dagiliūtė, 2021).

***BBC*** "Planet Earth" is the public service broadcaster in the United Kingdom, where Sir David Attenborough, with his distinctive voice, tells stories about plants and animals and explains the impact of human activities on climate. Years ago, the physicist Brian Cox presented the broadcast "Wonders of the Solar System," simply explaining natural complex phenomena. For example, he discusses how the Sun influences each planet in our Solar System, how the magnetosphere deflects solar winds, which are the factors that cause a borealis aurora and where such phenomena can be observed, which are the effects of Coriolis force in river bank erosion, signs of life in our solar system.

The ***Guardian*** is recognized worldwide as an independent and investigative British newspaper that occasionally publishes articles on STEM topics. For example, the "The Climate Crisis" series provides objective, extensive, and in-depth environmental coverage, presenting the points of view of all involved actors (lobbyists, governmental agencies, industries). ***National Geographic*** magazine publishes articles on science, nature and wildlife, history and culture, environment, travel, and exploration. The authors approach topics such as travelling to exotic places around the world, the ecosystems of deep oceans, the human body and mind, and how the body and mind are affected by different substances. They also present the animal world (from species migrations to mating rituals and the struggle for survival).

***Netflix*** is a streaming platform that delivers movies, series, documentaries, and original shows. The documentary "*Our Planet*" uniquely captures the beauty of Earth. This documentary covers the importance of bees for flower pollination, the life of fish and animals from ocean depths, Earth's structure, volcanoes and earthquakes and their effects during human history. This documentary aims to raise awareness of environmental issues and encourage people to take action to protect "Mother Nature" and to learn how to react to natural threats. It describes nature's greatness and draws attention to the dangers threatening the fragile balance of the Earth's planet. "*Cosmos: A Spacetime Odyssey*" is a documentary series that explores the universe, the history of science, and the fundamental concepts of astronomy and physics. "*The Mind Explained*" is a series that explores the human mind, including emotions, memory, anxiety, and other aspects of human psychology. "*Explained*" is a documentary series that explores topics ranging from science and technology to culture and politics, offering clear and concise explanations.

On the ***Amazon Prime Video*** streaming platform, many documentaries exploit STEM subjects. For example, in the series "*The Grand Tour*," Jeremy Clarkson, Richard Hammond, James May, and Andy Wilman test different cars in places where nature is spectacular (e.g., they filmed on Trasfagarasan road that crosses southern the Carpathian Mountain in Romania) and discuss the features of some powerful engines. "*Nova*" is a series of documentaries produced by the Public Broadcasting Service from the United States. These series of documentaries present scientific topics from astronomy and space, sciences and technology (artificial intelligence and robots, future sources of energy), biology and human health (pandemics, medical research, ageing, future medicine), archaeology and history (antic structures, the origin of human species and anthropology, lost civilizations, antic Egypt), climate change and environment (global warming, ocean depth exploring, pollution effects), and people who make things happen in science. It also explains some discoveries and innovations to the public. "*The Brain with David Eagleman*" tries to answer questions about the human brain, how it works, and the complexity of cognitive processes. The "*Breakthrough*," a series of scientific documentaries produced by National Geographic that explores major scientific innovations, talks about the cosmos, the origin of the universe, black holes, the discovery of fundamental particles at CERN and their possible use in medicine, the role of oceans for the planet, the eruption of volcanoes, the melting of glaciers in the context of global warming, the rise in water levels in some regions of the globe, extreme phenomena affecting humanity and the effect of pollution on ecosystems.

***HBO*** is a well-known streaming platform offering movies, series, shows, documentaries, and live sports programs. "*The Inventor: Out for Blood in Silicon Valley*" is an American documentary that describes the investigation of a huge scandal at Theranos, a biotech company. The founder of this company, Elizabeth Holmes, is an American biotechnology entrepreneur who promoted medical technologies that later became fraudulent. She promised that her company could conduct numerous tests based on a few drops of blood and a device. "*Going Clear: Scientology and the Prison of Belief*" approaches the issue of the Church of Scientology, the practices and influence of the group and its founder. Some ex-members tell the stories of the abuses and exploitation they have been subjected to. "*Food, Inc.*" investigates the US food industry, presenting the US food industry giants and their impact on health, the economy, and the environment, emphasizing some issues about food control. "*How to Survive a Pandemic*" is about the lessons learnt by humanity from the 1918 influenza pandemic, the response of the governments, and international efforts in developing the vaccine against COVID-19. "*Going Nuts: Tales from the Squirrel World*" is a documentary about Mother Nature's power to create and destroy, showcasing the behavior, biology, and adaptability of squirrels.

***Disney +*** is a streaming platform that offers movies, series, and documentaries from Disney, Pixar, Marvel, Star Wars, and National Geographic. "*The Imagineering Story*" explores the history and technology behind the Disney theme parks. This documentary miniseries offers a behind-the-scenes look at the creation and innovation process. "*Science Fair*" is a documentary that presents the journey of nine science-loving and dedicated teenagers who participated in a prestigious competition named Intel International Science and Engineering Fair. "*Marvel's Hero Project*" tells the inspiring stories of exceptional kids doing exceptional things that impact their local communities. "*One Strange Rock*" is a documentary series by National Geographic presented by Will Smith that offers information about unique conditions that allow the development and supporting life on the blue planet Earth.

***Apple TV***'s offering is more limited compared to other streaming platforms. It contains titles such as "*The Elephant Queen*," a story of a female elephant in Africa who successfully leads her herd during a drought in search of water sources. This documentary series explores the struggle for the survival of this species in a hostile and fragile ecosystem. "*Tiny World*" is an English documentary presenting details about tiny creatures' lives, the ecosystems they are part of, and what they do to survive. "*The Year the Earth Changed*" is about how the COVID-19 pandemic changed our world, helping us evaluate humanity's effect on nature. "*Becoming You*" follows 100 children in their first 2000 days of life.

Impressive videos, engaging and interactive lessons, practical demonstrations, and experiments are presented on channels like YouTube, TED-Ed, Khan Academy, Crash Course, and edX. Some of their plans are free, and others are paid.

On the ***YouTube*** platform, content creators offer videos, tutorials, experiments, and courses on different topics in English and other languages, and the pandemic has multiplied their number significantly. Short animated videos, lessons, and educational projects on various topics, including science, technology, history, art, and more, are uploaded on the ***TED-Ed*** platform. ***Khan Academy*** is an online platform that offers free courses on various topics. It includes tools for educators who can assess students' progress and can adapt their lessons according to their needs. It uses interactive exercises and engaging videos. The academy promotes skill mastery through personalized learning at the student's desired pace in and out of the classroom. ***Crash Course*** represents a series of educational video programs covering astronomy, climate and energy, film production, world history, science, literature and philosophy, government, and more. The ***edX*** platform hosts more than 4000 courses. The top two are about artificial intelligence (*Implications for Business Strategy* and *Introduction to Artificial Intelligence with Python*). The learning approach is based on experience, practice, and application.

## 4 How Important Is STEM Education for the Next Generations?

In a context where robots and artificial intelligence are becoming a reality, when there is so much talk about digitization, devices are getting smaller and smaller, and technology is evolving rapidly, educating future generations to become adaptable, understand, and answer the challenges of science and technology becomes a priority for any society. Countries' investments in education will make a difference between developed and underdeveloped countries.

A sound education has to maximize the potential of each new generation and lead to the development of critical thinking that helps the student search for, find, and confidently apply solutions to the problems faced. Education should also reduce the likelihood of manipulation in a world where artificial intelligence makes it increasingly difficult to distinguish between the real and imaginary world and social media propagates fake news.

STEM education helps pupils and students understand how to solve complex problems, keeps their curiosity alive, helps them understand abstract concepts, helps them analyze any problem from multiple points of view, and stimulates their creativity and imagination. In STEM activities, students observe, ask questions, formulate ideas, try new things, make mistakes from which they learn, apply what they discover, learn programming concepts, and acquire practical and theoretical skills. Using educational kits, students learn how to build robots or sensor networks. Students can make sensors that monitor different parameters (at home or in school laboratories) (Velea et al., 2023); these sensors can be incorporated into more complex devices or connected to independent sensor networks (IoT) (El Mghouchi et al., 2024).

The presentation in the mass media of some career models through inspirational stories among children and adolescents might influence them to pursue a STEM career. Also, the media can provide access to information and educational resources to students from isolated backgrounds, thus encouraging diversity and inclusion. Promoting STEM education in the mass media might influence students' career choices (Chen et al. 2023a, 2023b). Of course, the decision to pursue a STEM career is also influenced by many other factors, such as personal interests and individual abilities, socioeconomic factors, advice, guidance, and support from parents, teachers, and educational advisors. Students may be more likely to choose fields they feel drawn to and have natural abilities. The availability and accessibility of educational resources, financial support, and career opportunities play essential roles in students' decisions regarding their path. There is a significant gender imbalance in STEM jobs, which requires compliance with ethical criteria in companies with such vacant positions and in STEM education and research to encourage girls to develop such careers. This fact should not be neglected either (Brzozowy et al., 2017).

The link between communication and community is culture (Carey, 2008), and between science and society is popularization through media (Weingart, 1998). Humanity's tastes in culture, science, and technology are developed through education, exposure, practical experience, and an environment that stimulates them (Murphie & Potts, 2017).

## 5 Conclusions and Considerations Regarding Policy and Education

Science and media play an essential role in STEM education. Science contributes to a good understanding of the world and develops logical thinking. Media helps students from disadvantaged environments or with disabilities attend online classes or access information through TV educational programs that might help them take exams. Media can promote innovative and inclusive education for all ages. Together, science and the media educate society in the spirit of peace and cooperation.

STEM education develops well-informed and responsible citizens who are able to understand and approach each issue and actively contribute to protecting and preserving the quality of their environment. An interdisciplinary approach enhances students' understanding of the social and economic impacts of environmental problems, encouraging them to explore sustainable solutions. Technological progress is influenced by the development of emerging technologies (artificial intelligence, blockchain, biotech), changes in the labor market, and workforce diversity.

STEM education promotes behaviors that can improve health, well-being, and wealth, such as energy savings, reducing pollution, adopting a circular economy, using clean energy sources, and living an active lifestyle in harmony with nature. It contributes to an informed, responsible, and active generation of citizens who can make informed choices and positively influence the planet's future.

Even if the number of STEM graduates is lower today, technological progress is not necessarily restricted by this. Technological progress occurs because of international collaboration and knowledge exchange between researchers, engineers, and professionals in different fields.

Regarding policy considerations, there are a few recommendations:

1. Educational policies must ensure access and equity for all students, regardless of socioeconomic background, to high-quality STEM education and free digital resources.
2. Critical thinking and media literacy in STEM education help students analyze scientific claims in news and social media.
3. Governments, media, and tech companies have to collaborate to improve STEM education.
4. Policies must align STEM education with workforce needs and encourage media campaigns to promote careers in the STEM field.

The Next Generation Science Standards (NGSS) in the United States represent a successful policy of incorporating media and technology into STEM education (National Research Council, 2015; Pruitt, 2014).

There are a few observations about the educational implications of STEM education:

1. Using media-based STEM education in teaching students (XR experiences, gamification, interactive simulations, virtual experiments) makes learning more interactive and engaging for them.

2. Students should learn the difference between evidence-based science and misleading information during formal education.
3. Personalized learning: Learning from the pandemic experience and keeping digital platforms alive is necessary because they allow adaptive learning, helping students progress in their rhythm.
4. Educators need access to professional development to incorporate new technologies into teaching methods.

## References


Amankwah-Amoah, J., Khan, Z., Wood, G., & Knight, G. (2021). COVID-19 and digitalisation: The great acceleration. *Journal of Business Research, 136*, 602–611. https://doi.org/10.1016/j.jbusres.2021.08.011

Anua, S. M., Anwar, N. N. K., Zain, N. M., Abd Rahman, W. N. W., Hamzah, N. A., & Rahman, H. A. (2022). Reduce, reuse and recycle (3R) awareness programme to increase the knowledge, attitude and practice on 3R among primary school students. *International Journal of Academic Research in Business and Social Sciences, 12*(13), 62–74.

Arlt, D., Hoppe, I., & Wolling, J. (2011). Climate change and media usage: Effects on problem awareness and behavioural intentions. *International Communication Gazette, 73*(1-2), 45–63. https://doi.org/10.6007/IJARBSS/v12-i13/14144

Barashkova, A. L., Vorob'ev, I. V., Shavaev, A. A., & Zapolskaya, A. N. (2019, September). New methods of science popularisation in the social media: modern trends and communications. In *2019 International Conference "Quality Management, Transport and Information Security, Information Technologies" (IT&QM&IS)* (pp. 463–465). IEEE. https://doi.org/10.1109/ITQMIS.2019.8928354

Barkemeyer, R., Figge, F., Hoepner, A., Holt, D., Kraak, J. M., & Yu, P.-S. (2017). Media coverage of climate change: An international comparison. *Environment and Planning C: Politics and Space, 35*(6), 1029–1054. https://doi.org/10.1177/0263774X16680818

Bermudez, G. M. A., Pérez Mesa, R., & Ottogalli, M. E. (2022). Biodiversity knowledge and conceptions in Latin American: Towards an integrative new perspective for education research and practice. e-ISSN: 2147-611X.

Boni, A., & Calabuig, C. (2017). Education for global citizenship at universities: Potentialities of formal and informal learning spaces to foster cosmopolitanism. *Journal of Studies in International Education, 21*(1), 22–38. https://doi.org/10.1177/1028315315602926

Boothby, C., Murray, D., Waggy, A. P., Tsou, A., & Sugimoto, C. R. (2021). Credibility of scientific information on social media: Variation by platform, genre and presence of formal credibility cues. *Quantitative Science Studies, 2*(3), 845–863. https://doi.org/10.1162/qss_a_00151

Bradley, K. S., & Bradley, J. A. (2006). Teaching Students to Organise and Synthesise Information. *Journal of the International Association of Special Education, 7*(1).

Brumfiel, G. (2009). Supplanting the old media? Science journalism is in decline; science blogging is growing fast. But can the one replace the other. *Nature, 458*(7236), 274–278.

Brzozowy, M., Hołownicka, K., Bzdak, J., Tornese, P., Lupiañez-Villanueva, F., Vovk, N., et al. (2017). Making STEM education attractive for young people by presenting key scientific challenges and their impact on our life and career perspectives. In *INTED2017 Proceedings* (pp. 9948–9957). IATED.

Carey, J. W. (2008). *Communication as Culture, Revised Edition: Essays on Media and Society* (2nd ed.). Routledge. https://doi.org/10.4324/9780203928912

Cernicova-Buca, M., & Palea, A. (2021). An appraisal of communication practices demonstrated by Romanian district public health authorities at the outbreak of the COVID-19 pandemic. *Sustainability, 13*(5), 2500. https://doi.org/10.3390/su13052500

Chen, C., Hardjo, S., Sonnert, G., Hui, J., & Sadler, P. M. (2023a). The role of media in influencing students' STEM career interest. *International Journal of STEM Education, 10*(1), 56. https://doi.org/10.1186/s40594-023-00448-1

Chen, Y., Sherren, K., Smit, M., & Lee, K. Y. (2023b). Using social media images as data in social science research. *New Media & Society, 25*(4), 849–871. https://doi.org/10.1177/1461444821103876

De Sola, J. (2021). Science in the media: the scientific community's perception of the COVID-19 media coverage in Spain. *JCOM, 20*(02), A08. 10.22323/2.20020208.

Demirbilek, Y., Pehlivantürk, G., Özgüler, Z. Ö., & Meşe, E. A. (2020). COVID-19 outbreak control, example of ministry of health of Turkey. *Turkish Journal of Medical Sciences, 50*(9), 489–494. https://doi.org/10.3906/sag-2004-187

Dixon, G. N., & Clarke, C. E. (2013). Heightening Uncertainty Around Certain Science: Media Coverage, False Balance, and the Autism-Vaccine Controversy. *Science Communication, 35*(3), 358–382. https://doi.org/10.1177/1075547012458290

El Mghouchi, Y., Udristioiu, M. T., & Yildizhan, H. (2024). Multivariable Air-Quality Prediction and Modelling via Hybrid Machine Learning: A Case Study for Craiova. *Romania. Sensors, 24*(5), 1532. https://doi.org/10.3390/s24051532

Fox, D., Gouthro, M. B., Morakabati, Y., & Brackstone, J. (2014). *Doing events research: From theory to practice*. Routledge. https://doi.org/10.4324/9781315815084

Frankel, K. K., Deanna Brooks, M., & Learned, J. E. (2021). A meta-synthesis of qualitative research on reading intervention classes in secondary schools. *Teachers College Record, 123*(8), 31–58. https://doi.org/10.1177/01614681211048624

Gasco, L., Clavel, C., Asensio, C., & de Arcas, G. (2019). Beyond sound level monitoring: Exploitation of social media to gather citizens subjective response to noise. *Science of the Total Environment, 658*, 69–79. https://doi.org/10.1016/j.scitotenv.2018.12.071

Irfan, F. B., Minetti, R., Telford, B., Ahmed, F. S., Syed, A. Y., Hollon, N., Brauman, S. C., Cunningham, W., Awad, M. E., Saleh, K. J., Waljee, A. K., & Brusselaers, N. (2022 Aug). Coronavirus pandemic in the Nordic countries: Health policy and economy trade-off. *Journal of Global Health, 8*(12), 05017. https://doi.org/10.7189/jogh.12.05017

Jechow, A. (2019). Observing the impact of WWF earth hour on urban light pollution: A case study in Berlin 2018 using differential photometry. *Sustainability, 11*(3), 750. https://doi.org/10.3390/su11030750

Kardaras, N. (2016). *Glow Kids: How screen addiction is hijacking our kids-and how to break the trance*. Martin's Press.

Korfmann, F., Müller, S., Ehlert, S., & Haase, K. (2021). Students' perceptions, academic departments' image, and major-choice in business administration studies—The example of Hamburg Business School. *Higher Educ Q., 75*, 51–76. https://doi.org/10.1111/hequ.12258

Landrum, A. R., Olshansky, A., & Richards, O. (2019). Differential susceptibility to misleading flat earth arguments on YouTube. *Media Psychology, 24*(1), 136–165. https://doi.org/10.1080/15213269.2019.1669461

Le, H. V. (2021). An investigation into factors affecting concentration of university students. Journal of English Language Teaching and Applied. *Linguistics, 3*(6), 07–12. https://doi.org/10.32996/jeltal

Liobikienė, G., & Dagiliūtė, R. (2021). Do positive aspects of renewable energy contribute to the willingness to pay more for green energy? *Energy, 231*, 120817. https://doi.org/10.1016/j.energy.2021.120817

Lorente-Riverola, I., & Ruiz-Sánchez, J. (2018). In A. Bisello, D. Vettorato, P. Laconte, & S. Costa (Eds.), *Spatial footprints of context-aware digital services. Eventual self-regulated alignments of dating apps with the urban shape, in Smart and Sustainable Planning*

*for Cities and Regions* (pp. 133–150). Springer International Publishing. https://doi.org/10.1007/978-3-319-75774-2_10
Malecki, K. M., Keating, J. A., & Safdar, N. (2021). Crisis communication and public perception of COVID-19 risk in the era of social media. *Clinical Infectious Diseases, 72*(4), 697–702. https://doi.org/10.1093/cid/ciaa758
Muncie, E. C. (2024). Campaigning for the Future and the Future of Campaigning: A Case of Greenpeace International and the Emergence of New Environmental Communication Landscapes (Doctoral dissertation, University of Leicester).
Murphie, A., & Potts, J. (2017). *Culture and technology*. http://digital.casalini.it/9781137089380.
National Research Council, Board on Science Education, & Committee on Guidance on Implementing the Next Generation Science Standards. (2015). *Guide to implementing the next generation science standards.*
Nikolova, M., Angova, S., Valchanov, I., Valcheva, K., Zlatkova, V., Valkov, I., & Marinkova, B. (2022). Bulgarian public administration's social media communications strategies during COVID-19 crisis. Global issues: Disease control and pandemic. *Prevention, 24*, 55–62. https://doi.org/10.54941/ahfe1001359
O'connell, G., De Wilde, J., Haley, J., Shuler, K., Schafer, B., Sandercock, P., & Wardlaw, J. M. (2011). The brain, the science and the media: The legal, corporate, social and security implications of neuroimaging and the impact of media coverage. *EMBO Reports, 12*(7), 630–636. https://doi.org/10.1038/embor.2011.115
Pérez-Seijo, S., & Silva-Rodríguez, A. (2024). Innovation in Digital Media beyond Technology: The Audience-Centered Approach and Pending Challenges. *Journalism and Media., 5*(1), 311–324. https://doi.org/10.3390/journalmedia5010021
Peters, H. P. (2013). Gap between science and media revisited: Scientists as public communicators. *Proceedings of the National Academy of Sciences, 110*(supplement_3), 14102–14109. https://doi.org/10.1073/pnas.1212745110
Pruitt, S. L. (2014). The next generation science standards: The features and challenges. *Journal of Science Teacher Education, 25*(2), 145–156. https://doi.org/10.1007/s10972-014-9385-0
Raykova, Z., Janka, R. J., Udriştioiu, M. T., Yildizhan, H., Stoyanova, D., Tsokov, G., Stoyanov, S., Hruška, M., Pfefferová, M. S., Petrişor, I., Buligiu, I., Săraru, S. C., Etegan, C. M., Yilmaz, S. E., & Çelik, Y. (2023). *New Teaching And Learning Methods For The Post-Pandemic Time* (pp. 61–94). Akademisyen Yayınevi Kitabevi. https://doi.org/10.37609/akya.2497
Raykova, Z., Stoyanova, R., & D. J., Yildizhan, H., Udristioiu, M.T., Yilmaz, E., Petrisor, I. Buligiu, I., Sararu, S.C., Hruska, M., Pfefferova, M.S., Dudas, A., Skrinarova, J., Michalikova, A., Aksay, B. (2024). Some modern methods related to advanced technologies in teaching STEM specialists. *AIP Conf. Proc., 3181*, 040001. https://doi.org/10.1063/5.0215435
Reinemann, C., Stanyer, J., Aalberg, T., Esser, F., & de Vreese, C. H. (2019). *Communicating populism: Comparing actor perceptions, media coverage, and effects on citizens in Europe*. Routledge. https://doi.org/10.4324/9780429402067
Robertson, M. E., Chu, S. M., Cloutier, A., Mongeon, P., Driscoll, D. A., Heer, T., & Westwood, A. R. (2023). Interference in science: scientists' perspectives on their ability to communicate and conduct environmental research in Canada. *FACETS, 8*, 1–31. https://doi.org/10.1139/facets-2023-0005
Rosenthal, S. (2020). Media Literacy, Scientific Literacy, and Science Videos on the Internet. *Frontiers in Communication, 5*, 581585. https://doi.org/10.3389/fcomm.2020.581585
Sampson, H., & Turgo, N. N. (2018). Finding the way into a global industry: the usefulness of elite events to social science researchers. Journal of Organizational. *Ethnography, 7*(1), 2–15. ISSN: 2046-6749.
Schäfer, M. S. (2017). How changing media structures are affecting science news coverage. *The Oxford handbook of the science of science communication, 51*, 57.
Shchepina, E., & Surikov, A. (2022). Modeling the trajectories of interests and preferences of users in digital social systems. *Procedia Computer Science, 212*, 104–113. https://doi.org/10.1016/j.procs.2022.10.212

Slovic, P. (2013). *Risk, media and stigma: Understanding public challenges to modern science and technology*. Routledge.

Takahashi, B., & Tandoc, E. C. (2016). Media sources, credibility, and perceptions of science: Learning about how people learn about science. *Public Understanding of Science, 25*(6), 674–690. https://doi.org/10.1177/0963662515574986

Tsoy, D., Tirasawasdichai, T., & Kurpayanidi, K. I. (2021). Role of social media in shaping public risk perception during COVID-19 pandemic: A theoretical review. *International Journal of Management Science and Business Administration, 7*(2), 35–41. https://doi.org/10.18775/ijmsba.1849-5664-5419.2014.72.1005

Udristioiu, M. T., & Amarie, D. (2024). Stem Students' Progress in Two Romanian and United States of America Universities Before, During, and Beyond the COVID-19 Pandemic. In S. Puiu & S. O. Idowu (Eds.), *Online Education During COVID-19 and Beyond. CSR, Sustainability, Ethics & Governance*. Springer. https://doi.org/10.1007/978-3-031-49353-9_9

Vai, B., Cazzetta, S., Ghiglino, D., Parenti, L., Saibene, G., Toti, M., et al. (2020). Risk perception and media in shaping protective behaviors: insights from the early phase of COVID-19 Italian outbreak. *Frontiers in Psychology, 11*, 563426. https://doi.org/10.3389/fpsyg.2020.563426

Velea, L., Udriştioiu, M. T., Puiu, S., Motişan, R., & Amarie, D. (2023). A Community-Based Sensor Network for Monitoring the Air Quality in Urban Romania. *Atmosphere, 14*(5), 840. https://doi.org/10.3390/atmos14050840

Wang, L., Sheng, G., She, S., & Xu, J. (2023). Impact of empathy with nature on pro-environmental behaviour. *International Journal of Consumer Studies, 47*(2), 652–668. https://doi.org/10.1111/ijcs.12856

Weingart, P. (1998). Science and the media. *Research Policy, 27*(8), 869–879. https://doi.org/10.1016/S0048-7333(98)00096-1

Zsóka, Á., Szerényi, Z. M., Széchy, A., & Kocsis, T. (2013). Greening due to environmental education? Environmental knowledge, attitudes, consumer behavior and everyday pro-environmental activities of Hungarian high school and university students. *Journal of Cleaner Production, 48*, 126–138. https://doi.org/10.1016/j.jclepro.2012.11.030

**Mihaela Tinca Udristioiu** works as a lecturer for the Department of Physics, University of Craiova, Romania. She is a Fulbright Alumna at Georgia Southern University.She developed and implemented a project focused on STEM education, interactive physics teaching methods, and continuous assessment at the University of Craiova, Romania. For her, the standardised assessment is a reference system for quality assurance in education. Dr. Udristioiu was a member of or coordinated the organizing team of the European Researchers' Night at the University of Craiova for 5 years. This event brought science closer to the general public each year and gathered in the hall of this university between 2500 and 4500 children, parents, and grandparents within a few hours. She coordinated three volunteering projects focused on developing a network of air quality monitoring sensors in the Oltenia region (southwestern part of Romania). In the framework of these projects, her team organized public awareness campaigns in mass media about the importance of screening and prevention in detecting health issues and the importance of a clean environment for health. She promotes physics with her colleagues through experiments in the network of high schools in Oltenia region (SW Romania). Dr. Udristioiu participated in summer schools for children from disadvantaged backgrounds to convince them to consolidate their education. Her professional experience includes four EU projects as a project manager and more than ten as an expert.

# Index




M. Todorovic et al. (eds.), *Sustainability and Social Responsibility of the Media and in the Media*, CSR, Sustainability, Ethics & Governance,
https://doi.org/10.1007/978-3-032-00086-6

**D**

**V**



**W**

MIX
Papier aus verantwortungsvollen Quellen
Paper from responsible sources
FSC® C105338

If you have any concerns about our products,
you can contact us on
**ProductSafety@springernature.com**

In case Publisher is established outside the EU,
the EU authorized representative is:
**Springer Nature Customer Service Center GmbH**
**Europaplatz 3, 69115 Heidelberg, Germany**

Printed by Libri Plureos GmbH
in Hamburg, Germany